SOLVING PROBLEMS WITH PROJECTIONS

It is a curious fact that even notoriously difficult computational problems can be expressed in the form of a high-dimensional Venn diagram, where solutions lie in the overlap of a pair of remarkably simple sets, A and B. The simplicity of these sets enables operations called projections that locate the nearest point of A, or B, starting anywhere within the high-dimensional space.

This book introduces a novel method for tackling complex problems that exploit projections and the two-set structure, offering an effective alternative to traditional, gradient-based approaches. Beginning with phase retrieval, where A and B address the properties of an image and its Fourier transform, it progresses to more diverse challenges, such as sphere packing, origami design, sudoku and tiling puzzles, data dimension reduction, and neural network training. The text presents a detailed description of this powerful and original approach and is essential reading for physicists and applied mathematicians.

V E I T E L S E R, a professor of physics at Cornell University, is an expert in phase retrieval. His first major professional achievement was recognising that the atomic positions in quasicrystals could be understood as projections of points from a six-dimensional space into three dimensions. Projections continued to play a key role in his career, notably in 2001, when he discovered that the leading algorithm for solving the phase-retrieval problem was based on a pair of competing projections. Elser's work in physics was recognized by the William L. McMillan Prize.

SOLVING PROBLEMS WITH PROJECTIONS

From Phase Retrieval to Packing

VEIT ELSER

Cornell University

CAMBRIDGE
UNIVERSITY PRESS

Shaftesbury Road, Cambridge CB2 8EA, United Kingdom

One Liberty Plaza, 20th Floor, New York, NY 10006, USA

477 Williamstown Road, Port Melbourne, VIC 3207, Australia

314–321, 3rd Floor, Plot 3, Splendor Forum, Jasola District Centre,
New Delhi – 110025, India

103 Penang Road, #05–06/07, Visioncrest Commercial, Singapore 238467

Cambridge University Press is part of Cambridge University Press & Assessment,
a department of the University of Cambridge.

We share the University's mission to contribute to society through the pursuit of
education, learning and research at the highest international levels of excellence.

www.cambridge.org
Information on this title: www.cambridge.org/9781009475525

DOI: 10.1017/9781009475518

First published 2025

Cover image: Albrecht Dürer, *Melencolia I* (detail). 1514. Credit: Private Collection Photo
© AF Fotografie/Bridgeman Images.

A catalogue record for this publication is available from the British Library

A Cataloging-in-Publication data record for this book is available from the Library of Congress

ISBN 978-1-009-47552-5 Hardback

Cambridge University Press & Assessment has no responsibility for the persistence
or accuracy of URLs for external or third-party internet websites referred to in this
publication and does not guarantee that any content on such websites is, or will
remain, accurate or appropriate.

For EU product safety concerns, contact us at Calle de José Abascal, 56, 1°, 28003 Madrid,
Spain, or email eugpsr@cambridge.org

Contents

Preface

A proper account of the purpose of this book and the events that led to its creation would not fit in the space of a preface. For that you should read Chapter 1.

Even after reading Chapter 1, you might wonder, is this a textbook that explains things step-by-step, or is it more like a scholarly monograph, about something on the frontier of knowledge? My short answer is that I've tried to do both. All the material is self-contained. With a good command of undergraduate math, you should not have to look anything up. At the same time, the problem-solving method you will learn is new, experimental, and an active area of research.

With those important details out of the way, I will use the rest of this preface to thank the many people who contributed to the project.

By a large margin, the most active contributors were the talented and creative group of Cornell physics graduate students I have had the pleasure to work with over the past 20 years: (chronologically) Jean-Francois Carrier, Pierre Thibault, Ivan Rankenburg, Simon Gravel, Duane Loh, Yoav Kallus, Kartik Ayyer, Hyung Joo Park, Ti-Yen Lan, Yi Jiang, Sean Deyo, Avinash Mandaiya. That they were willing to take on projects starting from scratch, and work outside the established disciplines, speaks to their curiosity, confidence, and intellectual mettle.

My own career was launched when, as a postdoc at AT&T Bell Laboratories, I published a paper with the title "The diffraction pattern of projected structures" [34]. This book, which features both diffraction and projections, is the return on a long-term investment made back when industrial labs were still in the business of doing basic research. I thank my math friends from that time, Ron Graham, Jeff Lagarias, Andrew Odlyzko, and Neil Sloane, for welcoming a colleague from the other part of the building.

A nice feature of writing in the technical domain is that the experts are universally generous when pressed for details about their work – even when emailed by a complete stranger. Many thanks (to individuals who left a record in my inbox) go to Keith Ball, Tamir Bendori, Stephen Boyd, Henry Cohn, Dan Gordon, Johan Håstad, Nick Howgrave-Graham, Greg Kuperberg, Victor Miller, Amit Singer, Andrew C. Stuart, Mike Szydlo, Alex Vardy, Doug Weakley, and William Wootters.

Special thanks go to a group of people who have followed my work for apparently no reason other than they find it interesting: Heinz Bauschke, Joe Buhler, Persi Diaconis, Bill Gosper, David Mermin, Rick Millane, Cris Moore, Danny Sleator, and Jonathan Yedidia. Jon Borwein, who hosted the sabbatical at Simon Fraser University where I first dared to go beyond phase retrieval, would be first on this list but sadly is no longer with us. The encouragement expressed by these people over the past 20 years kept me going when the conventional sources of support came up short.

My wife Liz and sons Nick, Till, and Toby have been steadfast in their encouragement and provided the ideal work environment. At times, just knowing they might be peeking over my shoulder kept my meandering prose in check. They are thanked in the book by a certain kind of joke I am contractually obligated to produce.

My parents Lilo and Karl valued originality and creativity above all else. While they didn't live to see the finished product, it brought a smile to their faces when I explained a book of this kind had never been written before.

Were I limited to thanking just two people, it would be my seventh and eighth grade math teachers. Ariel "Ted" Juel was a Navy veteran from Iowa who resettled in California to study the "new math," which he embraced at Edwin Markham Junior High School in San Jose. We debated the meaning of infinity, and he showed me its practical side when proving $1^3 + 2^3 + \cdots + n^3 = (1 + 2 + \cdots + n)^2$ by a trick called induction. Ann Coffee, the daughter of a Greek immigrant father and pioneer midwestern mother, had a master's degree from San Jose State and legitimized my interest in "finding patterns" by gifting me books by Martin Gardner, George Pólya, and a slim treatise on continued fractions by C. D. Olds. This book is dedicated to the memory of these devoted and inspiring teachers.

Notes on mathematical notation

Much of the math in this book is about sets, and the reader should be familiar with the standard notation for set intersection, union, and so on. The "square cup" $\sqcup$ is used for union when the sets are disjoint. Less familiar may be the backward slash in $A \setminus B$, which is the set of elements of A that are not elements of B.

Sets are always set in upper case. Latin font tells you the set is a subset of Euclidean space. A and B are hands down the most popular set names. Finding elements of $A \cap B$ is what this book is all about. Script font is reserved for sets that are just collections of elements, like the antenna ($\mathcal{A}$) and broadcast ($\mathcal{B}$) nodes of a bipartite graph.

Vertical bars need context. Applied to a k-dimensional subset A of Euclidean space, $|A|$ is the k-volume of A. Applied to a discrete set $\mathcal{A}$, $|\mathcal{A}|$ is the number of elements (cardinality). A very different context is real and complex numbers, where $|z|$ is the absolute value or magnitude of z.

Variables, or the "unknowns," are always set in lower case Latin and x is hands down the most popular variable name. Solutions are in general not defined by equations but by statements of the form $x \in A \cap B$. Now is a good time for you to start thinking about solutions x as points in Euclidean space.

The symbol x doesn't reveal the shape of x. Most generally, x is a *list of lists*. At some level a list comprises ordinary high-school variables that take values from the real or complex numbers. For example, if $x_1, \ldots, x_7$ are ordinary variables, then

$$x = \Big((x_1, x_2, x_3), \big((x_4, x_5), (x_6, x_7) \big) \Big)$$

is a variable having the shape of a list comprising a vector of size 3 and a matrix of size 2×2. The elements of a list are its *factors* and live in a *product space*. The domains of variables are always products of $\mathbb{R}$ or $\mathbb{C}$. For the example above,

$$x \in \mathbb{R}^3 \times \mathbb{R}^{2 \times 2} \ .$$

Because the algorithm for finding elements of $A \cap B$ always works in some

Euclidean space, the definitions of the sets A and B always begin with a "declaration" of their factorized domains.

Matrix variables with symmetries come up frequently, and there is special notation for those domains. $\mathbb{S}^m$ is the set of $m \times m$ real-symmetric matrices, and $\mathbb{H}^m$ is the Hermitian counterpart. The symmetries are expressed via the two kinds of transpose: $x^\mathsf{T} = x$, for $x \in \mathbb{S}^m$, while $x^\dagger = x$, for $x \in \mathbb{H}^m$. For a matrix $x \in \mathbb{S}^m$ or $x \in \mathbb{H}^m$, the symbol $\mathrm{spec}(x)$ is the set of its m real eigenvalues.

Variables of the same shape are combined element-wise, no differently from vectors in a vector space. The Euclidean norm is also defined no differently. For the example above,

$$\|x\|^2 = x_1^2 + \cdots + x_7^2 \,.$$

In the complex case this is replaced by $|x_1|^2 + \cdots + |x_7|^2$. Sometimes the Euclidean norm has the modified form (real case)

$$\|x\|_g^2 = g_1 x_1^2 + \cdots + g_7 x_7^2 \,,$$

where the positive real numbers $g_1, \ldots, g_7$ are called *metric coefficients*.

When non-Euclidean norms make an appearance, it is mostly to highlight their faults.

From a typesetting perspective, a list of numbers (real or complex) should be a *row vector*. This is our convention and breaks with the standard (pre-Gutenberg?) column-vector convention. That explains why in this book $xx^\mathsf{T} = \|x\|^2$ is a number, while $x^\mathsf{T}x$ is a rank-1 matrix.

Since the typesetting of vectors defaults to rows, matrix-vector multiplication is written xM, not Mx. The capitalization of the matrix M brings up another break with convention: Capitalization is reserved for constant (non-variable) entities. As explained earlier, variables have diverse shapes, including matrices. So for consistency, matrix variables are treated no differently from vector variables and typeset in lower case. For example, a matrix variable $u \in \mathbb{R}^{m \times m}$ that satisfies $uu^\mathsf{T} = I$ is an orthogonal matrix (the constant identity matrix is the only thing capitalized).

The ˆ accent is reserved for Fourier transforms. Using the above convention for multiplication by a matrix, the Fourier transform of a vector x is written $xF = \hat{x}$. In this setting, x is always a vector, even when it represents pixels in a 2D image or voxel samples of a 3D density. The array structure that comes with the vector x is only there to reveal its translational symmetries. Note that the Fourier matrix F is capitalized because it is a constant. Of course in computations, $\hat{x}$ is obtained from x by the FFT algorithm, not explicit matrix multiplication.

The set of $m \times m$ unitary matrices is denoted differently, as $\mathcal{U}^m$, to remind us this is not a linear space. Similarly, O^m are the orthogonal matrices and the chiral variant, $\mathcal{O}^m$, is the subset with positive determinant. As explained earlier, you will not find these sets in the declarations of A and B.

The use of boldface is limited to the vector positions $\mathbf{p}$ of pixels/voxels that represent the contrast in microscopy or crystallography, or the vector positions $\mathbf{q}$ of detector pixels. Their non-geometric counterparts p and q are the indices on the vectors and matrices that arise in phase retrieval. The geometrical entities $\mathbf{p}$ and $\mathbf{q}$ are needed when defining the sizes and shapes of sets.

Instead of using functions (e.g., zero-norm) to express sparsity, we use the symbol $\mathbb{R}^m_k$ for the set of m-component k-sparse vectors (having at most k nonzero elements). This notation reminds us that the constraint set for such vectors is the union of $\binom{m}{k}$ Cartesian subspaces. Surely every student of "analytical geometry" is familiar with $\mathbb{R}^2_1$.

The expression

$$\arg\min_{x \in A} f(x)$$

denotes the set of elements x in a compact domain A that achieve the minimum of the function f. Almost everywhere this notation (or its companion $\arg\max$) is used, the set is a singleton and the symbol defaults to that one element.

1

Origins

1.1 Phase Retrieval

Like many new lines of research, the subject of this book was not the product of a systematic, well-argued plan of investigation. It came about as a complete accident.

In May 2001, a workshop on "New approaches to the phase problem for nonperiodic objects" [103] was held at the Lawrence Berkeley National Laboratory. I was invited for the flimsiest of reasons, in that I had some background in very famous nonperiodic objects called quasicrystals, discovered by Dan Shechtman in 1982.

For some background, we first turn to a 1986 meeting of the American Crystallographic Association (ACA), held at McMaster University. A representative sample of the quasicrystal data [97] that was hotly debated at the meeting is shown in Figure 1.1. The bright spots are a record of electrons (produced by an electron microscope) reflected into special directions by planes of atoms in the quasicrystal. Many scientists, including Linus Pauling, were bothered by the fact that the angles of the planes could not be reconciled, mathematically, with a material that was periodic [93]. My ACA presentation, where I argued that the spots could be explained by a generalization of periodicity that has six axes instead of the usual three (but still in three dimensions) [34], was overshadowed by Pauling's, in which he claimed the precision of the spot measurements was insufficient to rule out a periodic arrangement, albeit with a very large repeat distance [88]. In any case, the six-axis generalization of periodicity had nothing to do with the "nonperiodic objects" of the Berkeley workshop 15 years later. The objects of interest there had no periodicity whatsoever.

There was a recognition event at the 1986 ACA meeting that proved to be remarkably prescient. In the previous year, the crystallographers Herbert

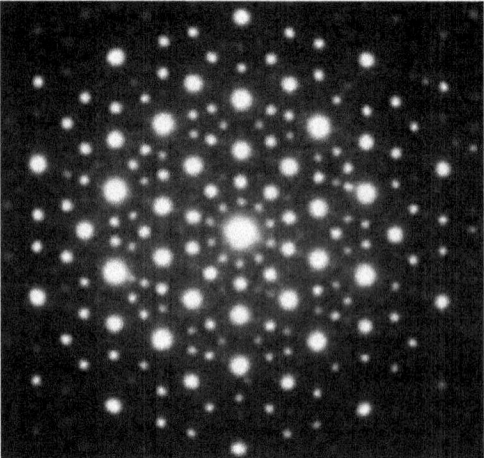

Figure 1.1 An early electron diffraction image of the AlMn quasicrystal [97]. The 10-fold arrangement of the spots challenged the orthodox model of crystalline order. Reprinted from Shechtman et al. (1984) with permission from © 1984 American Physical Society.

Hauptman and Jerome Karle had been awarded the Nobel Prize in Chemistry for having solved "the phase problem." The substance of this accomplishment can be explained even for the data shown in Figure 1.1, which after all only represents a break from orthodox periodicity.

From the physics of the electron reflection process, called diffraction, one can show that the spot intensities are directly interpretable as the coefficients of a Fourier series. By summing the series, called Fourier synthesis, one can reconstruct the microscopic "sources" of the electron reflection, a function in three dimensions commonly referred to as *contrast*. Hauptman and Karle were working with X-ray data where the contrast is the electron density in the material, while it is the electrostatic potential produced by those electrons (and nuclei as well) that the electron microscope data reveals. In either case, by doing the Fourier synthesis, one can recover not only the geometry of the reflecting planes, but the makeup of those planes by atoms.

But the Fourier synthesis dream runs up against a major obstacle. The Fourier coefficients are complex numbers, and the spot intensities only provide the magnitudes of those numbers. Fourier synthesis cannot be attempted without the phase angles of those coefficients, which go unmeasured. This is the phase problem.

Early (pre-1950) crystallographers were undeterred by the phase problem. Progress was made through *modeling* the contrast (arrangement of atoms), as in the iconic photo of Crick and Watson posing with their model of the double helix. Using strong constraints coming from chemistry, only a limited number of atomic arrangements had to be considered, and getting just the magnitudes of the Fourier coefficients to work out was usually enough to nail down the structure. At the time of the 1986 ACA meeting, the modeling approach was the leading method being applied to determine the atomic structures of quasicrystals.

Hauptman and Karle [60] are credited with solving the phase problem because they found a method that avoids modeling, making atomic structure discovery available to even chemically challenged crystallographers. Their method comprises both a principle and an algorithm. The principle is surprisingly simple. Consider a thought experiment where *random* phases are applied to the Fourier coefficients. In all except a fantastically lucky combination of phases, the resulting contrast (by Fourier synthesis) is a complete mess. Chemistry does not come up because the contrast does not even resemble atoms!

The simplicity of the principle behind the phase problem solution – conjuring up phases to get something that at least resembles atoms – is in sharp contrast with the complexity of exploiting that principle. The unknown phases of the Fourier coefficients are like the dials of an enormous combination lock. Only one combination produces a sensible contrast and unlocks the atomic structure. But finding that combination is hard when there are many dials/phases. The phase problem cannot be declared solved unless there is also a practical algorithm that can unlock the phases encoded in the Fourier magnitudes.

In 1986, I did not think of the phase problem in these terms. The code-breaking angle might have inspired me to dive deeper into the subject, but at the time my interests were still aligned with physics. The secret of quasicrystals, like the secret of life, had to be simple and elegant and not involve difficult computations! Still, I recall being in awe walking through the poster gallery at the ACA meeting, each one chronicling a phase problem success on a complex biomolecule. Unlike the simplicity I was counting on to make the quasicrystal phases amenable to modeling, the giant virus structures on display all owed their existence to an algorithm.

The 2001 Berkeley workshop was organized by the electron microscopist John Spence and was meant to launch a revolution in microscopy. Spence had invited a diverse group of X-ray scientists, electron microscopists, engineers, mathematicians, and physicists, some of whom (myself included) knew almost nothing about the phase problem or any form of cutting-edge microscopy. Those in the latter group were quickly brought up to speed, most notably with regard to the "abcdef" experiment [76].

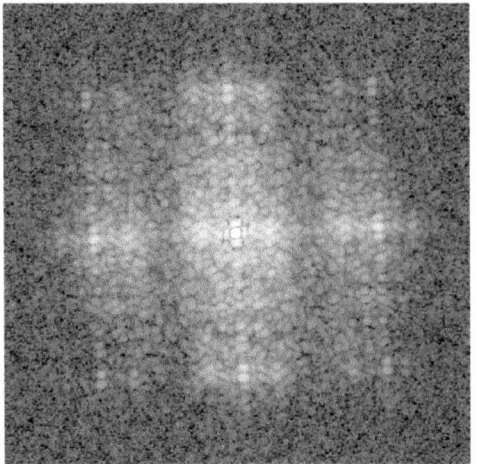

Figure 1.2 First demonstration of lensless microscopy with X-rays, by Miao et al. [76]. The diffraction intensities on the left were given to an algorithm that reconstructed the contrast shown on the right. Reprinted from Miao et al. (2000), with permission from Copyright © 2000, AIP Publishing.

In 1999, a group at SUNY Stony Brook published a proof-of-concept paper [76] demonstrating the algorithmic reconstruction of a nonperiodic contrast from just the magnitudes of its Fourier transform. The data, shown on the left in Figure 1.2, were obtained by shining a coherent (laser-like) beam of X-rays on a fabricated specimen of gold dots deposited on a thin membrane. The image on the right is their reconstruction but is practically indistinguishable from the ground truth, which was obtained using the higher resolving power of an electron microscope. In any case, no conventional, lens-based microscope using light (even X-ray light) was capable of doing what these researchers had managed to do with the help of an algorithm.

The intensity of the diffracted X-rays in the "abcdef" experiment was not concentrated in spots, as in the quasicrystal data. Instead of a Fourier series with unknown phases for the coefficients, in this data the measured and continuously distributed Fourier magnitude is accompanied by an unknown and similarly continuous phase *function*. Later I learned that this feature made the phase problem much easier than it was for periodic contrasts. At the workshop I also learned to think more optimistically about the phase problem, because the participants never used the word "problem" and instead chose to call the algorithmic task "phase retrieval."

Probably the most far-reaching implication of algorithmic phase retrieval of nonperiodic objects I learned about at the workshop was the possibility of reconstructing the shapes of complex biomolecules (proteins, viruses, etc.) without having to coax them into forming crystals first [81]. Most biomolecules are terrible at forming crystals, but structural biologists persist because only a crystalline specimen, comprising 10^{16} identical copies of the same object, can produce measurable diffraction data. A star-wars caliber X-ray source, with the power to form a diffraction image of a single biomolecule, would eliminate the need to grow crystals. That was in part the rationale behind the Linac Coherent Light Source, an X-ray laser under construction at SLAC. If this feat of engineering succeeded, determining the shapes of the molecules of life might become as easy as "abcdef."

My engagement at the workshop ratcheted up by several levels when the talks turned to the algorithms for phase retrieval. One algorithm stood out: Jim Fienup's *hybrid input-output* (HIO) algorithm [42]. HIO was wildly successful but also mysterious. It belonged to the class of algorithms that acted iteratively and deterministically on estimates of the contrast. Starting with an initial guess, which typically was random, a sequence of transformations was applied until the contrast stopped changing, hopefully after it had settled on the true contrast. My general impression was that the transformations would alternate between modifying the contrast and modifying its Fourier transform, of which only the phase function was free to change because the magnitude was fixed by the data. As far as I could tell, the transformations were doing reasonable things: Fixing what was obviously wrong in the contrast or restoring the known magnitude to the Fourier transform. In this respect HIO was not exceptional. The secret of its success seemed to lie in the precise way the transformations were combined.

A more complete theoretical understanding of HIO was clearly a worthwhile goal. Only HIO had been able to reconstruct "abcdef" in the final and computational stage of the new microscopy. If anyone understood HIO at the time of the workshop, it could only have been Jim Fienup. For the rest of us, and here I mostly refer to myself and three mathematicians [10] at the workshop, his "hybrid" construction looked ad hoc. It was a puzzle why the algorithm worked at all.

When I returned to Ithaca, I set myself the challenge of understanding HIO. This book, written 22 years later, is the outcome of that investigation. However, already within months of the workshop, the three mathematicians Bauschke, Combettes, and Luke [10], came up with a formal and more general statement of what HIO was doing that matched what I had come up with [35]. And as often happens in mathematics, Bauschke and coworkers realized the HIO "formula" had been written down nearly half a century earlier, by mathematicians Douglas and Rachford [30], in a completely different context: numerical methods for solving partial differential equations.

1.2 Nonconvexity

The context of the Douglas–Rachford formula plays a fundamental role in the nature of the algorithm. In symbols the formula reads

$$x \mapsto x + P_B(R_A(x)) - P_A(x) . \tag{1.1}$$

Here x is the thing being reconstructed, like the "abcdef" contrast or the solution of a partial differential equation. The building blocks of the algorithm are called *projections*, and P_A "projects to set A," while P_B "projects to set B." The third operation is called a *reflection*, and

$$R_A(x) = 2P_A(x) - x$$

"reflects in set A." We do not need to get into the motivation and roles for these operations here – that is the purpose of this book – to make a point about the Douglas–Rachford formula. This is that the formula makes sense for arbitrary sets A and B even though the context of the formula is normally limited to the case of *convex* sets. A set is convex if all the points between any pair of points in the set are also in the set. We will see that the "convexity context" of formula (1.1) plays an oversized role in the history of the subject.

When the Douglas–Rachford formula is used with convex sets, it is as though the algorithm never has to make decisions. This comes about because $P_A(x)$, a point in A that is closest to x, is always unique when A is convex. A simple example shows how this changes when A is nonconvex. Suppose x is just a single unknown (an image of one pixel) and A is the set of two elements $\{-1, +1\}$. Now if $x < 0$ then $P_A(x) = -1$, while $x > 0$ implies $P_A(x) = +1$. In both cases, the result of the projection is unique. However, when $x = 0$, the algorithm has to make a decision because both -1 and $+1$ have the same distance to $x = 0$. If $x = 0$ never comes up in the course of the Douglas–Rachford iterations, then one might argue that the algorithm is still not having to make decisions. But this is a very naive view if it turns out that the x values the algorithm encounters have a nonzero density near $x = 0$. In that scenario, some fraction of the $P_A(x)$ values will depend very sensitively on x. Though still formally deterministic (no actual decisions), the algorithm is better characterized as sometimes making *random* decisions. Thinking of the Douglas–Rachford formula as defining a dynamical system, nonconvexity of the sets A and B is a potential source of *chaos*.

1.3 Chaos and Ergodicity

The analysis of algorithms that have the potential of behaving chaotically is hopeless. That is the reason the Douglas–Rachford formula has traditionally been used only when the sets A and B are convex. With this restriction, and when a solution exists, one can prove the iterations converge to a fixed point and a solution to the problem is at hand [4]. But a simpler and more obvious algorithm, studied by John von Neumann and called alternating projections [8],

$$x \mapsto P_B(P_A(x)),$$

also has these good properties when the sets are convex. Fienup tried this in his phase retrieval experiments and rejected it because it always got stuck on nonsolutions. One of the sets in phase retrieval, the constraint on the Fourier magnitudes, is nonconvex.

After I had managed to express HIO as the general formula (1.1) and was in a position to analyze the algorithm geometrically, it seemed to me what was special about this algorithm was its ability to extricate itself from the traps that plagued alternating projections. Because there are many traps in a hard problem, the resulting dynamics will be very chaotic. Biased by my physics background, I liked the idea of exploiting chaos. Understandably, this perspective was not shared by the mathematics community. Already at the Berkeley workshop, a proposal was made to move HIO into safe territory by "convexifying" the Fourier magnitude constraint. Fienup obliged, changed one line in his code, and probably had already anticipated the result: HIO no longer worked.

In physics, it is fair to study things phenomenologically: make observations, form hypotheses, perform experiments, and so on. This applies even when the subject has been reduced to mathematical formulas. Some formulas are just too hard to yield to mathematical analysis! Over the past two decades, I have taken the phenomenological approach in studies of HIO/Douglas–Rachford. An example of this kind of study, singled out for its kinship to phase retrieval, is the problem of *bit retrieval* [37]. My first experiments were performed at Simon Fraser University, in a sabbatical hosted by Jon Borwein, an early champion of experimental mathematics.

In bit retrieval, one tries to reconstruct the contrast of a one-dimensional crystal from the magnitudes of its Fourier series coefficients (diffraction data). As the name suggests, "retrieval" is possible because the contrast is known to have only two values, say -1 and $+1$, at each pixel. In the same spirit as phase retrieval for physical crystals, where Fourier synthesis with the

Figure 1.3 Chaotic dynamics in bit retrieval [37], the reconstruction of a two-valued sequence from its Fourier magnitudes. Each row shows the time evolution of one bit in the sequence. After about 34 000 time steps (Douglas–Rachford iterations), the random initial pattern on the left arrives at the solution fixed point on the right.

wrong phases produces a nonatomic mess, the wrong phases in bit retrieval produce a contrast where the pixels deviate from the two special values.

The time evolution of the contrast in bit retrieval, by the Douglas–Rachford algorithm, can be rendered in the same way that is popular for one-dimensional cellular automata. Figure 1.3 shows the retrieval, left to right, of a sequence of 35 bits starting from a random contrast. The juxtaposition of chaos with stability, at the fixed point on the right, is striking.

It is important that formula (1.1) is able to recognize and converge on solutions when it finds them. The local convergence of Douglas–Rachford, near solutions where the sets A and B may be approximated as convex, is responsible for that. But just as important is the behavior prior to arrival at the fixed point, where the dynamics can only be characterized as chaotic. The two contrast values (± 1) are being sampled much like we think of the positions of molecules in a fluid. All configurations in the fluid, subject only to the conservation of energy, arise eventually if one is patient and waits long enough. Thanks to chaos introduced by the projections (when a contrast value is close to zero), the sampling of candidate contrasts in bit retrieval is similarly exhaustive.

The bit retrieval experiments also showed that the dynamics of formula (1.1) was very good at doing something else. Though there is no analog of energy, the stream of contrasts being generated were all *near*-solutions: Their Fourier magnitudes were close to the known values, and the contrast values themselves were close to either -1 or $+1$. Empirically I knew a very targeted search was taking place because solutions were being found with far, far fewer iterations than the $2^{35} \approx 3 \times 10^{10}$ possible bit sequences!

Might chaotic search be a competitive alternative to the systematic approach to search we are taught in computer science? The time spent in a successful chaotic search has an analog in physics called the *Poincaré recurrence time*. Technically this is the time needed by a finite system of particles to return to its initial configuration within some specified precision. Because water molecule configurations are not special in any obvious way, we expect all of them to be visited over the course of one recurrence time. If Kurt Vonnegut's ice-nine [111] did in fact exist as a (catastrophically) stable solid configuration of water molecules, it would be found within about one recurrence time! Fortunately, the recurrence time for even microscopic physical systems is astronomical, and the possibility of an ice-nine fixed point – should one exist – poses no danger to humankind!

The recurrence time of chaotic Douglas–Rachford dynamics is often comparable and sometimes much shorter than the time taken by a systematic branching search. Ice-nines are found routinely. The most detailed study is on the bit retrieval problem and includes estimates of the run/recurrence time and how this depends on a quantifiable hardness measure. The 35-bit instance shown in Figure 1.3 is the hardest for that size because all of its Fourier magnitudes are equal [37].[1] In 2002, an easier, average-case instance with 160 bits was solved in one week. This may still be the record for consummated Poincaré recurrence times.

There is really just one obstacle for chaotic search to be more widely adopted, and it is not the inherent randomness in the run time. Some initial points are luckier than others because the dynamics stumbles on a solution fixed point earlier. The distribution of run times is accurately described by the curve for radioactive decay, consistent with a loss of memory of the past (due to chaos) and there being a time-independent exposure to fixed points. But systematic branching searches are just as unpredictable, with some decision orderings luckier than others.

The unpredictability of the time required to arrive at a fixed point is not so much a concern as is the possibility that the dynamics simply fails to explore a significant fraction of the solution candidates. In physics, the property of visiting all configurations (subject to energy conservation) is called *ergodicity*. It is easy to see how sensitivity to initial conditions, as a result of chaotic dynamics, helps ergodicity. But that still falls short of a guarantee.

Fortunately, there is an important use-case that makes me optimistic: thermodynamics. Boltzmann faced a very similar challenge when he tried to derive the macroscopic equilibrium properties of gases from the complex dynamics of the constituent particles. The assumption that all configurations consistent with energy conservation were visited with exactly the same frequency gave the correct answers but seemed impossible to prove. Because the phenomenological model – thermodynamics – worked so well, Boltzmann formalized his assumption as the *ergodic hypothesis*. Today no one doubts the truth of the ergodic hypothesis for systems such as gases and fluids, even though the only established results are for simple "billiard" models [101].

The ergodic hypothesis is false for some physical systems, a fact we should be thankful for when contemplating our place in the solar system (and its repetitive and predictable dynamics). By analogy, we should not expect all applications of formula (1.1) to be as ergodic as bit retrieval. Over the years, I've encountered applications where instead of arriving at fixed points, whose existence is not in doubt, the dynamics often gets trapped in the analog of a solar system. Sometimes this can be resolved by defining the sets A and B differently or modifying the metric used to define the distance. In any case, the existence of bad actors is not relevant for the many applications in which ergodicity appears to be upheld.

1.4 Convergence and Logic

When I first started giving talks about (scandalously) applying Douglas–Rachford (DR) to nonconvex problems, I would invariably be asked, "... but does it converge?" I was completely comfortable knowing that DR was doing a very good job at something just as important – a thorough search – and that the mathematically provable convergence that happened in the last few iterations was simply what terminated the search. Though I could point to my experiments on bit retrieval, as a demonstration of the new kind of convergence, I got the sense that the people asking about convergence were looking for something more. Around 2005 a way to satisfy even these skeptics came from an unexpected source: sudoku.

Solving a problem by *logic* is usually construed as the process of increasing knowledge by the application of inference. Knowledge about the problem can only grow, culminating when the culprit is confidently unmasked (Colonel Mustard). Sudoku is a minimalist exercise in inference, something that surely is part of its appeal. Players learn (or discover on their own) a set of inference rules whose application increases the set of filled-in numbers. A popular rule is to look for two instances of the same number, say 5, in two 3×3 blocks that fall on a shared set of three rows or columns. The unknown location of a third 5 is now constrained not just by a block, but also by the row or column not already used by the two other 5's. Beginners and experts differ mostly in the number of inference rules they have amassed and the complexity of applying them. The size and complexity of the rule toolbox needed to solve a puzzle may also be the basis of the mysterious difficulty ratings.

Computer programs for solving sudoku fall into two groups. Either the program goes through a list of human-engineered inference rules (naked-quad, X-wing, ...), or more laboriously, systematically tries out all ways of completing the puzzle that are not in direct violation of the rules (with the promise that one of the many hypotheses is bound to work out). But these are really just extremes on a spectrum, the elaborate inference rules being just a highly creative alternative to "brute" search. The DR algorithm, if it could be applied to sudoku, would be doing something else entirely.

The first challenge in solving sudoku with DR was formulating the puzzle in terms of two easy constraints, A and B. This turned out to be not very hard at all[2]:

A : In each of the 3×3 blocks, numbers and cells are paired one-to-one.

B : Each number appears nine times, pairing rows with columns, one-to-one.

Filling in numbers consistent with just A or just B is easy, and provided the projections (nearest patterns) P_A and P_B are also easy to compute, the DR algorithm will do the rest. Though projections aren't covered until Chapter 4, the information in Figures 1.4 and 1.5 will help you understand what DR is doing. Both show the algorithm solving, in six iterations, a New York Times puzzle with difficulty rating "easy."

Figure 1.4 shows just the progress of the projection $p_A = P_A(x)$, so not the actual point x being updated. All the 3×3 blocks in each iteration are in compliance with rule A. The "given" numbers or "clues" are marked by the gray cells and never change. Figure 1.5 shows the progress of the projection $p_B = P_B(R_A(x))$. Think of these as nine overlapping number patterns, all having the same kind of one-to-one pairing of rows with columns as shown highlighted for the number 5. After six iterations (bottom-right panel) the two number patterns, p_A and p_B, are the same and the puzzle is solved.

The constraints A and B, as geometrical sets, are as nonconvex as possible: sets of isolated points. In spite of this, there is a kind of convergence not so different from the mathematically rigorous convergence when both sets are

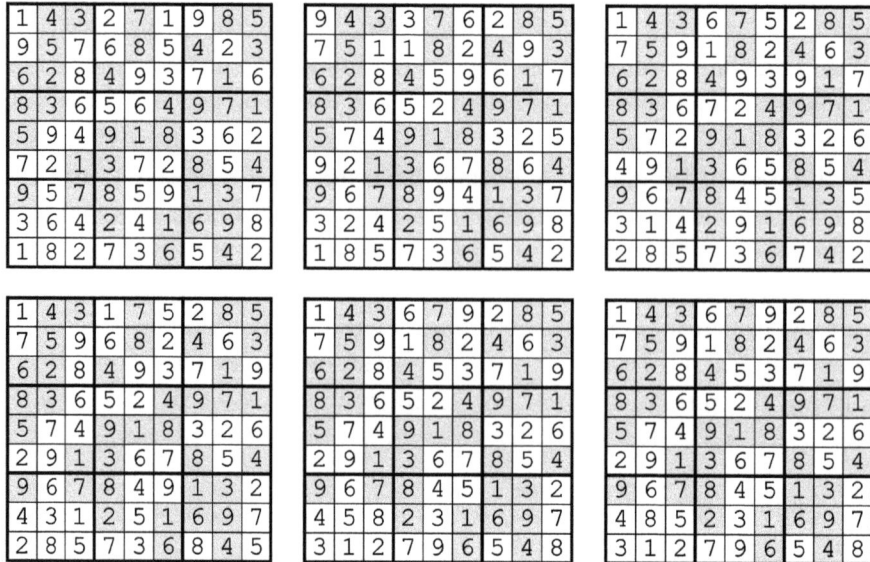

Figure 1.4 A sudoku puzzle being solved in six Douglas–Rachford iterations (top-left to bottom-right). These do not show the updating of the point x, just consistency of the projection $P_A(x)$ with sudoku rule A. The gray cells hold the given numbers (and never change).

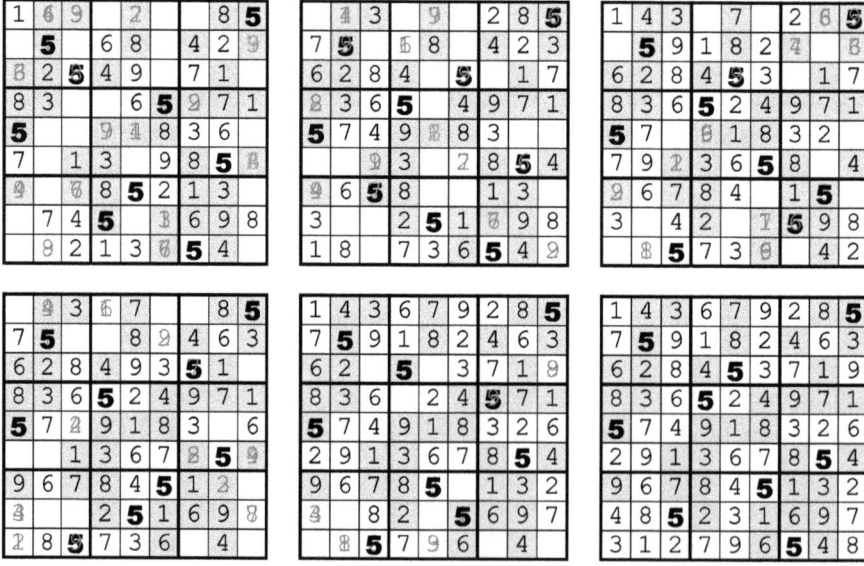

Figure 1.5 Same as Figure 1.4 but for rule B. All numbers have the relationship displayed by the nine highlighted 5's, a one-to-one pairing of rows with columns.

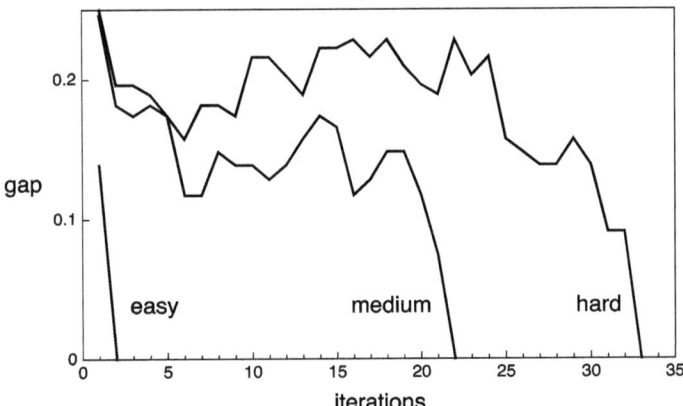

Figure 1.6 Typical behavior of the gap (distance between constraints A and B) when solving the three difficulty levels of New York Times sudoku puzzle.

convex. To convey this, we consider the *gap*, or the current distance between the points p_A and p_B. When the gap is zero, the two points are the same point, and this point is a solution to the puzzle (because both constraints are satisfied). The generalization of convergence in this very nonconvex application of DR is the behavior of the gap.

Figure 1.6 shows plots of the gap for three New York Times sudoku puzzles. In these runs, the clue data were implemented differently,[3] with the result that "easy" puzzles are always solved in just two iterations. In "medium" puzzles, it's harder to argue something analogous to inference is going on, because the gap is not always decreasing. A mostly decreasing gap for "hard" puzzles does not begin until after the algorithm has done some exploring (here about 20 iterations). This progression aligns with our expectations, but is also necessary if a widely believed technical property of sudoku is true.

In the study of computational complexity, sudoku belongs to a class of problems where (it's believed) long exhaustive searches can never be completely avoided. To realize the extremes of hardness, puzzles must be "diabolically" designed or generalized to 16×16 grids and higher. Puzzles for human consumption are neither so large nor diabolical to obviate a logical route to the solution. Interestingly, these puzzles also have the property that the Douglas–Rachford gap is mostly decreasing.

The sudoku episode of the nonconvex Douglas–Rachford story was important in three respects. First, it attracted far more interest than any accomplishment in phase retrieval could have achieved! Second, and strictly as

sudoku difficulty	gentle	moderate	tough	diabolical	extreme
average DR solution-time	3.8	6.8	9.3	12.9	15.6

Table 1.1 Growth in solution time with a logic-based difficulty scale [105].

empirical evidence, I could now point to the behavior of the gap as a generalization of convergence for even highly nonconvex problems. Third, and for me the most inspiring, was the discovery that DR seemed to be capable of generating its own brand of logic. Though my sudoku solver was built using just the elementary rules of the puzzle, upon execution, its progress toward the solution was apparently as systematic (a shrinking gap) as a solution by standard logic. And if the solution process for easy puzzles/problems had the hallmarks of logic, then DR was likely also doing something clever on the harder counterparts. A model to explain the cleverness is proposed in Section 5.5.6 where we study the "flow" of the algorithm and define the *solution time*, closely related to the number of iterations. Table 1.1 lists solution times for five classes of puzzles graded by the depth of logic required to solve them. These results were obtained by sampling Andrew Stuart's puzzle collection [105]. All puzzles require some amount of logic, but the number and complexity of inference rules needed to solve them grows substantially between "gentle" and "extreme."

1.5 Deconstruction

So far this narrative has betrayed my background as a physicist. Though a computer scientist may not view search as a dynamical system, when subject to chaos and limited by ergodicity, this perspective is certainly one that most physicists are comfortable with. On the other hand, one of the main objectives of this book would not be met if physicists were not also taken into unfamiliar territory.

Search algorithms are often characterized as "physical" if they mimic the dynamics of a physical system. I will give two examples: disk packing and neural network training.

Figure 1.7 shows a set of 14 disks being packed by an algorithm that simulates the process of compression. The enclosure, a circle, starts out large and shrinks opportunistically whenever allowed by the positions of the disks. In between compressions the disks make small random steps called Brownian motion. The idea is that through Brownian motion the disks can rearrange into configurations with more wiggle room and a more efficient use of space.

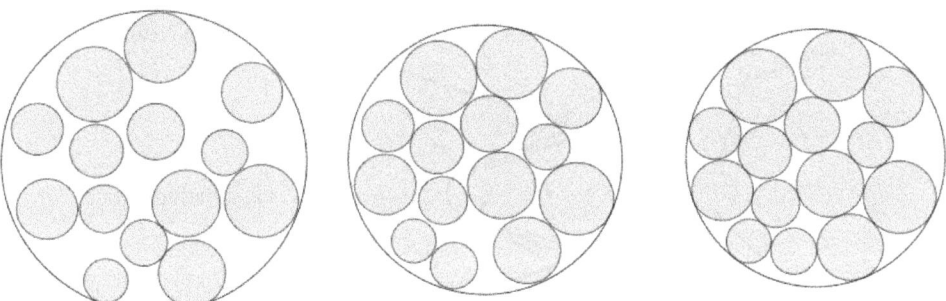

Figure 1.7 The compression of a system of 14 disks by a shrinking circular enclosure comes to a halt (frame on the right) when the disks become jammed.

Eventually there is no wiggle room left. When this happens, one says the configuration of disks is "jammed"; no further compression is possible.

The second example concerns the application that is pervasive in nearly all of modern technology: artificial neural networks. The core algorithm for the training of networks was set forth in "Learning internal representations by error propagation" by Rumelhart, Hinton, and Williams[4] (RHW) [91, 92] whose citation count increased even as you were reading this sentence. This algorithm is physics based in the sense that the discrepancy between the network's target output and its actual output is quantified by an energy function ("loss"), and the network's parameters ("weights") are adjusted to decrease the discrepancy/loss in much the same way a ball rolls down a hilly landscape so as to reduce its gravitational energy.

Though most of the RHW citations are for the back-propagation formula,[5] this paper also features several simple and revealing applications. The network for one of these, shown in Figure 1.8, is called an autoencoder. Here the network is tasked with exactly reproducing n input patterns in its outputs, both expressed on n nodes. What makes this challenging is a bottleneck of only $\log_2 n$ nodes that the information is forced to pass through. By choice of the nonlinear functions that generate the values on the bottleneck nodes, which smoothly interpolate between 0 and 1, the autoencoder is biased toward transmitting the information through the bottleneck as a binary code. Though the authors were thrilled that their error-propagation-trained autoencoder worked (outputs exactly matched inputs), they had to concede that often the codes appearing in the bottleneck were not strictly binary, because they included "intermediate values." It was a victory for self-taught data

Origins

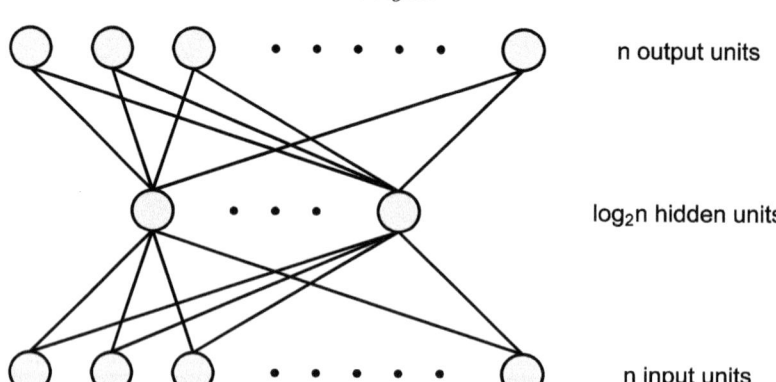

Figure 1.8 Autoencoder network from "Learning internal representations by error propagation" [91] used in the study of the binary encoding problem.

compression, but a mixed result for interpretability of the data representation.

Figure 1.9 shows the distribution of code values for the $n = 16$ autoencoder, using the same "sigmoid" activation functions used by RHW, but with a state-of-the-art training algorithm [62] that uses "momentum" in addition to energy to guide its course through the loss/energy landscape. Though the endpoint of training, a point with zero loss, is found faster than the method used by RHW, the results are no better: Most of the codes are nonbinary because of an admixture of intermediate values.

Were I to remain true to my physics training, I would view the jammed, nonoptimally packed disk configuration, and the not-quite-binary autoencoder as inevitable outcomes when systems cross some threshold of complexity. There is even a subdiscipline of physics, for disordered and glassy

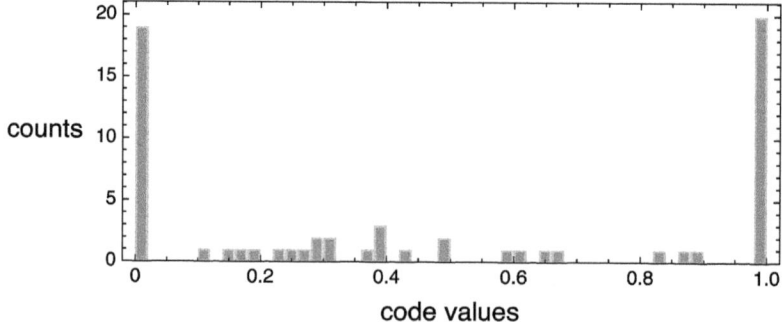

Figure 1.9 Distribution of the 16×4 code values found by the gradient-descent-trained autoencoder.

systems, where both of these applications have been welcomed with open arms [1, 59]. The point of this book, however, is to offer you an alternative.

The alternative is not a more powerful, physics-inspired algorithm. It goes deeper than that. A good name for the main idea is *deconstruction*. Before taking on a new problem, the first step is to deconstruct it into sets A and B. Phase retrieval was important in the development of the subject simply because A and B were fairly obvious. For most other problems, deconstruction into A and B is a new skill.

Let's deconstruct the disk packing problem. My former student Simon Gravel came up with a great name for this style of deconstruction: *divide and concur* [48]. Each of the 14 disks to be packed is "divided" into 13 versions, called "replicas." For example, disk 3-7 is the replica of disk 3 that "watches out for disk 7," while disk 7-3 is the replica of disk 7 that "watches out for disk 3." Set A corresponds to all such replica pairs satisfying the packing constraint (not intersecting). This is a very easy constraint to satisfy because the constraint on each replica pair is independent of the constraints on all the other replica pairs. If replicas 3-7 and 7-3 do not intersect, nothing needs to be done. Otherwise, to project to set A, the replicas are moved the minimum distance, which amounts to making them tangent. Set B implements "concur," or the constraint that replicas 3-1, 3-2, 3-4, ... , 3-14 are actually the same disk! The projection here is to replace their centers by the mean of their centers.

The packing problem can be deconstructed even further in a way that takes advantage of the nonuniform disk sizes. Replica 3-7 not only has a center (to be determined), but is given a variable radius. To keep replicas 3-7 and 7-3 from intersecting, the projection may shrink their radii in addition to moving their centers. Whichever combination of those actions involves the smallest sum-of-squares change is the one taken by P_A. The variable radii are addressed in concur, or P_B. After all the replicas of disk 1, disk 2, and so on are given a shared radius by averaging, the resulting set of 14 radii are minimally modified to match the radii of the original packing problem.

This is quite an elaborate deconstruction, but very much worth the effort. Three configurations encountered in the Douglas–Rachford (DR) dynamics are shown in Figure 1.10. An animation would show disks moving according to some strange laws of physics that allowed intersections and penetration of the enclosure. Even stranger, disks of different sizes occasionally appear to swap positions by teleportation (when they swap radii). But the exotic dynamics are not at all exotic in the deconstructed world. The figure shows only the dynamics of the 14 concurred centers and radii and cannot convey what all the replicas are doing!

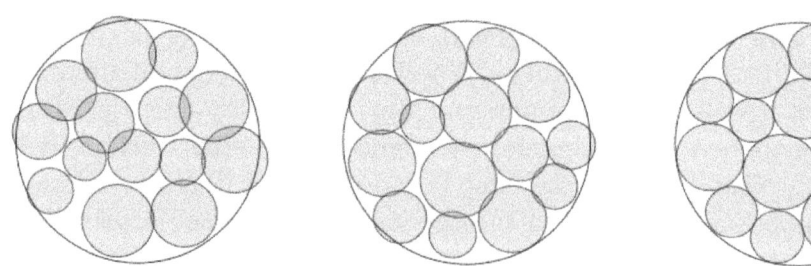

Figure 1.10 Douglas–Rachford iterations acting on the constraints A and B of the deconstructed packing problem (with a fixed circular enclosure) terminate when a true packing is found (right frame).

The 14 disk-radii in this packing puzzle were designed so there is a unique solution for the given enclosure.[6] The DR algorithm finds this solution in about the same time taken by the compression algorithm to find one of the many jammed nonsolutions (enclosures larger than optimal by several percent).

Let's now turn to the deconstruction of the neural network training problem. In constraint A each neuron is treated in isolation, independent of all the other neurons in the network. Every neuron has a vector of inputs, x, a single output, y, and a vector of weight parameters w. In the binary encoding/decoding network, or any other network where we want the neurons to learn dichotomies, the outputs may have only two values, say $y = 1$ or $y = -1$. The choice of output is determined from the value of the dot product $w \cdot x$. Projecting to the "neuron constraint" is actually quite easy and involves no transcendental functions (as in the sigmoid step-like activation function). At the end of projection P_A, all the neuron outputs are either 1 or -1, exactly. P_A also minimally modifies the inputs and weights for each neuron. Interestingly, x and w experience the same degree of change because they appear symmetrically in the constraint through their dot product. If you are impatient, the details of the neuron projection are spelled out in Section 7.7. Finally, the network's input nodes are special and for them P_A simply sets the y variables to the data values.

The connectivity of the network and the constraint coming from the output-side of the data is addressed in constraint B. Each neuron's output y should concur with the inputs x of the receiving neurons. A key part of the deconstruction is to realize that the input variables live on the edges of the network, not the nodes. That's because one neuron's output y gets sent to multiple neurons, and the replicas of that y, called x, align with the edges that connect the output node with the various input nodes. The complete concur projection therefore involves, at each neuron's output, a single y and the x values on all the edges from that output. Like the neuron constraint projection, P_B is very simple (replacing numbers by their average) and a local computation because independent sets of variables are involved. At the output nodes, where there are no outgoing x variables, P_B simply sets the y variables to the data values.

Neural networks are trained on batches of data and this continues to be the case in the deconstructed problem. In the $n = 16$ autoencoder, there are only 16 data and it is feasible to have the batches be the entire data. This is how the gradient-descent results shown in Figure 1.9 were obtained. In the deconstructed problem, the entire network of x, y, and w variables is replicated 16 times. For the most part, the variables in the different network replicas are independent, and each receives a different set of data that constrain the input and output nodes. But P_B now has an additional task: The weights w should concur across all the network replicas.

If this deconstruction also strikes you as elaborate, it just means you can use some training in the craft of deconstruction. Chapter 2 is a warm-up with simple examples; the more elaborate examples in Chapter 7 convey the scope of deconstruction. But already now you can see that the deconstructed neural-network-training problem is quite nice: many simple and local constraints that are easily satisfied (projected to). Most of the programming work is in the implementation of the two projections, and is comparable in complexity to back-propagation code. A single DR iteration, comprising one P_A and one P_B and little else, involves about as much computation as a single gradient step.

How can the progress of training be monitored when there is no loss function? In this and any application of DR, we can do exactly what we did for sudoku: Measure progress by the behavior of the gap, the currently achieved distance between a pair of points in sets A and B. Points $p_A \in A$ and $p_B \in B$ now correspond to an assignment of values to all the replicated x, y, and w variables. If you take a moment to review the deconstruction and what it means when $p_A = p_B$, you will realize that the problem has been solved: A concurring set of replicated variables that satisfy all the constraints at the neurons and the input/output nodes.

Figure 1.11 shows the progress of the gap for the $n = 32$ binary autoencoder problem. Unlike the loss in gradient descent, the gap does not decrease in every iteration. When it plummets toward zero, the training is complete. There is no point in plotting a distribution analogous to Figure 1.9 because

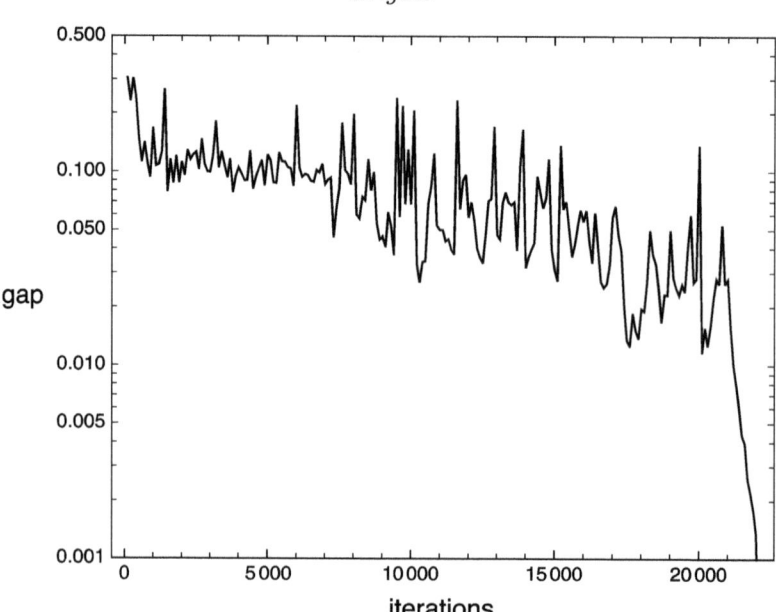

Figure 1.11 Progress of the gap in the training of the deconstructed auto-encoder by DR iterations.

zero-gap means all the concur projections (P_B) of the neuron outputs match the binary values dictated by the neuron constraints (P_A). Finally, if you review the DR formula (1.1), you will see that a gap of size zero means the dynamics has arrived at a fixed point.

I could also mention that training the deconstructed network using DR is significantly faster than gradient descent training, but that misses the more important point that even after reaching zero loss the gradient-descent trained autoencoder has not learned to encode all the data into binary codes. This is a serious shortcoming when the top decoder-half of the trained auto-encoder is to be used as a generative model. A generative model is supposed to generate the data when a simple code is sampled at the decoder's inputs. This fails when the codes generated in the autoencoder training have the distribution of Figure 1.9. Another benefit of the interpretability of the codes generated in the deconstructed problem is that the learned weights are interpretable as well, thereby eliminating, in this instance at least, the black-box nature of the data representation. To achieve this to the fullest extent, a norm constraint on the weights and a margin separating the neuron outputs must be imposed as well. These are minor tweaks in the definitions of A and B and covered in Section 7.7.

There is no evidence that either training method – by gradient descent on loss or DR iterations on the deconstructed problem – is relevant for *natural* neural networks. The method used by biology is probably none-of-the-above. On the other hand, the method that makes the most sense from an engineering perspective can and should be debated. But the gradient method is embedded so deeply in our technology, it is unrealistic to imagine it being supplanted any time soon. We can draw a parallel with the long reign of the Roman Empire. Similar to the many amenities provided by the main purveyors of machine learning, the Romans kept the roads paved and water in the viaducts flowing. However, they also imposed – and here my analogy is inspired by the binary encoding exercise – a rather poor system for representing numbers! Though 2023 may prove to be a pivotal year in the deployment of artificial intelligence, the core technology may in introspect be remembered as vintage MMXXIII.

The Roman system of representing numbers held sway for hundreds of years. If this depresses you, then you might find comfort in a story where "One small village of indomitable Gauls still holds out against the [Roman] invaders" [47]. The villagers owed their ability to resist assimilation, in part, to a magic potion that conferred superhuman powers. While I cannot promise you superhuman powers, and there is nothing magical about the skills you will learn, the new method of solving problems will at times seem miraculous.

Exercises

1.1 Though a whole chapter is devoted to the Douglas–Rachford formula (1.1) and an important generalization, now[7] is a good time for you to establish the very direct relationship between fixed points and solutions. If x^* is a fixed point, then

$$x^* = x^* + P_B(R_A(x^*)) - P_A(x^*) \,.$$

(a) Use this equation to write down an explicit expression for a solution $x_{\text{sol}} \in A \cap B$. Your expression will *involve* x^* but will not be as simple as $x_{\text{sol}} = x^*$.

(b) As a first example of the relationship between fixed points and solutions, consider a problem in the plane (two variables). For A take an infinite line and for B a single point on that line. The solution x_{sol} is obvious, but what about the fixed points x^*?

1.2 In the deconstruction of the disk packing problem, the radii were allowed to vary, and the projection P_B was tasked with restoring their true values (those of the original puzzle). Suppose the puzzle has three disks whose radii comprise the set $R = \{1, 2, 3\}$. Let (r_1, r_2, r_3) be the vector of radii that P_B will act on. In the complete problem P_B does a lot more, but in this exercise we just focus on the restoration of the disks' radii. For example, it might turn out that

$$P_B(r_1, r_2, r_3) = (2, 1, 3) \,,$$

or some other permutation of the numbers in R. You don't know which is the correct one until you are given the numbers r_1, r_2, and r_3. The projection $(2, 1, 3)$ is correct if the Euclidean distance, or its square,

$$(r_1 - 2)^2 + (r_2 - 1)^2 + (r_3 - 3)^2 \,,$$

is the smallest possible for the particular (r_1, r_2, r_3).

(a) Work out the projection

$$P_B(\, 0.5 \,,\ 2.1 \,,\ -0.5 \,) \,.$$

While it's true that a negative radius makes no sense, it's important that projections can handle arbitrary inputs. Notice that the reflection R_A in the Douglas–Rachford formula can turn positive numbers negative.

(b) With three disks, there are only $3! = 6$ cases to consider. Can you find an efficient algorithm that avoids trying all $n!$ cases (!) when the packing has n disks?

2

Bipartisanship

If deconstruction is the craft, then the principle behind the craft is *biparti-sanship*. The sets A and B in a deconstruction capture very different characteristics of the solution. In fact, the prospects of a bipartisan solution are best when the "partisans" are defined by quite different criteria, instead of simple disagreements on particular ones. A signal (or image) and its Fourier transform are well known to have a complementary relationship. And so it is fitting that in the debut application (phase retrieval), one constraint is imposed on an image while the other constrains its Fourier transform.

Still, the bipartisan scheme seems strange, and you probably suspect that it only applies to very special kinds of problems. The examples in this chapter are meant to convince you otherwise. Hopefully, by the end, you will see that bipartisanship forces you to think along creative lines that open up novel strategies for solving a problem.

The first example is almost too simple: anagrams. How should the letters in

UTSOONLI

be rearranged to form an English word? If you are an average reader, you will have paused to solve the puzzle ... and are now ready to learn how anagrams can be formulated in terms of A and B. We make a rare exception, in this instance, of not revealing the solution.

There are $8!/2! = 20\,160$ distinct ways of rearranging the letters of UT-SOONLI, and about $6\,000$ eight-letter common English words. If we call the set of rearrangements A, and the entries of the eight-letter dictionary B, then an anagram solution is any element x that belongs to both A and B. In mathematical notation, the anagram problem may be stated as

$$\text{find some} \quad x \in A \cap B. \tag{2.1}$$

All problems in this book will be formulated in this way. The two sets will always be written A and B and a solution, a single element, will always be written as x. Often the intersection of A and B will have multiple elements and the challenge is to find any one of them[1].

The anagram example is too simple because A and B manifestly lack the structure that the computations in our search algorithm require. In these computations, A and B must be specified as sets in space, where space has some number of dimensions d, usually much larger than three. The solution x is then a point in this space, and represented by a list of d real numbers.

Though A and B will always be geometrical sets, the algorithm is not limited to problems whose sets are explicitly geometric, say where A and B are planes, spheres, and so on. Discrete problems can often be formulated in terms of geometric sets as well. In our anagram problem, x would be an 8×26 table of numbers, one row for each of the eight letters. For example, to represent the word BIRTHDAY (an element of B but not A), the table would look like this:

$$
\begin{matrix}
0 & 1 & 0 & \cdots & 0 & 0 & 0 \\
\vdots & \vdots & \vdots & & \vdots & \vdots & \vdots \\
0 & 0 & 0 & \cdots & 0 & 1 & 0
\end{matrix}
$$

The set B (dictionary) for anagrams does not have a nice geometrical definition – just a collection of some 6 000 oddball points, all with 0 and 1 coordinates, in a space with $d = 8 \times 26 = 208$ dimensions.

When referring to A and B as geometrical sets, we made the assumption that there is a distance between any pair of points x and x' which has the Euclidean form:

$$
(\text{distance})^2 = \|x - x'\|^2 = \sum_{i=1}^{d}(x_i - x_i')^2. \tag{2.2}
$$

The core computations of the algorithm, the projections P_A and P_B, make use of the distance to find the nearest point in A or B from any given x. In the anagram problem, there is no better way to find the nearest dictionary word – the projection to B – than to try out all 6 000 word-points. We should not expect the algorithm to offer an advantage over exhaustive search when the projections themselves are similarly exhaustive. Fortunately, most of the sets we will need to formulate problems are defined by general geometric properties, and for such sets the projections can be computed very efficiently. For example, a much used discrete set is one where each of the coordinates corresponds to an independent binary choice having (conventional) values 0 and 1. Projecting to such a set involves nothing more than rounding each coordinate to the nearest bit.

The sudoku application you saw in Chapter 1 should amplify the remarks in the previous paragraph. Of the many bipartisan formulations of sudoku, the one featured in Chapter 1 is the most economical in terms of the number of dimensions: $d = 9^3 = 729$. Like the anagram problem, both A and B are sets of points with $0/1$ coordinates. But the number of points is astonishing: $|A| = |B| = (9!)^9 \approx 10^{50}$! [2] To be more precise, this is the size of these sets when there are no given numbers in the puzzle. In that case the solution is not unique, and in fact the precise number of solutions, $|A \cap B| \approx 6.7 \times 10^{21}$, has been worked out [69]. The given numbers in published puzzles guarantee that $|A \cap B| = 1$ and also reduce $|A|$ and $|B|$. For example, in a hard NYT puzzle, $|A| \approx |B| \approx 10^{29}$. But these are still far, far too large if finding the nearest element required trying out each one. Fortunately, the structure of these sets is such that finding the nearest element can be done very efficiently, though the computation is not quite as easy as rounding coordinates to the nearest bit. As described in Section 4.6, considerable cleverness is involved in what turns out to be a standard task in computer science. The tricky part of the projections P_A and P_B can therefore be relegated to black-box software. In contrast to the enormous sizes of the sets that are within the scope of the Douglas–Rachford method, the dimension d in which these sets live cannot also be fantastically large. That's because one needs to be able to store the coordinates of the points x, $P_A(x)$ and $P_B(x)$.

The numbers within x will come packaged in many different ways. They may be vectors, matrices, structures associated with graphs, and the numbers themselves might be complex valued. In all these cases, the distance always reduces to a sum of squares of real numbers. There is a standard mathematical "declaration" for the packaging. In the anagram example, it would be $x \in \mathbb{R}^{8 \times 26}$. When retrieving the phases of a Fourier-transformed image of 256×256 pixels, we declare $\hat{x} \in \mathbb{C}^{256 \times 256}$. In this last example, the distance between (Fourier-transformed) images $\hat{x}$ and $\hat{x}'$ takes the form

$$(\text{distance})^2 = \|\hat{x} - \hat{x}'\|^2 = \sum_{i=1}^{256} \sum_{j=1}^{256} |\hat{x}_{ij} - \hat{x}'_{ij}|^2 \,, \qquad (2.3)$$

where $|z|$ now denotes the magnitude of the complex number z.

We now begin a survey of bipartisan formulations, where both A and B have enough structure that the projections can be computed efficiently. Because phase retrieval launched the subject, we will begin with that.

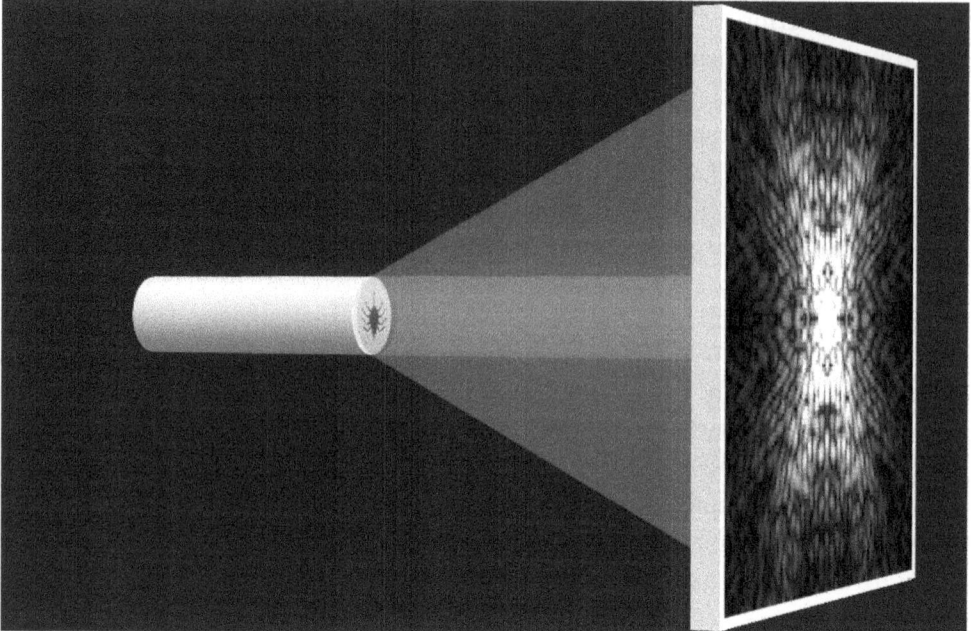

Figure 2.1 The wave microscope. Light with very pure wave content –
a laser beam – is incident on the specimen, in this case a spider. Waves
scattered by the specimen and recorded on a detector create very distinctive
patterns, called diffraction intensities.

2.1 Microscopy with Waves

In a conventional microscope, rays of light originating from points in the
specimen are refracted by lenses to create the appearance that the rays are
emanating from points that have a greater separation than in the specimen.
The size enhancement of the "virtual specimen" is the microscope's magnification. In conventional microscopes, the magnifying power has an upper
limit set by the wavelength of light. The notion of magnification, for features
of size approaching the wavelength, does not apply because the ray model of
light breaks down at that scale.

A very different kind of microscope design, that in principle can operate
all the way to the fundamental wavelength limit, is shown in Figure 2.1.
There are no lenses. All information comes from a record of the light waves
produced when light of very pure wave content, such as produced by a laser,
is scattered by the specimen. Thinking of light waves as the wave packets
of individual photons, the low/high spatial frequency features in the plane
of the specimen impart small/large sideways kicks to the incident photons.
The result is that the photons arriving at the center/edge of the detector

carry information about the low/high frequency content of the specimen. The record of detected photons over the plane of the detector is called the *diffraction intensity.*

The smallest features in the diffraction intensity, called "speckles," are easily resolved just by moving the detector far enough away from the specimen. But the pattern formed by the speckles is not a magnified image of the specimen at all. To complete the wave microscope, we need another stage, a purely computational one, that reconstructs the specimen's *contrast* from the diffraction pattern data. The symbol x in statement (2.1) now represents this contrast, and our task is to use the diffraction data to define the sets A and B.

Students of physics may notice a similarity between the speckled nature of the diffraction intensities produced by the spider, and the diffraction patterns formed when a laser shines on a screen with a complex aperture. In fact, if the specimen's opacity is modeled as two-valued (0% and 100% transmission), then by Babinet's principle,[3] the diffraction pattern recorded by the wave microscope would exactly match that of a spider-shaped aperture. Our wave microscope will handle a continuum of opacity values across the specimen, and it is this property that we identify with the contrast x. Mathematically, x is an $\ell \times \ell$ array of numbers, where ℓ defines the resolving power of the wave microscope. Since opacity is nonnegative, we always have the constraint

$$x \geq 0 . \tag{2.4}$$

This is standard notation when a property (nonnegativity) applies to every element of x.

The waves or photon wave packets arriving at any point on the detector have two characteristics: an amplitude and a phase (relative to the reference defined by the incident wave). These two pieces of information can be combined into an array of complex numbers, $\hat{x}$, called the *diffraction amplitudes*:

$$\hat{x} = |\hat{x}| e^{i\varphi} .$$

This is another instance of the compact notation: The array of magnitudes $|\hat{x}|$ is multiplied element-wise by the array of phasors $e^{i\varphi}$.

Physics students learn that the arrays x (contrast) and $\hat{x}$ (diffraction amplitudes) are related by the Fourier transform $\mathcal{F}$. Because we are interested in computations, we choose $\hat{x}$ also to be an $\ell \times \ell$ array. The two arrays then carry exactly the same information, because from one we can always obtain the other:

$$x \underset{\mathcal{F}^{-1}}{\overset{\mathcal{F}}{\rightleftharpoons}} \hat{x} . \tag{2.5}$$

In practice, it is the linear size of the detector, in pixels, that defines the array size ℓ. The contrast x, obtained computationally, lives on an array of that same size.

Equation (2.5) seems in conflict with one array being real valued and the other complex valued, nominally having twice as much information. This is resolved by a symmetry property of $\hat{x}$. Consider a pair of detector pixels with coordinates $\pm\mathbf{q}$, so diametrically opposite the detector center at $\mathbf{q} = (0,0)$. By a general property of the Fourier transform, and when x is real-valued, the magnitude and phase of $\hat{x}$ have the following symmetry properties:

$$|\hat{x}(-\mathbf{q})| = |\hat{x}(\mathbf{q})| \,, \tag{2.6a}$$
$$\varphi(-\mathbf{q}) = -\varphi(\mathbf{q}) \,. \tag{2.6b}$$

That is, between the pixels at $\pm\mathbf{q}$ there are just two, not four, independent real numbers.

The light arriving at the detector oscillates too fast to establish a phase φ (relative to the incident light) at each pixel. All that the detector can measure is the time-averaged property of the light intensity I or, equivalently, the arrival rate of photons at each pixel. Because I is proportional to $|\hat{x}|^2$, from the square root $\sqrt{I}$ one can obtain $|\hat{x}|$ at each pixel. However, without $\varphi = \arg\hat{x}$, the Fourier transform $\hat{x}$ is incomplete and cannot be inverse-transformed to produce x (the spider contrast).

By examining properties of the diffraction amplitudes on a very fine scale, corresponding to impractically tiny detector pixels, we can see that some information about the phase is present in the intensity itself. Figure 2.2 shows both the magnitude (left) and phase (right) of the spider diffraction amplitudes near the center of the detector. The phase rendering is confusing because of the sharp black/white discontinuities wherever φ crosses over from $+\pi$ to $-\pi$. These rendering artifacts should be ignored, if possible, with the exception of some special points where the discontinuity curves terminate. These points coincide with the crossings of two sets of contour curves superposed on the magnitude plot. One set of contours shows where the real part of $\hat{x}$ is zero, the other where the imaginary part is zero. You should convince yourself that both the magnitude and phase rendered in Figure 2.2 are consistent with the symmetries expressed by equations (2.6).

At the crossings of contours, both real and imaginary parts of $\hat{x}$ are zero, and therefore $|\hat{x}|^2$ is zero. Something special happens with φ at all the places in the detector plane where the intensity is zero. That's because φ switches between 0 and π on the $\text{Im}(\hat{x}) = 0$ contour right where it crosses the $\text{Re}(\hat{x}) = 0$ contour. Similarly, φ switches between $\pi/2$ and $-\pi/2$ on the

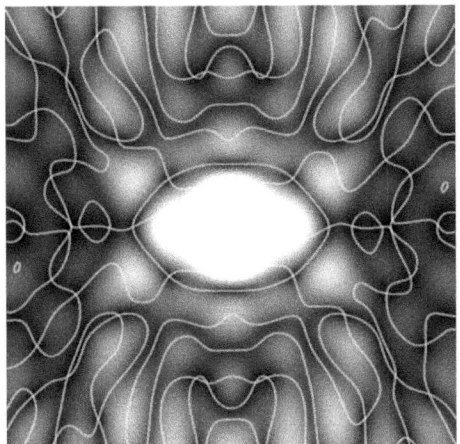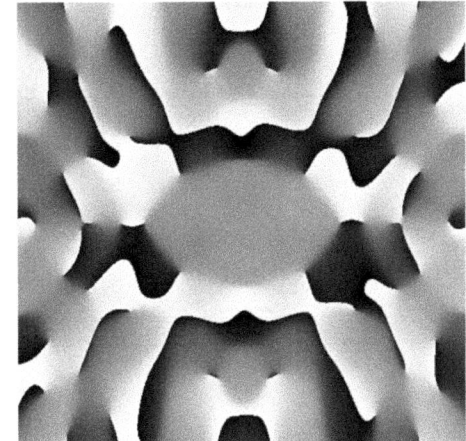

Figure 2.2 Magnitude (left) and phase (right) rendering of the spider diffraction amplitudes $\hat{x}$ near the center of the detector. Superposed on the magnitude are contours corresponding to $\mathrm{Re}(\hat{x}) = 0$ and $\mathrm{Im}(\hat{x}) = 0$.

$\mathrm{Re}(\hat{x}) = 0$ contour right where it crosses the $\mathrm{Im}(\hat{x}) = 0$ contour. This is neatly summarized by the statement that φ goes around a complete circle (clockwise or counterclockwise) when $\hat{x}$ is sampled on a path that winds around one of the special points where $I = 0$. The 2π phase winding is seen more directly in the phase rendering on the right, though without being able to see the coincident vanishing of $|\hat{x}|$. We are dwelling on these phase windings because it is a large effect (2π) that happens in arbitrarily small regions of the detector plane.

To understand the consequences of the large phase changes at the zero-intensity points, we turn to the simple one-dimensional example shown in Figure 2.3. For simplicity, we consider a constrast x that is symmetric about the origin, so that $\hat{x}$ is real valued and φ is either 0 or π. The top panel shows how $\hat{x}$ is obtained from the product of the magnitude $|\hat{x}|$ (solid curve) and the phasor $e^{i\varphi}$, which because of the symmetry is either $+1$ or -1 (dashed curve). Notice that $|\hat{x}|$ has cusps wherever the intensity $I \propto |\hat{x}|^2$ is zero. When multiplied by $e^{i\varphi}$ the resulting $\hat{x}$ is a much smoother curve. In the two-dimensional spider diffraction pattern, the 2π rotation of φ at $I = 0$ points accomplishes exactly the same thing: a very smooth $\hat{x}$ (in both its real and imaginary parts).

Figure 2.4 shows what goes wrong when the magnitude $|\hat{x}|$ is multiplied by the wrong $e^{i\varphi}$. Symmetry does not dictate where $e^{i\varphi}$ switches sign, and if

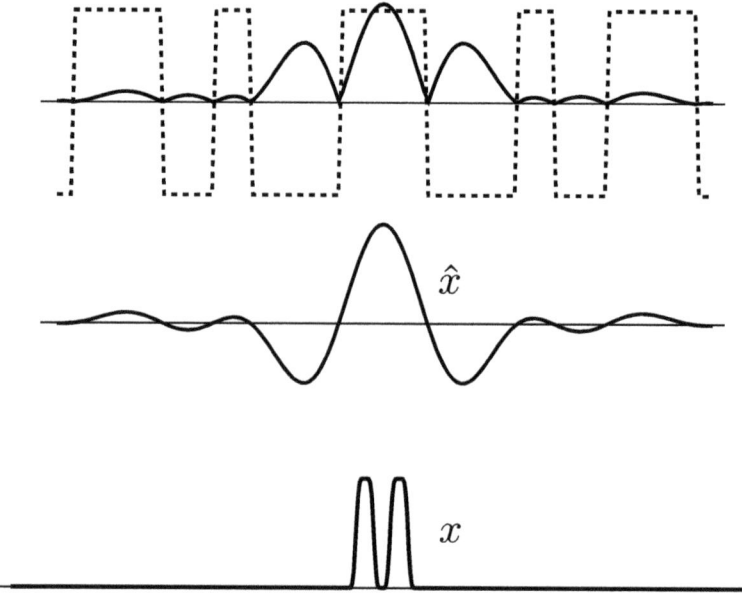

Figure 2.3 A simple one-dimensional $\hat{x}$ (middle) and how it is obtained from the product of $|\hat{x}|$ and $e^{i\varphi}$ (top). The Fourier transform of the smooth $\hat{x}$ is shown at the bottom.

the sign switches do not coincide with the cusps in $|\hat{x}|$, the resulting product, $\hat{x}$, will not be very smooth at all.

"Smoothness of the Fourier transform" seems like a strong constraint that could be used to determine the unknown φ. But in two dimensions, and without centro-symmetry to make $\hat{x}$ real valued, this is easier said than done! Instead of a two valued and real $e^{i\varphi}$ that flips in sign at all the zero-intensity points, the precise way that φ continuously rotates by 2π at these points must be specified in detail. Missing also is a strategy for stitching together the φ pattern from one special point to the next.

A bipartisan strategy should be considered when a direct approach – make $\hat{x}$ smooth by conjuring a magic φ – seems too hard. The bottom panels in the previous two figures show the contrast x that results from Fourier transforming the smooth (correct) and discontinuous (wrong) $\hat{x}$. These differ qualitatively in a way that relates directly to the smoothness of $\hat{x}$. The correct $\hat{x}$ has very little high-frequency content, and so its (inverse) Fourier transform, x, is confined to the region of low frequencies (at the center of the plot). But the discontinuities in the wrong $\hat{x}$ are high-frequency features, and we are not surprised to see that its Fourier transform is broadly distributed. The technical term for the region in which a signal (such as x) is nonzero

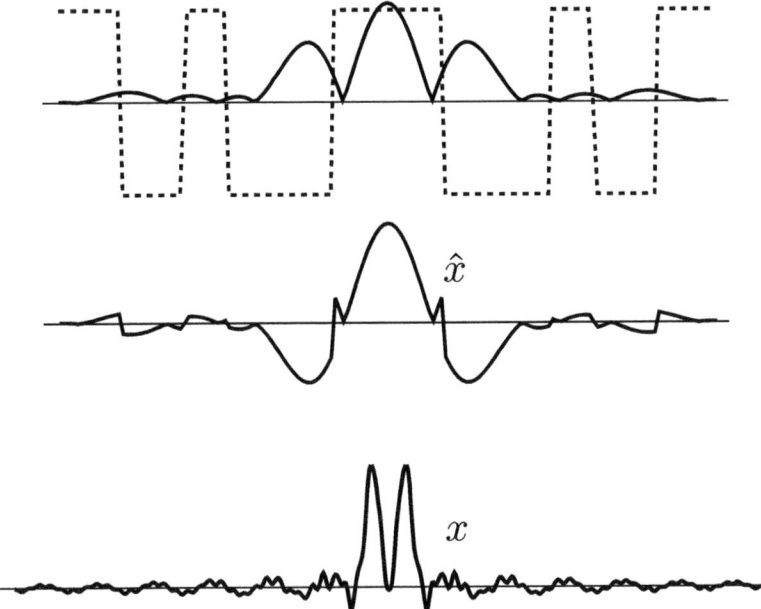

Figure 2.4 Same as Figure 2.3 but with the wrong φ. The resulting $\hat{x}$ is not smooth, and when Fourier transformed, does not have a small support.

is *support*. From the example we learned that the difference between the correct and wrong $\hat{x}$ is something that is more easily expressed in terms of x: The correct $\hat{x}$ is special in that the corresponding x has an unusually small support!

The support-size insight provided by the one-dimensional example suggests the following bipartisan scheme: Let A recklessly define any contrast consistent with the intensity data. The bad contrasts this includes, when the magnitudes $|\hat{x}|$ are combined with the wrong phases, is something that will be handled by constraint B. In fact, a constraint on the support size of x seems like a very good candidate for B.

Using mathematical notation, here is the proposed definition of set A:

$$A = \left\{ x \in \mathbb{R}^{\ell \times \ell} : |\hat{x}| = \sqrt{I} \right\}. \tag{2.7}$$

We first see the declaration that the contrast x (and therefore set A) lives in a space of ℓ^2 dimensions (the number of contrast/diffraction pattern pixels). As in most applications, this number is much too large to render A for human inspection. We will need another approach to develop an intuitive grasp of this set.

It is easy to create elements of A. For example, we might wonder whether the unmeasured phase angles matter at all. To test this, we can construct contrasts x from the known Fourier magnitudes and phases φ of our choice:

$$x(\varphi) = \mathcal{F}^{-1}\left(\sqrt{I}\exp(i\varphi)\right).$$

The phases φ can be random, taken from the Fourier transforms of our favorite bugs, or derived from some other source. The only restriction, so the resulting contrast is real valued, is the symmetry property (2.6b). Three elements of set A constructed in this way are shown in Figure 2.5, where the first (top row) is formed with the true phases of the spider.

It will not have escaped the reader's notice that the two examples of non-spider phases produce contrasts that cover most of the image, much like we saw in one dimension (Figure 2.4). The true phases are special in producing zero contrast ($x = 0$) over much of the $\ell \times \ell$ image, that is, a contrast with small support. We can reexpress this observation in terms that bear directly on the wave microscope. For suppose the spider was caught in a literal web of weakly scattering contrast. In that case the correct phases would need to reproduce that noisy background, no matter how peripheral to the object of interest. It therefore needs to be stated, unequivocally, that the microscopist has gone to great lengths to ensure the data is recorded while the spider is in a pristine, zero-contrast environment. Not only that, the size ℓ of the detector/image grid has to be set at a sufficiently large value that the zero-contrast surrounding the spider can be imposed as a constraint. The power of this last statement, which may seem like an innocent mathematical detail, cannot be overstated.

Consider what is involved to make x zero on a large number of pixels. To form x from $\hat{x}$ by Fourier synthesis, ℓ^2 waves are combined with a Fourier coefficient derived at each detector pixel from a known magnitude and an unknown phase. But conjuring phases for the Fourier synthesis to produce perfect cancellation ($x = 0$) at all those pixels surrounding the spider seems like a highly nontrivial task!

Formalizing the wholesale cancellation phenomenon defines the set B. Instead of data, this set is defined by a region $\mathcal{S}$ called the *support*, not because the spider is a support animal, but because this term defines the region where the contrast is allowed to be nonzero. The support is a compactly shaped subset of the $\ell \times \ell$ image pixels that the microscopist is confident will contain the contrast of the specimen. As a small detail, we also include in the definition of set B the property (2.4) that spider contrast is never negative. Using $\mathbf{p}$ for a general image pixel, here is the definition of set B:

$$B = \left\{x \in \mathbb{R}^{\ell \times \ell} : x(\mathbf{p}) \geq 0 \text{ if } \mathbf{p} \in \mathcal{S}, \text{ otherwise } x(\mathbf{p}) = 0\right\}. \qquad (2.8)$$

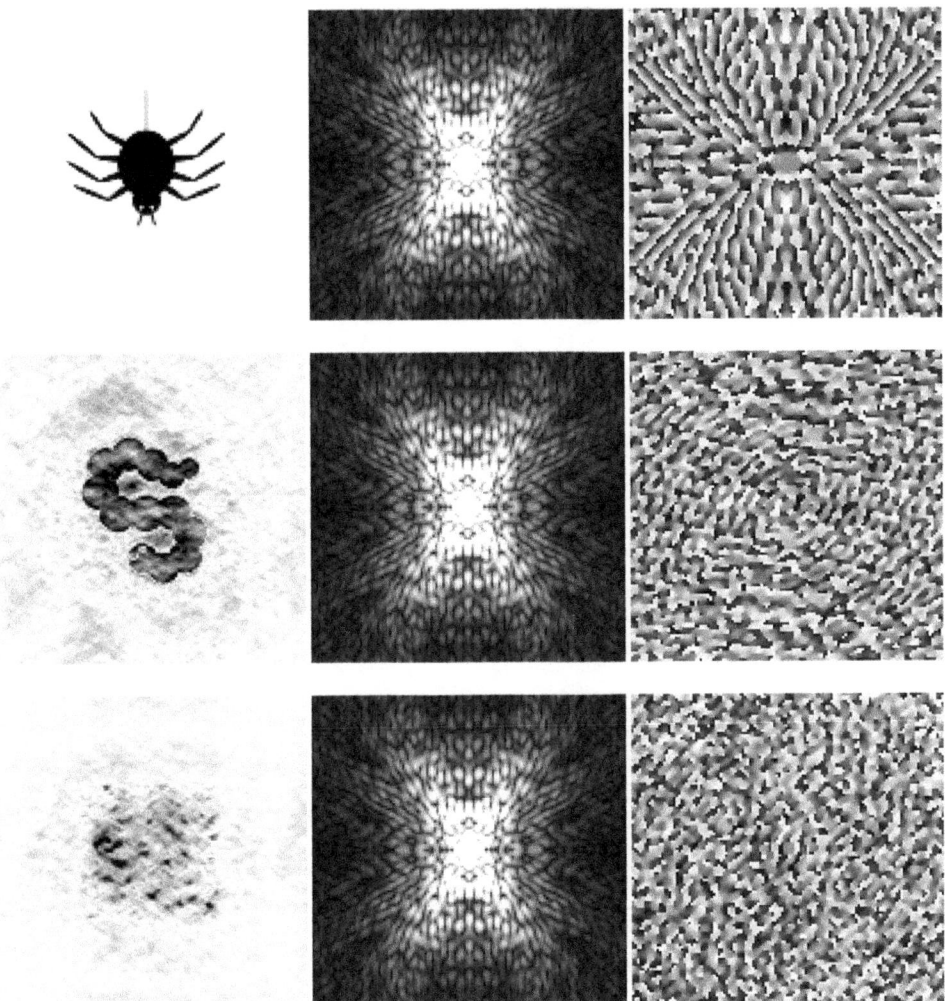

Figure 2.5 A sampling of three points or elements of the set A. With each contrast x (left) is shown the Fourier magnitude $|\hat{x}|$ (middle) and phase φ (right). Set A is defined by the fact that the magnitudes match the wave microscope data. On the other hand, set A knows nothing about the true Fourier phases. A spider results when the spider's phases are combined with the data magnitudes (top row), and mysterious contrasts are produced when other phase choices are used (other rows). The spider contrast is special in that it is not surrounded by a gray haze that fills the entire image.

Set B is easier to grasp than set A, but we repeat what we did with set A, now in Figure 2.6. In set B there is a restriction only on the contrast x being nonnegative and confined to the support region $\mathcal{S}$, shown with a dashed

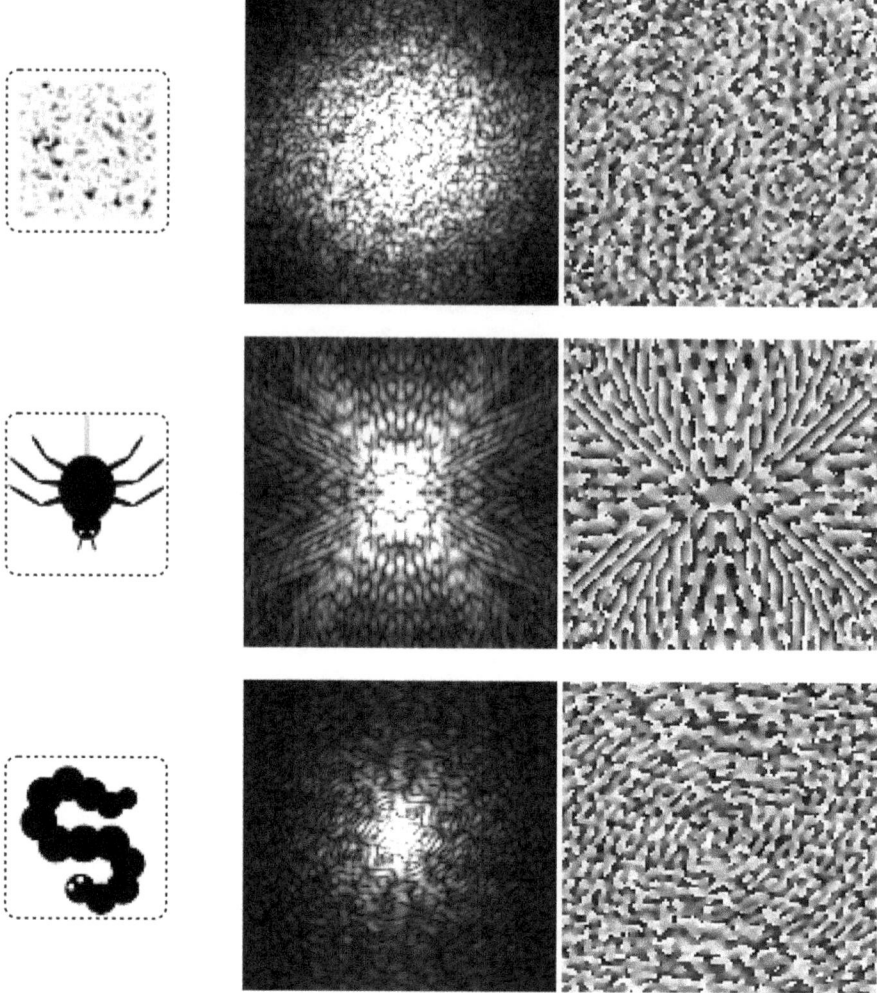

Figure 2.6 A sampling of three points or elements of the set B. This set is defined by the property that the contrast x (left column) is zero outside a support region $\mathcal{S}$ (dashed outline) and is nonnegative inside $\mathcal{S}$. Set B includes spiders, other bugs, and completely random patterns. Notice that the six-limbed spider has a slightly different Fourier magnitude $|\hat{x}|$ (middle column) than the eight-limbed one in Figure 2.5, and therefore this point is not an element of set A.

outline. The three samples all have Fourier magnitudes $|\hat{x}|$ that differ from those of the eight-limbed spider (Figure 2.5) and therefore are not elements of set A.

From our modeling of the wave microscope and the definitions of the sets A and B, solutions x of (2.1) would be very strong candidates of the specimen contrast. Even without knowing at this stage how to solve (2.1), or what can be said about the uniqueness of those solutions, it is clear that the mode of operation of a wave microscope is very different from a conventional microscope. Only the diffraction magnitudes are produced by a physical process and these bear no likeness to the specimen. Any hope of obtaining the latter requires an algorithm that can solve (2.1). Seen as an instrument, a wave microscope comprises a physical and a computational stage. The computational stage is known as *phase retrieval* because the reconstruction of x, given only the magnitudes $|\hat{x}|$, also reconstructs the phases $\arg(\hat{x})$.

We can now see why wave microscopy is such a great example of bipartisanship. While both constraints apply to $\ell \times \ell$ tables of numbers, in one we are obsessed about their magnitudes (as complex numbers), in the other we just want most of them to be zero. That such different properties apply to nominally the same thing (a spider) is possible because Fourier-transform pairs (x and $\hat{x}$) capture very different facets of that thing.

It would be premature to declare our sets A and B a valid bipartisan formulation without checking one important detail. If you haven't already, now is a good time to ask why we didn't simply define the set

$$C = \left\{ x \in \mathbb{R}^{\ell \times \ell} : |\hat{x}| = \sqrt{I}, \text{ and} \right.$$

$$\left. x(\mathbf{p}) \geq 0 \text{ if } \mathbf{p} \in \mathcal{S}, \text{ otherwise } x(\mathbf{p}) = 0 \right\}.$$

Pick any element of C and we have a solution (properly supported contrast having the correct diffraction intensities). The problem with this is that unlike sets A and B, it is very difficult to "just pick" an element of C. Being able to easily pick random elements is important, because calculating projections is at least as hard. Recall that the projections P_A and P_B are the basis of the algorithm that we will use to solve (2.1).

It's easy to see that there's an efficient projection to set B. This is a constraint on the contrast pixels, and the constraints are independent. If $\mathbf{p} \notin \mathcal{S}$ and $x(\mathbf{p}) \neq 0$, then P_B sets this pixel's value to zero. On the other hand, if $\mathbf{p} \in \mathcal{S}$, then the constraint is violated only if $x(\mathbf{p}) < 0$. The distance-minimizing change to restore the constraint in this case is also to set the pixel's value to zero (a positive value, while also allowed, would involve a greater distance).

The constraints in set A are also independent, now on the pixels of the Fourier transform $\hat{x}$. But these are all nonconvex constraints and as explained in Chapter 1, projections to nonconvex constraints can be a source of chaos. Chaos helps when a diverse variety of phase combinations must be searched to find the one that unlocks the contrast when combined with the magnitudes. Without chaos our algorithm might get stuck on the image of a six-limbed spider even if the data imposed in the A constraint was collected from an eight-limbed spider!

A wonderful feature of the Fourier transform (Parseval's theorem) is that if a change to $\hat{x}$ is distance minimizing for the distance (2.3), then the change in the corresponding x is distance minimizing for its sum-of-squares distance. To work out projections in phase retrieval it therefore doesn't matter whether we compute with x or $\hat{x}$. Because set A constrains $\hat{x}$, we will be minimizing distance in the complex plane for this projection.

Suppose constraint A is not satisfied at detector pixel $\mathbf{q}$:

$$|\hat{x}(\mathbf{q})| \neq \sqrt{I(\mathbf{q})} \ .$$

The projection P_A finds the complex number $\hat{x}'(\mathbf{q})$ that satisfies

$$|\hat{x}'(\mathbf{q})| = \sqrt{I(\mathbf{q})} \tag{2.9}$$

and also minimizes the change, that is, the distance

$$|\hat{x}'(\mathbf{q}) - \hat{x}(\mathbf{q})| \ .$$

This is an easy problem in plane geometry: For an arbitrary input point $\hat{x}(\mathbf{q})$, find the nearest point $\hat{x}'(\mathbf{q})$ on the circle defined by (2.9). The solution is

$$\hat{x}'(\mathbf{q}) = \left(\frac{\sqrt{I(\mathbf{q})}}{|\hat{x}(\mathbf{q})|} \right) \hat{x}(\mathbf{q}) \ .$$

We can see that the change is especially sensitive to the projection input, $\hat{x}(\mathbf{q})$, when this has a small magnitude and $\sqrt{I(\mathbf{q})}$ is not also small. The output $\hat{x}'(\mathbf{q})$ is then a point on a large circle whose angle is random because it was inherited from the largely meaningless phase angle on the small circle.

This completes the bipartisan formulation of wave microscopy, a special case of the phase retrieval problem. You might object that nothing has been said about solving (2.1), and whether the solutions will be unique – an important practical consideration! While these are important questions and addressed in later chapters, a significant part of the problem was solved simply by coming up with the definitions (2.7) and (2.8).

2.2 Hadamard Matrices

The Hadamard matrix problem should, conceptually, have started the chapter and was preceded by phase retrieval only to reflect the history of the subject. Whereas formulating a problem in the form (2.1) in many cases is very direct and natural, only recently was this characterization deemed useful when A or B is nonconvex. Phase retrieval served as the stimulus because the only successful algorithms for solving that nonconvex problem were algorithms that solved the problem when expressed in bipartisan terms.

A Hadamard matrix of order n is an orthogonal $n \times n$ matrix with only ± 1 elements. This one-sentence definition is an instance of (2.1) if we simply let A be the set of $n \times n$ orthogonal matrices, rows and columns with squared-norm n, and let B be the set of $n \times n$ matrices with ± 1 elements:

$$A = \left\{ x \in \mathbb{R}^{n \times n} : xx^{\mathsf{T}} = nI \right\}, \tag{2.10a}$$

$$B = \left\{ x \in \mathbb{R}^{n \times n} : x \in \{-1, 1\}^{n \times n} \right\}. \tag{2.10b}$$

A Hadamard matrix of order 28 is shown in Figure 2.7. Up to symmetries, there are just 487 such matrices [82]. This one was found by iterating the Douglas–Rachford rule (1.1) for the sets A and B above starting with a random initial x. More precisely, to find this Hadamard and (reliably) matrices of order $n > 8$, we had to use the small-time-step modification of Douglas–Rachford:

$$x \mapsto x + \beta \Big(P_B(R_A(x)) - P_A(x) \Big). \tag{2.11}$$

For the Hadamard in Figure 2.7 we used $\beta = 0.002$. The rationale behind taking small steps is explained in Section 5.5.2.

Whereas any element $x \in A \cap B$ is a Hadamard matrix, our bipartisan formulation is incomplete without the assurance that the projections P_A and P_B can be computed efficiently. In Chapter 1 we saw the B projection applied to vectors in the bit retrieval problem. The distance minimizing operation, for both shapes of x, is to round all elements to either -1 or $+1$ depending on sign.

The math behind the efficient projection to set A is more sophisticated and covered in Section 4.4.3. For now, you will have to be satisfied to know that at least it is easy to sample the elements of A. Start by generating a random vector of real numbers. After normalizing it to have squared-length n, it becomes the first row of x. Do the same thing for the second row, except that before normalizing, orthogonalize with respect to the first vector. Continue doing this for all the other rows, orthogonalizing with respect to all the previous rows. Orthogonalizing a new vector with respect to a set of vectors that are already orthogonal, known as the Gram–Schmidt process, is an easy computation. You should now be convinced that A is not a terribly

Figure 2.7 A random Hadamard matrix of order 28. Take any pair of rows or columns and check that they have exactly 14 matching and 14 contrasting elements.

n	1	2	4	8	12	16	20	24	28	32
number	1	1	1	1	1	5	3	60	487	13710027

Table 2.1 Number of inequivalent Hadamard matrices up to the highest order where this is known [82].

complicated set and because of that it is quite likely that an element $P_A(x)$ nearest to x can also be computed efficiently.

Unlike set A, the set of Hadamard matrices, $A \cap B$, is complicated and mysterious. Phase retrieval is often described as a *reconstruction* problem. Assuming the mathematical model is a good approximation of the physics and noise in the data is small, a solution consistent with the known information, as expressed by constraint sets A and B, is guaranteed to exists. The same thing cannot be said about Hadamard matrices. There is nothing in the definition of A and B that guarantees the existence of solutions. Rather than reconstructing something known to exist, the task is to construct something that is only conjectured to exist. The Hadamard matrix conjecture [87] asserts these matrices exist for all orders n divisible by 4, with $n = 668$ currently the smallest unresolved case [85].

The unknown existence of Hadamard matrices of large order stands in sharp contrast with what is known about low orders. Table 2.1 documents the explosive growth in the number of inequivalent Hadamard matrices for small n. Matrices are considered equivalent if they can be related by row/column permutations and sign flips of whole rows and columns. As no one has come up with an argument why the numbers should go back down, it is embarrassing that for $n = 668$ we do not have even one example!

One explanation of the grave international shortage of Hadamard matrices of large order is the nature of the known methods for constructing them. These all exploit special combinatorial structures, and are therefore limited by the imaginations of the mathematicians who come up with these structures. When combined, this zoo of constructions does not even establish a positive asymptotic density of orders where Hadamards exist [26]!

Though the bipartisan formulation (2.10) and the Douglas–Rachford rule for finding elements of $A \cap B$ will never resolve the Hadamard matrix conjecture, this method appears to be uniquely qualified for sampling the entire set of Hadamards, that is, without the restrictions imposed by combinatorial structures.

The fact that x lives in a continuous space, and is represented with floating point numbers in computations, should not detract from the applicability of the method to discrete problems (Hadamard matrices, bit retrieval, etc.). When the gap $\|x_A - x_B\|$ separating points $x_A \in A$ and $x_B \in B$ in the current Douglas–Rachford iteration falls below a small threshold, one can check directly whether x_B, with its exact ± 1 coordinates, is a proper Hadamard matrix (which will be highly likely when the gap is small).

2.3 Kissing Spheres

Given the geometric nature of the problem statement (2.1), in terms of a pair of sets in Euclidean space, and the distance-minimizing projection operations at the core of the search algorithm, we should expect geometric problems to be among the most natural applications. Geometric problems with a combinatorial character, where some form of exhaustive search is called for, might be especially interesting. As the variables are not discrete, a complete exploration by branching is not possible for such problems. But as we have hinted repeatedly, chaotic dynamics in a continuous space can serve the same purpose.

Figure 2.8 shows arrangements of "kissing spheres" in two and three dimensions. We use the term "sphere" in all dimensions, even when "disk" is available in two. Six spheres can "kiss" a central sphere in two dimensions with no room to spare under the constraint that they do not intersect. In this version of the problem, all spheres have the same diameter. The situation in three dimensions is qualitatively different. The 12 kissing spheres are

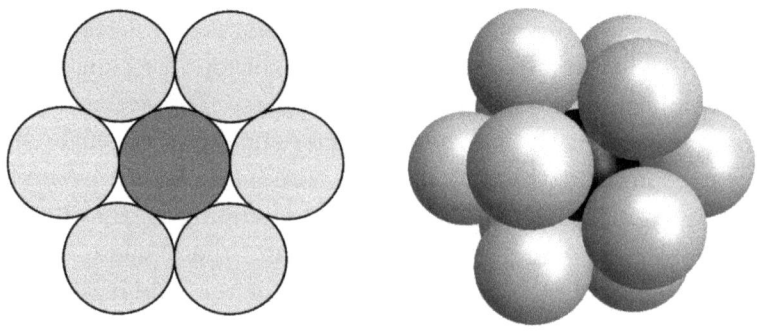

Figure 2.8 Kissing spheres (or disks) in two and three dimensions.

free to move continuously and there is even sufficient room for them to swap
positions while maintaining contact with the central sphere [25].

Isaac Newton and David Gregory had a friendly dispute about kissing
spheres in three dimensions [21]. Newton was convinced that 12 kissing
spheres was the maximum possible while Gregory thought there was room
for a 13th. It turned out that Newton was correct but this fact was not
established rigorously until 1952 [94]. Based on these results for the *kissing number* (maximum number and configuration) in dimensions $d = 2$ and
$d = 3$, what should we expect for general d?

A bipartisan formulation (2.1) for the kissing spheres problem is not quite
as obvious as it was for the previous two applications. The process of arriving
at such a formulation in nonobvious cases is "deconstruction." Let's start with
mathematical statements of the constraints that apply to the position vectors
$v_1, \ldots, v_n$ of the n kissing spheres. All spheres, including the central one, have
radius $1/2$, so the kissing sphere centers lie on a sphere of unit radius. Placing
the central sphere at the origin, the position vectors all satisfy

$$i = j: \quad v_i \cdot v_j = 1 \,. \tag{2.12}$$

Intersections are avoided provided distinct spheres satisfy

$$i \neq j: \quad (v_i - v_j) \cdot (v_i - v_j) \geq 1 \,.$$

With the help of (2.12), this simplifies to

$$i \neq j: \quad v_i \cdot v_j \leq \frac{1}{2} .$$

(2.13)

To arrive at a bipartisan formulation, we notice that all the constraints above are compactly expressed in terms of an $n \times n$ matrix of inner products, or *Gram matrix*, with entries

$$x_{ij} = v_i \cdot v_j .$$

If we let $\mathbb{S}^n$ denote the set of $n \times n$ real and symmetric matrices, then a natural candidate for set A is

$$A = \left\{ x \in \mathbb{S}^n : x_{ij} = 1 \text{ for } i = j, \ x_{ij} \leq \tfrac{1}{2} \text{ otherwise} \right\} .$$

Clearly, we are not done because the dimension d has not appeared anywhere! Also, how do we extract the n position vectors v from the matrix x?

To resolve the missing-d problem, we express the position vectors more explicitly, in terms of their coordinates in d dimensions, as an $n \times d$ matrix v:

$$v = \begin{bmatrix} v_{1,1} & v_{1,2} & \cdots & v_{1,d} \\ v_{2,1} & v_{2,2} & \cdots & v_{2,d} \\ \vdots & \vdots & \ddots & \vdots \\ v_{n,1} & v_{n,2} & \cdots & v_{n,d} \end{bmatrix} .$$

In terms of v, the Gram matrix x has the following product structure:

$$x = vv^{\mathsf{T}} .$$

This shows that not only is x symmetric, it is also positive semi-definite because $yxy^{\mathsf{T}} = (yv)(yv)^{\mathsf{T}} \geq 0$ for any $y \in \mathbb{R}^n$. Moreover, since $f(y) = yv$ is a linear map from $\mathbb{R}^n$ to $\mathbb{R}^d$, the rank of the linear map $g(y) = yx = yvv^{\mathsf{T}}$ is bounded by d (assuming $d \leq n$), thereby bounding the rank of x to d. The Gram matrix therefore has two important properties missed by constraint set A that we will use to define set B:

$$B = \left\{ x \in \mathbb{S}^n : x \succeq 0, \ \text{rank}(x) \leq d \right\} .$$

Here $x \succeq 0$ indicates x is positive semi-definite.

Does any solution of (2.1) for the A and B defined above provide a solution to the kissing spheres problem? Unlike the previous applications, here we need to do some work to establish this fact. Let's start with constraint B. A standard matrix decomposition exists for positive semi-definite $x \in \mathbb{S}^n$ called the Choleski decomposition:

$$x = bb^{\mathsf{T}} .$$

Here b is the Gram–Schmidt basis for (Gram matrix) x. To construct this basis, we arbitrarily choose to align b_1 with the 1st coordinate axis. This means only the first element of b_1 is nonzero and has magnitude $\sqrt{x_{11}}$. We are free to give this element a positive sign. From x_{12} we then infer the projection of b_2 on the first coordinate axis and choose to align the orthogonal complement, whose magnitude we infer from x_{22}, along the 2nd coordinate axis, and so on. Since x has rank bounded by d, that number of coordinate axes (rows) will suffice for expressing b. Finally, since the x matrix elements used to construct b also satisfy constraint A, the rows of b are an explicit solution for the sphere center vectors v. Seen as a matrix, b is lower-triangular.

From the problem statement in geometrical terms, we know solutions are never unique: arbitrary rotations and reflections of space can be applied to the position vectors with no effect on the properties (2.12) and (2.13). However, this nonuniqueness is not reflected in the Gram matrix x. Consider the case $d = 2$ where the 6×6 matrix x has 1's on its diagonal and $\pm 1/2$ everywhere else. What has happened to the continuous nonuniqueness?

The arbitrariness of the Gram basis construction resolves the puzzle about solution uniqueness. An arbitrary orthogonal matrix u (with normalization $uu^{\mathsf{T}} = I$) applied to the Gram basis b has no effect on the inner products and defines the general solution:

$$v = bu .$$

By formulating the kissing spheres problem in terms of the Gram matrix x, instead of the sphere positions v, we eliminated the nonuniqueness arising from trivial symmetries (rotations and reflections of space). *Nontrivial* nonuniqueness can still occur and in fact should not be eliminated. In particular, in the case $d = 3$ it is important that solutions reveal continuous variations in the inner products $v_i \cdot v_j$ because the configurations are not rigid, as in $d = 2$.

To certify our bipartisan formulation of the kissing spheres problem, we still need to check that our sets A and B have easy projections. Projecting to A is not that different from projecting to the B constraint of Hadamard matrices. The elements of x are processed independently, setting diagonal elements to 1, and changing the off-diagonal elements minimally, when necessary, to satisfy the inequality constraint. "Minimal" now means setting to the equality case of the inequality, $1/2$, when the latter is violated.

Projecting to the nearest symmetric, positive semi-definite low-rank matrix (set B), like the projection to orthogonal matrices, requires the more sophisticated math you will see in Chapter 4. You have already seen how

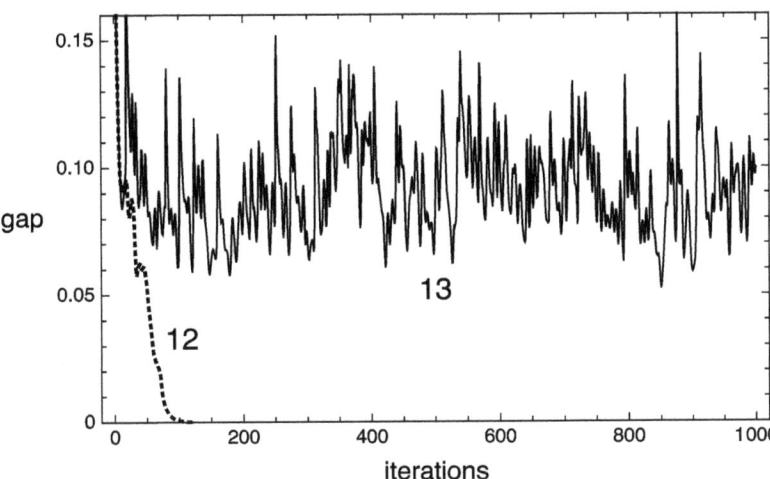

Figure 2.9 Progress of the gap when searching for 12 (dashed) and 13 kissing spheres in three dimensions.

easy it is to sample random elements of B: Generate n random vectors without regard to their inner products and let the latter define the matrix $x \in B$.

Searching for bipartisan solutions with Douglas–Rachford iterations is only conclusive when solutions are found. When the gap shows no sign of heading toward zero, it could mean a solution doesn't exist, the Poincaré recurrence time is very long, or the dynamics is beset with nonergodicity and gets stuck on nonsolutions. Trying to resolve the Newton–Gregory debate by chaotic search would look something like the plot in Figure 2.9. Whereas the gap with 12 spheres quickly goes to zero (and the fixed-point sphere configuration can be checked to be correct), the gap with 13 spheres tells a different story. The large amplitude chaotic swings suggest the search is not getting stuck, casting doubt on a 13-sphere solution the longer the algorithm is run.

Space as we experience it, the case $d = 3$, appears to be an outlier for kissing spheres because the known winning configurations (highest kissing number) in other dimensions always have discrete values in their Gram matrices [25]. That these discrete values may include irrational numbers first came to light with Simon Gravel's discovery [40], in a bipartisan search, of a 378-sphere configuration in 10 dimensions.[4] Rational Gram matrices are not surprising because winning kissing configurations are usually derived from winning (densest) packings of spheres in all of space, in which spheres are centered on the points of highly symmetrical lattices. Because the latter also serve as error correcting codes, it is no surprise their points can be expressed

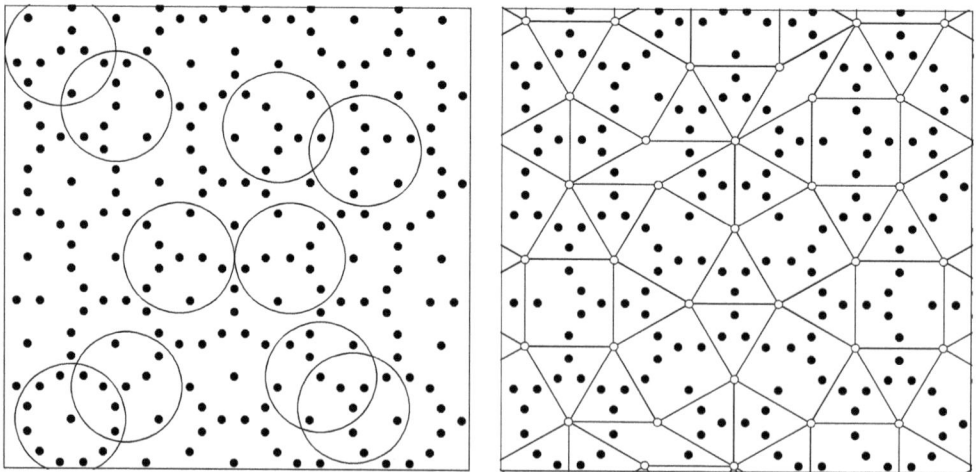

Figure 2.10 Two-dimensional shadow of the 10-dimensional nonperiodic sphere packing discovered by Simon Gravel [40]. The left panel shows just the shadows of the sphere centers and pairs of spheres that are tangent when their positions in the other eight dimensions are taken into account. A subset of the centers (right panel) form the vertices of a nonperiodic tiling. Reprinted from Elser et al. (2008) with permission from © 2008, Springer Science Business Media, LLC.

with all-integer coordinates. The vectors describing kissing sphere pairs will then also have integer coordinates and rational inner products.

Gravel's configuration of 378 spheres in 10 dimensions, which included the inner products

$$\frac{3 \pm \sqrt{3}}{12},$$

was explained by a *non*-lattice packing of spheres in 10 dimensions [40]. Figure 2.10 shows a shadow of the packing in a special two-dimensional plane where it is nonperiodic (it is periodic in the other eight dimensions). The shadows of the sphere centers form a nonperiodic tiling of triangles, squares, and rhombi. Though the symmetry is 12-fold, unlike the 10-fold patterns that accidentally got the author mixed up with "the phase problem for non-periodic objects," there is poetic justice in that the discovery was made by the phase-retrieval method.

Curiously, the three very different applications we have looked at so far involve computations on square arrays of numbers. This pattern is broken in the next application.

2.4 The Hardest Problem

In 1972 Richard Karp published a list of 21 diverse combinatorial problems, all believed to be exceptionally hard, and proved in a technical and also practically important sense, that these problems were all the same problem [61]. The 21 problems are in the class NP (Non-deterministic Polynomial time), the technical designation for problems whose solutions can easily be checked. Finding cliques, filling knapsacks, ... are in a sense the same problem because Karp showed how each one could be reformulated as an instance of the others. Now if there is a breakthrough on any one of the 21 problems – a polynomial-time algorithm – all the others can be solved efficiently by that same algorithm. Computing the reformulations, of course, should also be possible in polynomial time. This single "hardest problem" is better known as a computational complexity class, called NP-complete.

If the only algorithms worth developing are ones that run in a time that grows as a polynomial in the size of the problem input, then one should probably not bother with any NP-complete problems. Most experts believe a polynomial time algorithm for these hardest problems does not exist. But NP-complete problems come up all the time and need to be solved.

If N is the input size, and the best available algorithm runs in time $\exp(cN)$, there is an incentive for reducing c even by a modest amount since it will extend the utility of the algorithm significantly when N is large. In this section, we formulate the sixth problem in Karp's list with sets A and B, thereby demonstrating that even the hardest problem is not off the table for this method. We should not be too upset if the recurrence time of the (hopefully ergodic) dynamics of the algorithm grows exponentially with N.

There is no purer expression of bipartisanship than a bipartite graph, and Karp's sixth problem, called *set-cover*, is best expressed in the language of this kind of graph. A simple instance is the graph $\mathcal{G}$ shown in Figure 2.11. The two parts of $\mathcal{G}$ are called *antenna* and *broadcast* nodes, and denoted $\mathcal{A}$ and $\mathcal{B}$. All edges $(a, b) \in \mathcal{E}$ of $\mathcal{G}$ join a node $a \in \mathcal{A}$ with a node $b \in \mathcal{B}$.

A set-cover solution is a subset $\mathcal{E}^* \subset \mathcal{E}$ of the edges that corresponds to the following scenario. A subset $\mathcal{B}^* \subset \mathcal{B}$ of the broadcast nodes "beam out signal" on all their incident edges and these edges define the subset $\mathcal{E}^*$. Not any subset of broadcasters will do, because there is another constraint at the antennas. The constraint there is that the edges $\mathcal{E}^*$ are incident on every node of $\mathcal{A}$. In a solution we say the selected broadcasters "cover" all the antennas (of the listening area). What makes this problem hard is that there is an upper bound on the number of selected broadcast nodes, $|\mathcal{B}^*| \leq c$, and

antenna nodes

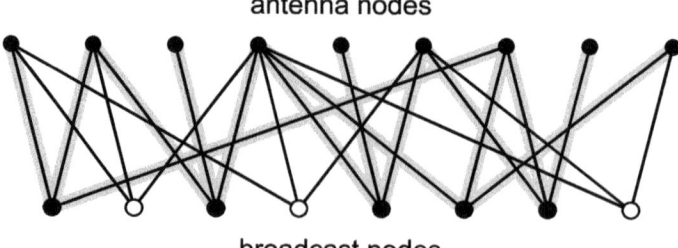

broadcast nodes

Figure 2.11 Bipartite graph (thin edges) for a set-cover instance and solution (gray highlighted edges). In a solution, a selection of the broadcast
nodes (black) have all their incident edges selected (highlighted) and the
selected edges cover all the antenna nodes. The number of selected broadcast nodes is the cover size c, and $c = 5$ is the smallest possible for this
graph.

the *cover size* c might be significantly smaller than the number of broadcast
nodes, $|\mathcal{B}|$.

Deciding whether a cover of size c exists is Karp's sixth problem. In practice, one is always trying to find the smallest c for which a solution exists.
The highlighted edges in Figure 2.11 form a cover of size $c = 5$, the smallest
possible. Problems on bipartite graphs are good candidates for bipartisan
formulations because the two sides of the graph can be associated with a
pair of constraints called A and B.

The most straightforward formulation of set cover, like most problems
on graphs, uses indicator variables. Indicator variables take value 1 when
something is present, 0 when it is absent. Our formulation has an edge-
indicator variable x_{ab} for every edge $(a, b) \in \mathcal{E}$, where $x_{ab} = 1$ means that
$(a, b) \in \mathcal{E}^*$. The reader will probably have realized that the variables in this
problem, x_{ab}, do not have the structure of a rectangular table (or matrix) as
they did in all the previous applications.

The many constraints (equations, inequalities) on the variables x_{ab} need
to be partitioned into sets A and B to complete the formulation. Though
there is no single correct way of doing this, it is important to keep in mind
the following two general criteria:

equivalence: The satisfaction of constraints (equations and inequalities) in
both A and B should be equivalent to having solved the original problem.
efficiency: The projections to the two constraints individually should have
efficient computations.

Though projections are the topic of Chapter 4, we get a preview here because
the second item above is such a key consideration.

In our bipartisan formulation, the constraints associated with antenna nodes will belong to A, while those relating to the broadcast nodes fall into B. We choose to make A discrete and keep set B continuous. We start with the constraints of set A:

$$A = \Big\{ \quad x \in \mathbb{R}^{|\mathcal{E}|} :$$
$$\forall (a,b) \in \mathcal{E} , \quad x_{ab} \in \{0,1\}$$
$$\forall a \in \mathcal{A} , \quad \sum_{(a,b)\in\mathcal{E}(a)} x_{ab} > 0 \quad \Big\} .$$

The specification of A begins with a shorthand declaration: one real variable for each of the $|\mathcal{E}|$ edges. This is followed by a statement that forces each edge to be absent or present in $\mathcal{E}^*$. Set A is now discrete. The statement in the third line ensures that at least one edge indicator, incident on every antenna node, equals 1 so that the edges $\mathcal{E}^*$ at least cover all the antennas. In the sum, we use the symbol $\mathcal{E}(a)$ for the set of edges incident on node a. We have not tried to make the selected edges $\mathcal{E}^*$ fit the broadcast pattern. This will be the job of the constraints in set B.

Projecting to set A is very easy. Starting from any values on the edge indicator variables, first round them to either 0 or 1. This is clearly distance minimizing but may not satisfy the last statement above. To fix things, we survey each of the antenna nodes in turn. If the inequality is satisfied at node a, we need do nothing more. Otherwise we review the values of x_{ab}, $(a,b) \in \mathcal{E}(a)$, before rounding, and select the one with the largest value. Flipping the rounded value of this variable, from 0 to 1, incurs the smallest added distance and should be our choice for restoring the covering property at node a.

Set B comprises several linear equality constraints and one linear inequality:

$$B = \Big\{ \quad x \in \mathbb{R}^{|\mathcal{E}|} :$$
$$\forall (a,b) \in \mathcal{E} , \quad x_{ab} = \bar{x}_b , \quad \sum_{b\in\mathcal{B}} \bar{x}_b \le c \quad \Big\} .$$

The first statement in the second line forces all indicators incident on a broadcast node b to be equal, and the equality value is given the symbol $\bar{x}_b$. The values $\bar{x}_b$ are arbitrary except that they must satisfy the inequality that follows. When the constraints of set A are also satisfied, and all indicators are 0 or 1, the values $\bar{x}_b$ will also be 0 or 1 and the inequality in set B is just the bound on the size of the cover.

This completes the check that any point $x \in A \cap B$ corresponds to a solution of the set cover problem. Since we have already shown that projecting to A is an easy computation, what remains is to check this for the projection to B. Because the constraints used in set B are so pervasive, the reader should study this projection very carefully.

Ignoring for the moment the cover size inequality, we have independent equality constraints at each node b. We focus on a particular b, drop the b index, and denote the incident indicators as x_i, where i ranges between 1 and $m = |\mathcal{E}(b)|$. We are interested in finding the equality value $\bar{x}$ that minimizes the squared distance

$$\sum_{i=1}^{m} (\bar{x} - x_i)^2 = m\,\bar{x}^2 - 2\bar{x} \sum_{i=1}^{m} x_i + \cdots$$

$$= m \left(\bar{x} - \frac{1}{m} \sum_{i=1}^{m} x_i \right)^2 + \cdots$$

where $\cdots$ are terms that do not involve $\bar{x}$. We see that the distance is minimized when $\bar{x}$ equals the arithmetic mean of the values of the variables participating in the equality constraint. Restoring the b index,

$$\bar{x}_b = \frac{1}{|\mathcal{E}(b)|} \sum_{(a,b) \in \mathcal{E}(b)} x_{ab} \,.$$

If these equality values at the broadcast nodes satisfy the cover inequality, we are done. If not, to minimize the projection distance we should try to achieve the equality case of the cover inequality. In Section 4.1.7 we will see, as a consequence of a general theorem, this involves modifying the equality values, $\bar{x}_b \to \widetilde{x}_b$, to satisfy

$$\sum_{b \in \mathcal{B}} \widetilde{x}_b = c \,,$$

while minimizing the distance[5]

$$\sum_{b \in \mathcal{B}} (\widetilde{x}_b - \bar{x}_b)^2 \,.$$

Geometrically, this is the problem of finding the nearest point $\widetilde{x}$ on a hyperplane (in $|\mathcal{B}|$ dimensions) from a given point $\bar{x}$. The solution is to change $\bar{x}$ along an axis orthogonal to the hyperplane, which in this case means each $\bar{x}_b$ changes by the same amount. If

$$\Delta = \sum_{b \in \mathcal{B}} \bar{x}_b - c > 0$$

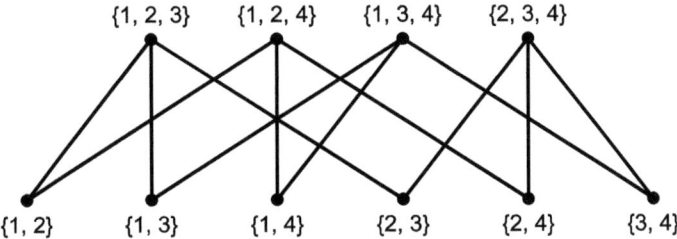

Figure 2.12 The bipartite graph $\mathcal{G}(4,3,2)$. Its antenna and broadcast nodes are, respectively, all the 3-subsets and all the 2-subsets of $\{1,2,3,4\}$. Because the smallest cover has size 2, the Turán number $T(4,3,2)$ equals 2.

is the deficit in meeting the cover bound, the required change is

$$\widetilde{x}_b = \bar{x}_b - \frac{\Delta}{|\mathcal{B}|} \ .$$

As a diversion while interned in the Hungarian labour service during World War II [108], Pál Turán took up the study of a very symmetrical class of set cover problems [107]. His idea was to take as the universe the set $\{1, 2, \ldots, n\}$, defined by an integer n, and then take all the subsets of size k, called k-subsets, as the antenna nodes. The broadcast nodes are defined likewise, as all the r-subsets, where $r < k$. Edges are defined by the containment of r-subsets in the larger k-subsets. We will refer to the corresponding bipartite graph as $\mathcal{G}(n,k,r)$. Figure 2.12 shows $\mathcal{G}(4,3,2)$. There is an edge between nodes $a = \{1,2,3\}$ and $b = \{1,2\}$ because, as sets, $b \subset a$.

The size of the smallest cover of $\mathcal{G}(n,k,r)$ is the Turán number $T(n,k,r)$. Using Figure 2.12 you can check that $T(4,3,2) = 2$. Because of their symmetrical definition, one might expect to find explicit formulas for the Turán numbers, and for small k and r this is indeed the case. But already for $(k,r) = (4,3)$, the Turán numbers are the subject of active research. Here is a conjectured formula when n is a multiple of three [100]:

$$T^*(3(m+1), 4, 3) = m(m+1)(2m+1) \ . \tag{2.14}$$

Because these graphs grow rapidly in size with n, and finding small covers is challenging, they provide a useful test case for the bipartisan approach to solving problems in the NP-complete class.

Table 2.2 shows the growth in the number of Douglas–Rachford iterations required to find covers of the conjectured sizes. The growth is steady but not alarmingly so. To add some perspective, consider the fact that the cover sizes up to $n = 12$ are known to be the best possible. If the $n = 12$ instance were run with a cover size of 83, smaller by one than formula (2.14), it would

n	6	9	12	15	18
$|\mathcal{E}|$	60	504	1 980	5 460	12 240
$T^*(n,4,3)$	6	30	84	180	330
iterations	37	420	2 500	13 100	66 000

Table 2.2 Growth in the number of Douglas–Rachford iterations to find covers for the conjectured Turán numbers $T^*(n,4,3)$.

never terminate (find a fixed point) because $A \cap B$ is known to be the empty set. The behavior with the first unknown case, $n = 15$, is similar. When the constant c in the cover size constraint is set at 180, the algorithm consistently finds solutions. Our faith in the conjecture is bolstered when the algorithm finds no solutions for covers of size 179, even when allowed a hundred times as many iterations.

A different kind of refinement of the method, and not as simple as shrinking the time step (2.11), was used to obtain the results in Table 2.2. This involved introducing a mechanism for adjusting the definition of the distance used by the projections P_A and P_B. As we have seen, these are easily computed for the isotropic distance (2.2). But with almost no extra computational effort, the projections can be generalized for the distance

$$\|x - x'\|_g^2 = \sum_{(a,b)\in\mathcal{E}} g_{ab}(x_{ab} - x'_{ab})^2 \,, \tag{2.15}$$

where $g_{ab} > 0$ are arbitrary *metric coefficients*.

Our first reaction to the nonisotropic-metric proposal, given the symmetry of Turán's graphs, is that this is a waste of time. But this overlooks the fact that symmetry is broken already by the choice of the initial point x. As x evolves by the Douglas–Rachford dynamics, the constraints across the antenna and broadcast nodes face different degrees of difficulty. Edge-variables incident on hard-to-satisfy nodes will fluctuate a lot, while those incident on easy nodes will hardly fluctuate at all. This observation is the basis of a heuristic for tuning the metric parameters in a way that tries to redistribute the difficulty of the constraints more evenly.

When the metric refinement is turned off for the $n = 12$ instance, the average number of iterations is increased by a factor of seven. The increase is far from uniform across starting points. There are lucky starting points where the isotropic metric quickly leads to a solution, but for the majority of starting points this metric is a poor choice. In fact, the dynamics gets trapped on nonergodic "cycles" in about 4% of the runs. The metric-tuning heuristic is described in Section 6.3.2.

2.5 Spin Glasses

Though "optimization" is construed in many ways, often it just comes down to the problem of finding the maximum of a function. The most challenging optimizations are for nonconvex functions of many variables. Simply going uphill is not a viable strategy for nonconvex functions. Still, "hill climbing" combined with random restarts, when the climber arrives at the top of a small hill with nowhere higher to go, is a much-used method.

There is a class of optimization problems where one should consider the possibility of a better approach. Suppose the climber has knowledge of the existence of a Mount Higherest uniquely towering over the expanse of minor peaks. To be sure, the knowledge of something existing is not the same as knowing how to find it. And indeed, in simple hill climbing, the existence of a Mount Higherest is of no help at all. The local nature of this strategy is like a dense fog that shrouds the landscape, including anomalies like a Mount Higherest.

To break free of wandering aimlessly among the foothills, we can try *deconstructing the landscape*. Bipartisanship will be our inspiration for how this deconstruction might work. Figure 2.13 tries to convey the idea as a cartoon. As in all two-dimensional depictions of high dimensional objects, we should not be too literal when interpreting the cartoon. On the left is shown the instance of optimization: a function $f(\theta)$ of a single periodic variable. Mount Higherest, the peak at the center, is surrounded by many minor peaks. The myopic hill climber does not know whether Mount Higherest is toward the right or left. An arduous trek appears inevitable.

The right panel of Figure 2.13 imagines the possibility that the complex $f(\theta)$ landscape is a circular section of a somewhat simpler landscape in one higher dimension. The elevation contours suggest a single peak located well inside the circle marked B. If $g(x_1, x_2)$ is the elevation in the deconstructed landscape, and B has radius 1, then

$$f(\theta) = g(\cos\theta, \sin\theta) \, .$$

We are not interested in finding the maximum of the function g. Finding the highest contour that reaches B is really all we care about, since that translates to finding the maximum of f. More formally, suppose we have advance knowledge that the elevation of Mount Higherest exceeds E. We can then define the set

$$A = \left\{ (x_1, x_2) \in \mathbb{R}^2 : g(x_1, x_2) > E \right\} \, ,$$

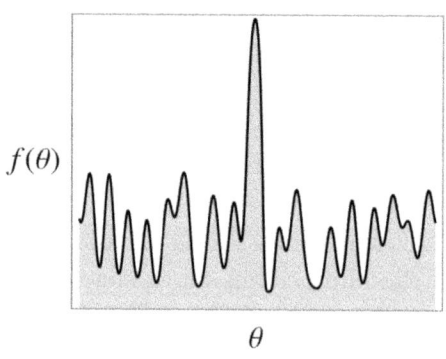

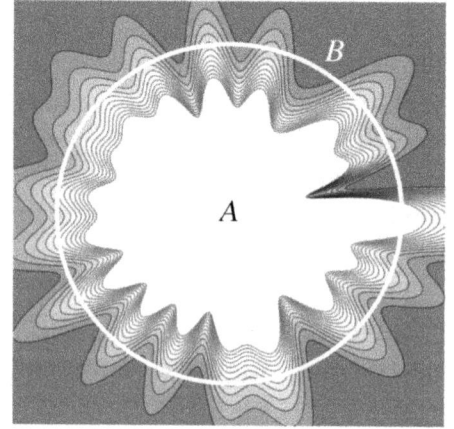

Figure 2.13 Deconstruction of a one-dimensional landscape $f(\theta)$, on the left, into a two-dimensional landscape $g(x_1, x_2)$, on the right. Set A is the white region above elevation E; set B is a circle. Finding the prominent maximum of f corresponds to finding points of $A \cap B$ in the deconstructed landscape.

shown as the white region in the Figure. With B defined as

$$B = \left\{ (x_1, x_2) \in \mathbb{R}^2 : x_1^2 + x_2^2 = 1 \right\},$$

the location of Mount Higherest, at least those parts above elevation E, corresponds to $A \cap B$.

The prospects of finding Mount Higherest by deconstructing the landscape crucially hinge on the feasibility of computing the projection P_A, something beyond the scope of the cartoon. For a taste of that question, we turn to a much studied optimization problem: spin glasses [3]. We will see that the spin-glass landscape can be deconstructed and that one of the constraints has advance knowledge of a bound on the maximum being sought, a key piece of information *not* used in hill climbing.

The *Ising spin glass* is an energy model for "spin" variables $s \in \{-1, 1\}^m$ given by the Hamiltonian function

$$H(s) = sJs^{\mathsf{T}},$$

where J is a symmetric matrix. Since H does not depend on the diagonal elements of J, because $(-1)^2 = (+1)^2$, we assume these are zero. When used as a toy model for the low temperature properties of a glass, the couplings J are treated as random variables and one is interested in low energy spin configurations. Keeping with the hill climbing convention, and replacing J by $-J$, we will be interested in s for which $H(s)$ is large.

Because the problem of deciding whether there exists an s such that $H(s) > E$ is known to be NP-complete [49], we can cast aside hopes that a bipartisan formulation will deliver an efficient algorithm for general J. On the other hand, if J has the property that there is a discrete analog of a Mount Higherest, then a search for a point in $A \cap B$ may be superior to hill climbing since it can take advantage of an exceptionally large bound E in one of its projections.

Our deconstruction is very generous, in that the higher-dimensional landscape is defined by variables $x \in \mathbb{S}^m$, the space of $m \times m$ symmetric matrices. The constraint analogous to the circle in Figure 2.13, which embeds the spin configurations, is the set

$$B = \left\{ x \in \mathbb{S}^m \colon \operatorname{rank}(x) = 1 \right\} .$$

This set is continuous and $x \in B$ only ensures that elements can be expressed as

$$x = s^\mathsf{T} s , \tag{2.16}$$

where $s \in \mathbb{R}^m$. To complete the bipartisan formulation, we define another set:

$$A = \left\{ x \in \mathbb{S}^m \colon \operatorname{Tr}(xJ) > E , \ x_{11} = \cdots = x_{mm} = 1 \right\} . \tag{2.17}$$

This set, defined by a few linear equations and one linear inequality, is also continuous. On account of (2.16), points $x \in A \cap B$ have the following constraint on their diagonal elements:

$$1 = x_{ii} = (s_i)^2 ,$$

with solution

$$s_i = \pm 1 .$$

Finally, because

$$\operatorname{Tr}(s^\mathsf{T} s \, J) = H(s) ,$$

we see that points $x \in A \cap B$ correspond to spin configurations with energy greater than E.

We are now in a position to answer the question left unanswered by the cartoon (Figure 2.13): Can the two projections be performed efficiently and take advantage of the bound E? The answer is a definite "yes." We encountered low-rank projections in the kissing spheres problem and, as a rule, linear equality/inequality constraints are not a concern because they are convex. Details on the projection computations are provided in Chapter 4. Most importantly, deconstruction is validated by the E-dependence of projection P_A.

In Chapter 3, we study *designed instances* of the set intersection problem more generally. The case of a Mount Higherest in an energy landscape falls in a special class, where A and B arise from the deconstruction of an optimization problem. Reconstruction problems, like wave microscopy, are also designed, but not associated with optimization. There the sets A and B are designed through the process of collecting data. Based on the great success of the Douglas–Rachford algorithm in wave microscopy, the bipartisan method is attractive for reconstruction problems more generally, as well as deconstructed optimization problems when there is some guarantee of a Mount Higherest. Among the latter, an obvious candidate is protein folding. That's because natural selection has evolved proteins to have unique, exceptionally deep minima in their energy landscapes. However, deconstructing protein landscapes to take advantage of the prior knowledge of a deep minimum is not as straightforward as it was in the spin glass model. In Section 7.4 the method is applied to a protein folding game, where it works quite well.

Exercises

2.1 For the bipartisan formulation of sudoku in Section 1.4, determine the set sizes $|A|$ and $|B|$ when you have the following two pieces of information about the given numbers (clues):

There are m_i givens in 3×3 block $i = 1, \ldots, 9$.
There are n_j instances of the given-number $j = 1, \ldots, 9$.

2.2 This exercise summarizes everything you need to know about the Fourier transform for wave microscopy (and a bit more). When we use the term "Fourier transform," we always mean "discrete Fourier transform," the only practical way of using this transformation in computations.

Applying the Fourier transform $\mathcal{F}$ to a vector of contrast values x to get the diffraction pattern $\hat{x}$ is equivalent to multiplying the former by a square Fourier matrix F to get the latter:

$$ xF = \hat{x} \ . $$

Most generally, the contrast x is sampled on a d-dimensional periodic grid with integer periods $(\ell_1, \ldots, \ell_d)$, and the corresponding diffraction pattern has the same grid structure. Indexing the grid points of x with integer d-tuples $p = (p_1, \ldots, p_d)$, and the grid points of $\hat{x}$ with integer d-tuples $q = (q_1, \ldots, q_d)$, the Fourier matrix has elements

$$F_{pq} = C \exp\left(2\pi i \left(\frac{p_1 q_1}{\ell_1} + \cdots + \frac{p_d q_d}{\ell_d}\right)\right).$$

(a) Explain why this F is well defined, given the periodicity assumptions on x and $\hat{x}$.

(b) Under what circumstances can the contrast grid be allowed to be periodic, given that wave microscopy specimens are not periodic? Your explanation should use the word "support."

(c) Under what circumstances can the diffraction pattern grid be allowed to be periodic, given that in an actual experiment the diffraction intensities are not periodic? Your explanation should mention resolution and the behavior of diffraction intensities at large angles.

(d) What is the size of the square matrix F ?

(e) For what positive value C is the Fourier matrix unitary, that is, $FF^\dagger = I$?

(f) Up to periodicity (which you've already dealt with), the coordinates $\mathbf{p}$ of image pixels have the same grid structure as the grid of integer indices p. The same relationship applies to the detector-pixel coordinates $\mathbf{q}$ and the integer grid of q indices. In both cases, the integer grid is related to the coordinate grid by a scale factor. Using the Fourier matrix (with real C), show that the diffraction pattern has the property

$$\hat{x}(-\mathbf{q}) = \hat{x}(\mathbf{q})^*$$

when the contrast x is real valued.

(g) Show that Parseval's theorem

$$\|x\|^2 = \|\hat{x}\|^2$$

follows from the unitarity of F. This relationship is very useful when computing projections to constraints. While some constraints are more directly expressed in terms of x, others in terms of $\hat{x}$, we are free to work with either because what is distance minimizing for one is also distance minimizing for the other.

2.3 Consider the following limiting case of spider contrast:

$$x(\mathbf{p}) = \begin{cases} 1, & \mathbf{p} \in \text{spider}, \\ 0, & \mathbf{p} \notin \text{spider}. \end{cases}$$

In words, the contrast has the constant value 1 over all the pixels that make up the spider. Next, suppose that the spider is sufficiently opaque to the incident light that the only pixels that have light passing through in the "exit wave" are those that are *not* part of the spider. The exit wave, an array of numbers in the plane of the specimen, is then the array $y = e - x$, where e is the constant array of all 1 elements. This also shows that for our simple opacity model of the contrast, the array x also has an exit wave interpretation: the wave exiting a spider-shaped aperture.

Using a general property of the Fourier matrix, show that

$$\forall\, \mathbf{q} \neq (0,0)\colon \hat{x}(\mathbf{q}) = -\hat{y}(\mathbf{q})\;,$$

that is, a sign flip everywhere except the pixel with grid indices $q = (0,0)$. Because the intensity at the center of the detector, $\mathbf{q} = (0,0)$, is too intense to be measured, the diffraction intensity of the spider, $|\hat{y}|^2$, and a spider-aperture, $|\hat{x}|^2$, are indistinguishable. This is Babinet's principle.

2.4 Bit retrieval [37] is a special case of phase retrieval where the elements of the contrast vector x are known to be two valued. As in wave microscopy, the Fourier magnitudes $|\hat{x}| = \sqrt{I}$ are known, but unlike wave microscopy, there is no support information. Instead, you know that $x(\mathbf{p}) \in \{-1, 1\}$ at all pixels $\mathbf{p}$. Define sets A and B for bit retrieval, checking that the two projections can be computed efficiently.

2.5 In wave microscopy, the contrast x is nonzero only at pixels $\mathbf{p}$ in the support region $\mathcal{S}$. The Fourier transform, when written as multiplication by the Fourier matrix, therefore involves a smaller sum than it would otherwise:

$$\hat{x}_q = \sum_{p \in \mathcal{S}} x_p\, F_{pq}\;.$$

Here, we have switched from pixel coordinates to their corresponding integer indices. Suppose there are m contrast pixels in $\mathcal{S}$ and n detector pixels, where $m < n$. We can write the Fourier transform more simply as

$$\hat{x} = x\widetilde{F}\;,$$

with the understanding that the vectors x and $\hat{x}$ have respectively m and n components, and $\widetilde{F}$ is an $m \times n$ submatrix of the $n \times n$ Fourier matrix F.

In *generalized phase retrieval* [20], the rectangular matrix $\widetilde{F}$ is replaced by a general *sensing matrix* $G \in \mathbb{C}^{m \times n}$. The unknown variables shift from x to $\hat{x}$, which we call y in the new setting, and the contrast x is allowed to be complex. Instances of generalized phase retrieval are completely specified by the sensing matrix G and a vector of magnitudes (using component-wise notation):

$$ w = |y| = |xG| \, . $$

The vector $x \in \mathbb{C}^m$ does not have an interpretation other than "whatever needs to be multiplied by G to produce a vector y having the given magnitudes." The task in generalized phase retrieval is to reconstruct all of y given just its magnitudes w and the sensing matrix G. Though there is nothing analogous to a support, this is implicit in the inequality $m < n$.

Give a bipartisan formulation of generalized phase retrieval for suitable constraint sets A and B:

$$ \text{find some} \quad y \in A \cap B \, . $$

To avoid confusion with Fourier transform pairs, we make a rare exception here of using the symbol y for the unknown. Because the Fourier transform has no special status in generalized phase retrieval, you should not use the symbol $\hat{y}$. When defining set B, you should instead use x as an auxiliary variable to express the most general relationship between y and G. Auxiliary variables in the definition of sets came up in set B of the set-cover problem ($\bar{x}$). Remember that set definitions always begin with a declaration.

2.6 Give a bipartisan formulation for *complex Hadamard matrices*, a generalization of ordinary Hadamard matrices where the matrix elements may be complex and unitarity replaces orthogonality. Drawing inspiration from phase retrieval, show that $n \times n$ complex Hadamard matrices exist for all orders n.

2.7 A complex generalization of kissing spheres, called *complex equiangular lines*, is important in quantum information theory where it is known by the endearing terms *symmetric, informationally complete, positive operator-valued measures*, or SIC-POVMs [89]. Just as complex Hadamard matrices are more plentiful than their real counterparts (they exist for all orders), "complex kissing spheres" appear to enjoy more regularity than real kissing spheres. Zauner's conjecture [113], verified up to dimension $d = 67$ [95], states that there exist vectors $v_i \in \mathbb{C}^d$,

$i = 1, \ldots, d^2$, such that

$$|v_i \cdot v_j^*|^2 = \begin{cases} 1, & i = j, \\ \frac{1}{d+1}, & i \neq j. \end{cases}$$

The "complex lines" are defined by points v that lie on them and have unit distance from the origin ($v_i \cdot v_i^* = 1$).

(a) Give a bipartisan formulation of complex equiangular lines whose sets A and B are closely analogous to the sets for the kissing spheres problem.

(b) If you have already studied quantum mechanics and can recall the three-dimensional Bloch sphere (for a $d = 2$ Hilbert space), construct the most symmetrical configuration of $d^2 = 4$ "lines" on this sphere and verify that it is equiangular in the above sense.

2.8 Karp's sixth problem, set-cover, makes no direct reference to a bipartite graph. An instance involves a set of items, say

$$\mathcal{U} = \{h, o, r, t, e, n, s, i, a\},$$

and a collection of its subsets,

$$\mathcal{C} = \Big\{ \{a, s, h\}, \{h, a, t\}, \{h, i, t\}, \{o, a, t\},$$

$$\{r, o, t\}, \{s, e, t\}, \{s, o, n\}, \{t, o, e\} \Big\}.$$

The task is to select a limited number of the subsets in $\mathcal{C}$ so that their union is $\mathcal{U}$.

(a) Can you solve this instance of set-cover with five subsets?

(b) Describe the bipartite graph associated with a set-cover instance. What are the antenna nodes and broadcast nodes?

2.9 Karp's 15th problem, called *hitting-set*, is closely related to set-cover. Using the example of the previous problem, the task is to find a subset $\mathcal{H} \subset \mathcal{U}$, a hitting-set, with the property that it intersects (hits) each subset in $\mathcal{C}$ in exactly one item:

$$\forall \, \mathcal{S} \in \mathcal{C} : \quad |\mathcal{H} \cap \mathcal{S}| = 1 .$$

(a) Can you find a subset of $\mathcal{U}$ in the previous problem that hits all the subsets in $\mathcal{C}$?

(b) Give a bipartisan formulation of hitting-set in terms of a bipartite graph with antenna nodes and broadcast nodes. The sets A and B will be slightly different from what they were in set-cover as there is no counterpart of the cover size.

2.10 Karp's first problem, *Boolean satisfiability*, served as a hub for showing NP-completeness of all the other problems on the list. Like set-cover and hitting-set, it too has a bipartite graph structure that makes it a natural candidate for a bipartisan formulation.

   We are interested in Boolean formulas that have the *conjuctive normal form* (CNF): a conjunction of *clauses* (combined with AND = $\wedge$), where each clause is the disjunction of *literals* (combined with OR = $\vee$). A literal is a Boolean variable $x_1, x_2, \ldots$ or its negation $\neg x_1, \neg x_2, \ldots$. Here is a CNF formula comprising four clauses on three variables:

$$(x_1 \vee x_2 \vee x_3) \wedge (x_1 \vee \neg x_2) \wedge (x_2 \vee \neg x_3) \wedge (x_3 \vee \neg x_1).$$

(a) Can you find a TRUE $(= 1)$ / FALSE $(= 0)$ assignment to the variables that makes the formula equal 1?

   The nodes of the bipartite graph for Boolean satisfiability also have the broadcast and antenna interpretations. There is a broadcast node b for each Boolean variable x_b, whose value is broadcast via edges (a, b) to all the clauses a in which it appears. Each clause node a is an antenna that listens to the truth values of its constituent Boolean variables. Negations on the variables add a new feature, a function $f : (a, b) \mapsto \{0, 1\}$. Variable x_b is negated in clause a when $f(a, b) = 0$ and is not negated when $f(a, b) = 1$. According to this convention, clause a is 1 if the broadcast value of one of the variables it listens to matches the f value of the edge to that variable.

(b) Give a bipartisan formulation of Boolean satisfiabilty that closely follows the formulation for set-cover and hitting-set. Instances are defined by a bipartite graph with edges $\mathcal{E}$ and the function $f : \mathcal{E} \to \{0, 1\}$. Variables live on $\mathcal{E}$ and are denoted x_{ab}. When defining the sets A and B, you may want to use the notation $\mathcal{E}(a)$ for the edges incident on a (and similarly for broadcast nodes b). There is no counterpart of the cover size c.

2.11 There are less obvious bipartisan formulations of Hadamard matrices than the one given in Section 2.2. In most, A and B impose orthogonality of at least the rows or columns of the matrix. Your task in this exercise is to devise a formulation that makes no reference to rows and columns at all! Instead, Hadamards will emerge as a satisfying assignment for a kind of logic problem, where the "clauses" are all the 2×2 submatrices of the matrix.

(a) Start by establishing some simple facts about $n \times n$ matrices x with ± 1 elements, where n is even. First notice that all the 2×2 submatrices of x are of two kinds, "good" and "bad." The good submatrices have full rank and an odd number of elements of either sign. The bad submatrices have rank 1 and an even number of elements of either sign.

(b) There are $\binom{n}{2}$ submatrices in any pair of rows or columns of x. Show that at most $(n/2)^2$ of these can be good, and exactly this number is attained when the rows or columns are orthogonal. It follows that Hadamard matrices are the "best possible" ± 1 matrices, having the greatest number of good 2×2 submatrices. Call this number g and show that $g = \binom{n}{2}(n/2)^2$.

(c) As in Boolean satisfiability, the variables live on the edges of a bipartite graph with the broadcast and antenna interpretations. The broadcast nodes are the n^2 elements of the matrix x, while the antenna nodes are its 2×2 submatrices, of which there are $\binom{n}{2}^2$. Each broadcast node sends out a value to all the submatrices that have that matrix element, while each antenna node listens to its four constituent matrix elements. Work out the number of edges $|\mathcal{E}|$ in the bipartite graph.

(d) Define the sets A and B in close analogy with Boolean satisfiability. Keep B, the constraint at the broadcast nodes, continuous, and let A impose the discrete, ± 1, values on the edges. Provide sufficient detail for the projection to the A constraint that confirms its efficiency.[6]

2.12 Is there a way to define a complex energy landscape without the many random parameters we saw in the spin glass model (the J matrix)? This exercise is about one such landscape [29] and its deconstruction.

The landscape or "universe" is the set of all scaled $n \times n$ orthogonal matrices:

$$U = \left\{ u \in \mathbb{R}^{n \times n} : uu^\mathsf{T} = nI \right\} .$$

Start by becoming familiar with the tangent spaces of this continuous landscape. You can do this by exploring what kinds of changes to a $u \in U$, $u' = u + x$, preserve orthogonality to first order in x:

$$u'(u')^\mathsf{T} = nI + O(x^2) .$$

Do this calculation, but to make life easier use the parameterization $u' = u + yu$ (which is possible because for any x, the corresponding y equals xu^T).

(a) Show that the only restriction on y is that it is antisymmetric. What does this tell you about the number of dimensions of the landscape?

(b) Here is the energy function (Hamiltonian):

$$H(u) = \sum_{ij} |u_{ij}| \ .$$

Apply the generalized mean inequality to the elements of u to show that

$$H(u) \leq n^2$$

and the equality case corresponds to all elements having the same magnitude (the value 1). The highest peaks in our landscape are therefore all the Hadamard matrices! If you are a mathematician and run into a physicist in a bar, or vice versa, you now can share an impossible dream.

(c) Show that the proposal to find Hadamard matrices by gradient ascent is doomed. Do this using the tangent space parameterization, that is, identify all points u_0 in the landscape with the property

$$H(u_0 + yu_0) = H(u_0) + O(y^2)$$

for all antisymmetric y. You will find that there is a peak (vanishing gradient) at all u_0 with the following symmetry property:

$$u_0^{\mathsf{T}} \operatorname{sgn}(u_0) = \operatorname{sgn}(u_0)^{\mathsf{T}} u_0 \ .$$

Since $\operatorname{sgn}(u) = u$ when u is Hadamard, this agrees with what you already found out using the generalized mean inequality. Unfortunately, there are also many, many non-Hadamard matrices with this symmetry property!

(d) Deconstruct this energy landscape using U for set A since you were promised that there is an efficient projection from the larger universe of all real $n \times n$ matrices. The obvious choice for set B is

$$B = \left\{ x \in \mathbb{R}^{n \times n} : H(x) = n^2 \right\} \ . \tag{2.18}$$

If it were not for the absolute values in the formula for H, projecting to B would be as simple as projecting to a very symmetrical hyperplane. As it is, the projection is to one of 2^{n^2} hyperplanes. With that as a hint, describe in detail an efficient projection to B.

(e) Though the discrete B (2.10b) we defined originally is probably superior to the continuous B (2.18) for the business of finding Hadamard matrices, when n is not a multiple of four there is nothing to be found with (2.10b) since orthogonal ± 1 matrices do not exist for those orders. Show how (2.18) can be modified to at least find a sequence of bounds on the energy for orders that are not multiples of four.

2.13 The great puzzle inventor Henry Dudeney asked his readers to place 16
pawns on a chessboard so no three fall on the same line [31]. Here is
one solution:

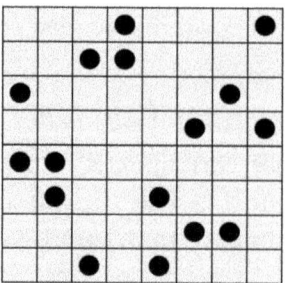

(a) What about larger chessboards? Argue that the number of no-three-
in-line pawns on the $n \times n$ board cannot exceed $2n$.

(b) To test whether $2n$ can be attained, work out a bipartisan formulation
for no-three-in-line. Use variables on the edges of a bipartite graph.
There are $n^2 = |\mathcal{B}|$ broadcast nodes and the set of antenna nodes $\mathcal{A}$
corresponds to the lines on the board that can hold three pawns. The
B constraint at the broadcast nodes will be almost identical to the
B constraint for set-cover. What is the A constraint at the antenna
nodes?

(c) To prove that you understand this problem, add the missing entries
in the table below that lists the number of antenna nodes in the $n \times n$
puzzle. Since you may be asked in the future to solve a no-three-in-
line puzzle, it makes sense for you to write a program that generates
all the elements of $\mathcal{A}$.

n	1	2	3	4	5	6	7	8
$|\mathcal{A}|$	0	0	8	14	?	?	?	?

3

Conspiracy Theory

A bipartisan formulation is *feasible* if solutions exist, *infeasible* when they do not. Solutions can be unique, or unique up to the symmetries of the problem. The feasibility-infeasibility transition is the obvious place to look for uniqueness. Even more interesting is the case where infeasibility is expected, and yet solutions exist. This may happen because the constraints were designed, or some other conspiracy is at work. In this chapter we develop a general theory for such phenomena because inevitably it will add to your understanding of a problem.

Figure 3.1 shows examples of the four scenarios just described in the highest number of dimensions humans can visualize. A and B are planes and lines, or *affine* sets of dimension 2 and 1. The *codimension* of a set defined by equations is the minimum number of equations required to define it. Planes have codimension 1, lines have codimension 2. In scenario (a) both sets are planes and $A \cap B$ is a line. The intersection comprises all points that satisfy two equations and has codimension 2. This problem is always feasible because the intersection (for infinite planes) is never empty, though without the solution being unique (a single point). There is one exception: the planes might not intersect when they are parallel. We can exclude this case by saying the two sets should be "in general position" – an informal way to assert our interest is in problems where the intersection property is stable with respect to perturbations. Two sets will be in general position when the equations that define them have noise, or when A and B are as different as apples and bananas.

In scenario (b) both sets are lines. When these are in general position the problem is infeasible. A solution point, having 3 coordinates, would have to satisfy 4 equations. We can define a *virtual dimension* even for nonintersections using codimensions, or equivalently, the number of equations. Subtracting the codimension from the number of dimensions of the ambient space

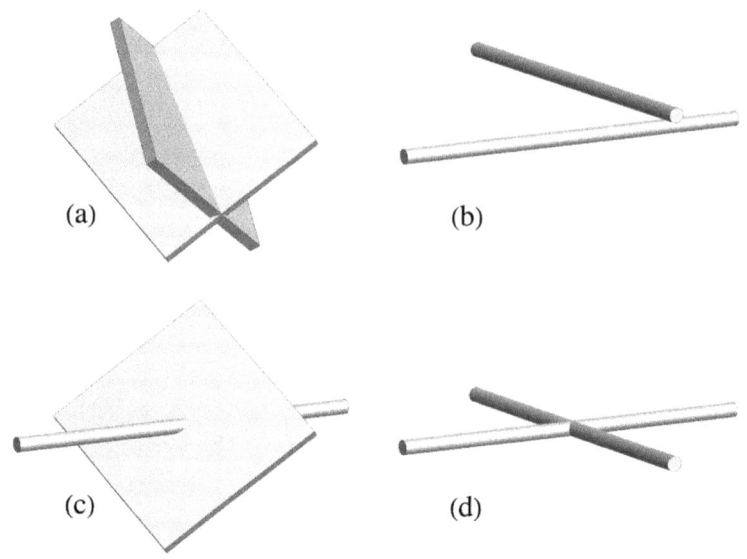

Figure 3.1 Four scenarios for $A \cap B$ in three dimensions when the two sets are planes and lines. (a) Feasible but nonunique, (b) infeasible, (c) feasible and unique, and (d) a conspiracy.

gives the dimension of the set. Using the total number of equations (instead of codimensions), the virtual dimension of scenario (b) is $d^* = 3 - 2 - 2 = -1$.

When a plane intersects a line, in scenario (c), the virtual dimension is $d^* = 3 - 2 - 1 = 0$, which is also the dimension of a point. In general position this problem is feasible and the solution, a single point, is unique. This bipartisan scenario would appear to be the best possible when one is interested in unique solutions. But here is a good dose of cold water we can throw on that tidy conclusion. In the applications of practical interest, the number of variables, or the dimension of the ambient space, will be some number in the hundreds, or even millions. The rule that the sum of two similarly large numbers – the codimensions of A and B – *exactly* matches this number cannot be a very robust criterion for uniqueness! The scenario that is superior in this respect is the one we discuss next.

Scenario (d) is the same as (b), but where the two lines are not in general position. We could invoke a conspiracy, but often problems with a negative virtual dimension have solutions for reasons that are not ominous at all. An important case is reconstruction problems, where the constraints are derived from experimental measurements, and these are usually redundant in some sense. Consider the support constraint, B, of wave microscopy. Suppose m symmetry-unrelated intensity measurements are above the level of noise, so that A is a set of m dimensions, one per unknown phase angle. To bring the virtual dimension of $A \cap B$ down to zero (and retrieve the phases), we need m equations. These are provided by B, in which we constrain m contrast pixels outside the support to be zero. In a well-designed experiment, the number of pixels where the contrast is guaranteed to be zero, say n, is much larger than m. So while the correct choice of the m phases will set m constrast pixels to zero, it will do much more: an additional $n - m$ contrast pixels will also, miraculously, be zero. This is not a true miracle, of course, because the data in the A constraint, the Fourier magnitudes, was derived from a specimen confined to a support that ensured n contrast pixels would be zero. The redundancy of the constraints in B completely explains the conspiracy.

The (negative) virtual dimension in a reconstruction problem can be large. In wave microscopy, $-d^*$ is a significant fraction of the number of image pixels. Because of noise, the constraint sets are in general position and the reconstruction problem is not strictly feasible. On the other hand, the non-solutions are *near-solutions* by nothing short of a conspiracy. Near-solutions are characterized by a small gap, the distance $\|x_A - x_B\|$ between a point $x_A \in A$ and a point $x_B \in B$. In a wave microscopy near-solution, because x_B is exactly zero at all the non-support pixels, x_A would have to be near zero on that same set of pixels. That n numbers can all be made small, given a significantly smaller number of free parameters (the m phase angles), is no less of a conspiracy than the case of a true solution (where the n numbers are exactly zero).

Conspiracy theory does not address particular instances of a problem but whole ensembles of instances. Ensembles are defined by a few parameters or properties. We will see in Section 3.1 that the wave microscopy ensembles are defined by just one property: the shape of the support. Ensembles are also key in the conspiracy theory of discrete problems. Instead of a virtual dimension, for these problems one can define a growth rate γ for the expected number of solutions in the ensemble as a discrete size parameter m is increased. When appropriately defined, the expected number of solutions behaves as $\exp(\gamma m)$. A positive γ means solutions are far from unique, while a negative γ means the solution probability, of a typical instance, is very small. Just like negative virtual dimensions, negative growth rates are interesting because instances with solutions can be designed. A designed solution is usually unique when the ensemble's γ is negative.

When A and B are continuous and defined by equations, the virtual dimen-

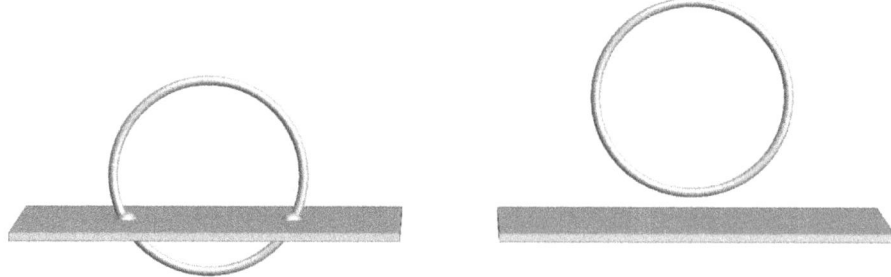

Figure 3.2 When one or both of the sets is nonaffine, the virtual dimension does not fully characterize the set of solutions, even when the sets are in general position.

sion calculus applies to the linearization of the equations, or equivalently, the locally affine approximations of the sets. One must be careful, in the global non-linear setting, not to assume the intersection $A \cap B$ behaves as it does for affine sets, even when these are in general position. Figure 3.2 shows what can happen in three dimensions when one of the sets is a circle. The virtual dimension $d^* = 3 - 1 - 2 = 0$ tells us the dimension of the solution set but nothing about the number of solutions with that dimension. This question comes up in Section 7.3, where we try to determine the maximum number of equidistant lines in four dimensions. Though the virtual dimension for 11 lines turns out to be 1, we had to settle for 10-line solutions whose virtual dimension is 5. It goes without saying that "looking" at the sets to see if they intersect (Figure 3.2) is not an option when the number of dimensions is large.

Edge-matching tile puzzles are mathematical recreations that have also been the subject of serious study. Deciding whether a solution exists was shown to be an NP-complete problem by Erik and Martin Demaine [27]. Figure 3.3 shows the six kinds of tile that can be made with two "colors" (all four rotations of a tile are the same tile). If six copies are made of each of these tiles, how hard is it to assemble them into a 6 × 6 square with all edge-colors matching? Figure 3.4 shows three solutions. Notice that the 6 × 6 square is actually a torus because the edge colors also match along the right/left and top/bottom edges of the solution.

The edge-matching-tiles ensemble we study in Section 3.6 is defined by two integers: the size n of the $n \times n$ assembly, and the number of colors k that may randomly be assigned to them. A problem instance corresponds to any

 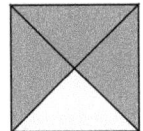 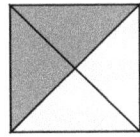 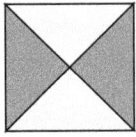 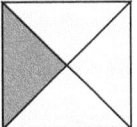 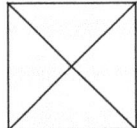

Figure 3.3 The six types of tile when there are just two colors.

set of $m = n^2$ tiles with a choice of k colors on their edges. Not surprisingly, for $(n, k) = (6, 2)$ the expected number of solutions is enormous, about 10^{40}, corresponding to a growth rate of $\gamma = 2.58$ when m is used for the size parameter.

The growth rate in the expected number of puzzle solutions first becomes negative at $k = 8$. For the same 6×6 size, the expected number of solutions in the eight-color ensemble is only 10^{-3}. In practical terms, this means one would have to try out about one thousand random tile sets before finding one that has a solution! Figure 3.5 shows one such "lucky" tile set, where the eight colors are replaced by parallel lines of two colors (black/white) and four directions. Of course, as the figure shows, the tile set was not found by trial and error. It was designed by first assembling *blank* tiles into a 6×6 "solution," then coloring them edge matched (optionally with tromino motifs) while they have the "correct" positions and rotations! Undoing the solution, by scrambling the tiles, gives the tile set of the actual puzzle. The uniqueness of the solution (up to symmetries of the torus) was checked by an exhaustive search.

In this chapter we develop the skill of locating the problem parameters where the virtual dimension d^* or the growth rate γ changes sign. The conspiratorial case, where these are negative, have the greatest practical importance. Uniqueness is critical in reconstruction problems and requires a

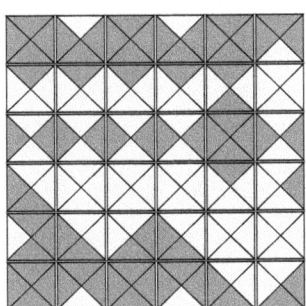 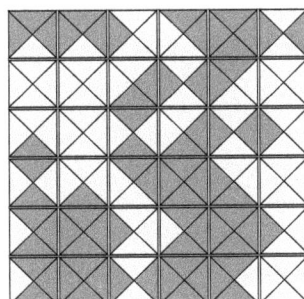 

Figure 3.4 Three edge-matched solutions formed from six copies of each of the tiles in Figure 3.3.

Figure 3.5 An eight-color edge-matched solution being designed on the right, and scrambled to make a puzzle on the left. The eight colors are rendered as four directions of slanted lines that are either black or white.

negative d^*. A virtual dimension that is also large (negatively) offsets noise in the data. If solutions persist even when the number of equations greatly exceeds the number of variables, one may interpret some fraction of the equations as validating the constraints imposed by a smaller, sufficient set of equations. Constraint redundancy keeps reconstructions from being sensitive to noise in the data that define the constraints.

The same considerations come up in discrete problems, and a sufficiently negative γ not only improves the chances a designed puzzle solution is unique, but allows, for example, cryptographic keys to be recovered when there is contamination by noise. Problems in the NP-complete class are decision problems, and deciding whether or not a solution exists is easier both when the expected number of solutions is astronomically large or astronomically small. A sufficiently negative γ, for uniqueness, is therefore also good from the perspective of tractability.

Problem ensembles with positive d^* or γ are often identified as *under-constrained* and their counterparts, where these numbers are negative, are *overconstrained*. That the counting of variables and equations mostly continues to apply when the equations are nonlinear is formalized in the *constant rank theorem* of multivariable calculus. We will always count real variables and real equations when computing d^*, even when the natural variables are complex valued.

The remainder of this chapter features various problem ensembles for in-depth study. For the continuous case, we first turn to phase retrieval. This is followed by a modern-day cousin called quantum state tomography. Continuing with quantum states, but from a geometrical perspective, we study a complex generalization of kissing spheres called complex lines. Boolean satisfiability and edge-matching-tile puzzles are featured in the very different method of analysis for discrete problems. When solutions have a small group of symmetries, like edge-matching tiles, the presence of the symmetry has a negligible effect on the transition between over- and underconstrained. Sometimes, however, the symmetry group is large and cannot be neglected when computing the transition. The complex line problem highlights this phenomenon.

3.1 Phase Retrieval with Known Support

3.1.1 Autocorrelation

For simplicity we represent the contrast x on square grids, comprising $\ell \times \ell$ pixels. The integer ℓ defines the resolving power of the microscope. The pixels in the support-set $\mathcal{S}$ cover the entire specimen, which we assume is isolated (surrounded by an infinite region of zero contrast). Before we get to the main topic, the counting of variables and equations, we need to clarify the relationship between $\mathcal{S}$ and the size ℓ of the grid.

Because we use the discrete Fourier transform when computing the diffraction pattern $\hat{x}$, the grid over which the intensity constraints on $|\hat{x}|^2$ are imposed is also $\ell \times \ell$. But knowledge of the intensity I over the diffraction grid is equivalent to knowing its Fourier transform $\hat{I}$ on a grid of the same size. How can $\hat{I}$ be at all useful when it is really the Fourier transform of $\hat{x}$ – the contrast x – that we care about?

Let's recall our notation for the Fourier transform:

$$\hat{x}_q = \sum_{p \in \Lambda} x_p \, F_{pq} \, ,$$

$$x_p = \sum_{q \in \Lambda} F_{pq}^* \, \hat{x}_q \, .$$

The grid Λ comprises all 2-tuples of integers $0, \ldots, \ell - 1$ and the Fourier matrix

$$F_{pq} = \frac{1}{\sqrt{|\Lambda|}} \exp\left(2\pi i \, p \cdot q \, / \ell\right)$$

is periodic in both p and q (in each coordinate) with periodicity ℓ. With our normalization, F is unitary in the usual sense,

$$\sum_{q\in\Lambda} F_{pq} F^*_{qp'} = \delta_{pp'} \, ,$$

where $\delta_{pp'} = 1$ when $p = p'$ (equal 2-tuples) and is zero otherwise.

Here is what the intensity constrains, expressed in terms of x with the help of the Fourier transform:

$$|\hat{x}_q|^2 = \frac{1}{|\Lambda|} \sum_{p\in\Lambda} \sum_{p'\in\Lambda} \exp\left(2\pi i(p - p') \cdot q / \ell\right) x_p \, x_{p'} \, .$$

When taking the complex conjugate, we used the assumption that x is real. Introducing the difference $k = p - p'$, this is reexpressed as

$$|\hat{x}_q|^2 = \sum_{k\in\Lambda} \exp\left(2\pi i\, k \cdot q / \ell\right) a_k \, , \tag{3.1}$$

where

$$a_k = \frac{1}{|\Lambda|} \sum_{p\in\Lambda} \sum_{p'\in\Lambda} \delta_{k\,(p-p')} x_p \, x_{p'} \tag{3.2}$$

is the contrast *autocorrelation*. Like p, k is a 2-tuple of mod-ℓ integers so that the autocorrelation also lives on an $\ell \times \ell$ periodic grid.

Before interpreting the Fourier relationship (3.1) between $|\hat{x}_q|^2$ and a_k, let's establish some basic facts about the autocorrelation. First, by interchanging the summation variables p and p' in (3.2), we see that $a_{-k} = a_k$. This implies the support of the autocorrelation, which we will denote $\mathcal{A}$, is a centro-symmetric set of pixels.

Whereas the size and shape of the support $\mathcal{S}$ both play a key role in defining the problem ensemble, the size ℓ of the grid is largely irrelevant as long as it exceeds a certain size $\ell_{\min}$. This point is explained in Figure 3.6, which shows the computation of a_k for a particular k in graphical terms, when the support has the shape of an isosceles right triangle. Shown on the right is the computation on a large grid, where the result of the computation is no different from what it would be on an infinite grid, a premise of wave microscopy. On the left we see what changes when $\ell < \ell_{\min}$: There are additional pairs of pixels separated by k that take advantage of the ℓ-periodicity of the grid. The extra pixel pairs would be appropriate if our specimen were truly periodic, like a crystal of spiders! However, the data in wave microscopy is taken from an isolated specimen, so there should be no such pairs. Wrap-around artifacts for the triangular support are avoided when $\ell/2$ is greater than the triangle's horizontal/vertical size.

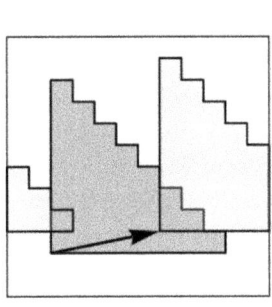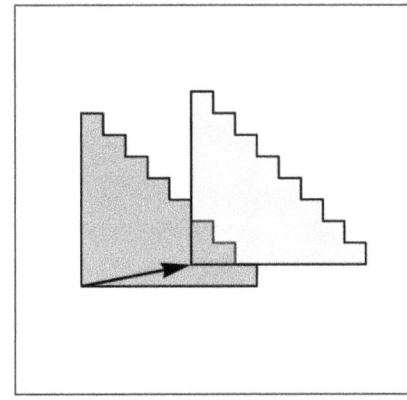

Figure 3.6 Wrap-around artifacts (left) in the autocorrelation are avoided when $\ell > \ell_{\min}$ (right), for which the autocorrelation computation for pixel separation k (arrow) gives the same result as on an infinite grid.

3.1.2 System of Quadratic Equations

The grid size is irrelevant when $\ell > \ell_{\min}$ because the Fourier transform relation (3.1) shows that the Fourier magnitude (intensity) data is represented in its entirety by the set of ℓ-independent autocorrelation numbers. Since the autocorrelation support is centered at $k = (0,0)$, its Fourier transform, the intensity, is a low-pass filtered function. Increasing ℓ only makes the intensity vary more smoothly over the pixels of the detector without adding new information. A smoothly varying intensity is important when modeling the detector, since its pixels record "point-samples" (single numbers). This is an additional reason for keeping ℓ large.

In addition to being real valued, the opacity model of the contrast ensures that x is nonnegative. This is useful because it eliminates the possibility of cancellations in the autocorrelation. Even better is the case where x is strictly positive everywhere in $\mathcal{S}$, as then the autocorrelation support $\mathcal{A}$ is explicitly determined by $\mathcal{S}$. In a wave microscope, this is realized by an aperture of shape $\mathcal{S}$ in an opaque screen. With a specimen placed in the aperture, the contrast will be positive over all of $\mathcal{S}$, provided the specimen opacity is nowhere 100%.

What we have learned about the two constraints of positive-contrast wave microscopy is summarized in the two panels of Figure 3.7, again for the ensemble defined by a right-isosceles-triangle support $\mathcal{S}$. Recall that the support, shown on the left, corresponds to the B constraint. We may interpret this as establishing the true "unknowns" in the reconstruction: one number

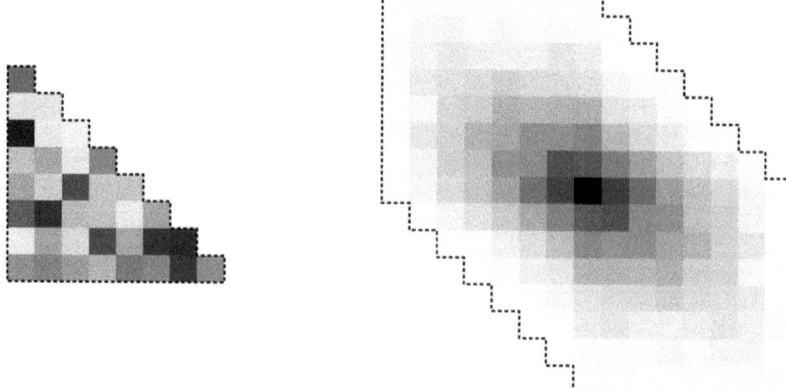

Figure 3.7 *Left*: Positive contrast in a triangular support. *Right*: The corresponding autocorrelation. The outlines of the two supports, $\mathcal{S}$ and $\mathcal{A}$, are shown dashed. Because $\mathcal{A}$ is generated by vectors k having one end at one of the three corners of $\mathcal{S}$, $\mathcal{A}$ is the union of three copies of $\mathcal{S}$ and three copies of $-\mathcal{S}$.

for each pixel in $\mathcal{S}$. On the right we see the autocorrelation support $\mathcal{A}$, which we construct from all possible separation vectors k in $\mathcal{S}$. The A constraint, which restores the measured I values to the Fourier magnitudes, we can alternatively interpret as imposing the Fourier transformed values $\hat{I}$ on all the autocorrelations in $\mathcal{A}$. From the definition (3.2) of the autocorrelation function, we see that when expressed explicitly in terms of the unknown contrast values (pixels in $\mathcal{S}$), phase retrieval corresponds to solving a very special system of quadratic equations.

3.1.3 Counting Variables and Equations

The number of quadratic equations is $(|\mathcal{A}|-1)/2+1$, since the equations for a_k and a_{-k} are the same equation. In the case of a triangular support, for any $k \neq (0,0)$ there exists a unique pair of support pixels, either $(p, p+k)$ or $(p, p-k)$, where p is one of the triangle's corner pixels. From this we see that $\mathcal{A}$ is the union of three copies of $\mathcal{S}$ and three copies of $-\mathcal{S}$ arranged to form a hexagonal parallelogon. Figure 3.7 shows this relationship when $\mathcal{S}$ is a right isosceles triangle. In the small-pixel limit $|\mathcal{A}| = 6|\mathcal{S}|$, so that even taking the $\pm k$ redundancy into account, there are three times as many equations as unknowns.

After eliminating the $\pm k$ redundancy, the equations in our system are in a sense linearly independent, because they have no common monomials. Using this standard for counting independent equations, the virtual dimension is

$$d^* = |\mathcal{S}| - (|\mathcal{A}| + 1)/2 \ .$$

As the case of the isosceles-right-triangle support shows, this can be a large negative number and we expect solutions only when the autocorrelation data is generated conspiratorially – via physical measurements. Though we confirm this for the triangular support in the next section, in the meantime we accept the legitimacy of the virtual dimension criterion and reexpress it in geometrical terms.

Here is a formal definition of the autocorrelation support, now for reconstructing contrast in any number of dimensions d:

$$\mathcal{A} = \left\{ p - p' : p \in \mathcal{S}, \ p' \in \mathcal{S} \right\} \ .$$

This is often written compactly as the *Minkowski sum* $\mathcal{S} - \mathcal{S}$. The ratio [41]

$$\Omega = \frac{|\mathcal{S} - \mathcal{S}|}{2|\mathcal{S}|} \tag{3.3}$$

corresponds to the number of linearly independent equations per unknown in the limit where $|\mathcal{S}|$ is large. The shape dependence of Ω, in the small-pixel limit where we interpret $\mathcal{S}$ as a subset of space, is interesting. Much of what is relevant for phase retrieval is captured by the inequality

$$\Omega \geq 2^{d-1} \ . \tag{3.4}$$

This follows [73] from a more general inequality attributed to the two Hermanns: H. Brunn [16] and H. Minkowski [78]. As long as $d > 1$, this tells us that the number of equations exceeds the number of variables by a factor greater than 1, so that the virtual dimension is negative and scales with the size of the support. The outlook for phase retrieval uniqueness, at least in two dimensions and above, looks good!

The worst (equality) case of (3.4) is realized only when the support $\mathcal{S}$ is both convex and centrosymmetric ($\mathcal{S} = -\mathcal{S}$). Wave microscopy ($d = 2$) with rectangular or circular supports appears to be safely unique, with $\Omega = 2$. And as we have seen, a triangular $\mathcal{S}$ gives a substantial improvement, with $\Omega = 3$. Triangles are apparently superior to squares by not being centrosymmetric!

3.1.4 Translational Nonuniqueness

The case of the isosceles-right-triangle support, and contrast that is strictly positive (partially transparent specimen in an aperture), turns out to be

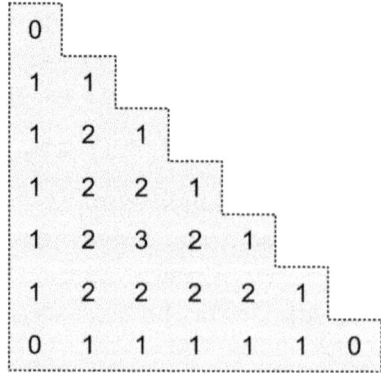

Figure 3.8 When the support has the shape above, the system of quadratic equations for the autocorrelations can be solved sequentially for the contrast pixels when these are ordered as shown. Pixels marked 0 are solved first, followed by the 1-pixels, and so on. The sequential solving method only works when all the autocorrelations are positive, which is true when the contrast is positive throughout the support.

one where the system of quadratic equations can be solved explicitly. The trick is to order the solving of variables according to the hierarchy shown in Figure 3.8. There are three k's whose autocorrelations involve just the pixels marked 0. These are solved first, and the solution always exists and is unique by positivity of the contrast. For all the pixels marked 1, there exists a k such that in the equation for a_k the pixel in question appears as the only new unknown in a linear equation. By using just these equations, all the pixels marked 1 can be solved for. Exactly the same procedure applies to the pixels marked 2, and continuing to the pixels at the top of the hierarchy. One of the exercises in this chapter provides more detail.

In the solving procedure just outlined, the number of equations used exactly equals the number of unknowns (pixels in the support). The equations for all the unused k's provide a very large number of checks. Or that is another way of thinking about the conspiratorial character of the autocorrelation data. The sequential outside–in solving scheme is probably a bad idea when there is noise that can get amplified. The Douglas–Rachford method should be used even in this directly solvable case. That's because each iteration gets the benefit of *all* the autocorrelation/intensity data through the projection P_A.

The direct solution example is instructive also when it breaks down. Suppose we are *not* using a triangular aperture to frame the specimen in an opaque screen, because there is no screen at all. Instead, we just have knowl-

edge of a triangular support region that contains the specimen. While we
know the contrast outside the triangle is zero, there are zero contrast pixels
inside the support as well. The direct solution procedure above breaks down
because we would have to divide by zero, starting with the many zero pixels
at the edge of the support.

The confounding zero pixels surrounding the specimen would be avoided
if the specimen were squeezed into the support with no room to spare. A
kinder approach (for our spider) is to accept the presence of zero pixels as
unavoidable along with a minor qualification of the criterion for uniqueness.
When the support is specified conservatively, the specimen can be translated
right/left and up/down by some number of pixels while keeping it inside the
support. Because these operations preserve the support constraint (B) and
also do not change the autocorrelation (Fourier intensity constraint A), they
represent a benign form of solution nonuniquness. This particular nonuniqueness can be viewed as a small adjustment of the virtual dimension.

If the translated contrast is defined in terms of the original contrast by
$x'_p = x_{p+r}$, where r is the amount of translation (by pixels along the two
axes), the Fourier transform has the following change:

$$
\begin{aligned}
\hat{x}'_q &= \sum_{p \in \Lambda} x'_p \, F_{pq} \\
&= \sum_{p \in \Lambda} x_{p+r} \, F_{pq} \\
&= \sum_{p' \in \Lambda} x_{p'} \, F_{(p'-r)\,q} \\
&= \exp\left(-2\pi i \, r \cdot q \,/\ell\right) \, \hat{x}_q \, .
\end{aligned}
$$

We see that the changed phase (in the new phase retrieval solution) is a
continuous function of r and well defined even when r involves fractions of
pixels. This makes sense when we think of the contrast as defined by its
Fourier transform,

$$
x_p = \frac{1}{\sqrt{|\Lambda|}} \sum_{q \in \Lambda} \exp\left(-2\pi i \, p \cdot q \,/\ell\right) \, \hat{x}_q \, ,
$$

where allowing p to vary continuously, called Fourier interpolation, gives
a representation of the contrast in a continuous space. But this is the appropriate representation for the objects of wave microscopy. The discrete r
translational nonuniqueness is therefore better described as a two-parameter
continuous symmetry. Continuous symmetries enter the virtual dimension
calculus negatively, reducing the effective number of variables. We will en-

counter continuous symmetries later in a big way, but here they are too few
to change the criterion for uniqueness.

3.1.5 One Dimension and Polynomials

The case $d = 1$ arises when the dimension is time and we wish to reconstruct
a temporal signal from its frequency spectrum magnitudes. From a mathematical perspective, this case is interesting because it is the simplest that
can be completely analyzed [15]. There is only one kind of convex support
in one dimension, an interval (of time, say), for which the equality case of
inequality (3.4) gives $\Omega = 1$. We can see this directly by noting that $n + 1$
equally spaced time samples have exactly $n + 1$ unique autocorrelations, including the self-correlation. Since the number of equations exactly matches
the number of unknowns ($d^* = 0$), the solution set will have no continuous
dimensions and comprise a discrete set. However, uniqueness is still at risk
because it may turn out that the number of these discrete solutions grows
rapidly with the length of the signal, $n + 1$.

Help in one-dimensional phase retrieval comes from an unexpected source:
the fundamental theorem of algebra. One-dimensional signals can be represented by polynomials:

$$f(z) = x_0 + x_1 z + \cdots + x_n z^n \ . \tag{3.5}$$

The equally spaced (in time, say) signal values are the coefficients $x_0, \ldots, x_n$,
which we assume are real numbers. By the fundamental theorem of algebra,
$f(z)$ has a unique factorization,

$$f(z) = F\, (z - \alpha_1)(z - \alpha_2) \cdots (z - \beta_1)(z - \beta_1^*)(z - \beta_2)(z - \beta_2^*) \cdots \tag{3.6}$$

where F is real, $\alpha_1, \alpha_2, \ldots$ are real roots, and $\beta_1, \beta_2 \ldots$ are complex roots
that come in complex-conjugate pairs. If r is the number of real roots and
c the number of conjugate pairs, the total number of roots is $r + 2c = n$. A
different number, $r + c = m$, will turn out to be important.

To keep the analysis as simple as possible, we assume the factorization
avoids two special symmetries. The first is that there are no multiple roots,
that is, the numbers $\alpha_1, \alpha_2, \ldots$ and $\beta_1, \beta_2, \ldots$ are distinct. The second is the
property that the operation of taking the reciprocals of the roots produces
a completely different set of n numbers. This means that $1/\alpha_i$ does not
equal any of the real roots, and $1/\beta_i$ does not equal any of the complex
roots. In particular, this eliminates factorizations where some of the roots
lie on the unit circle in the complex plane. These no-symmetry assumptions
are reasonable given that the roots of $f(z)$ are continuous functions of the

contrast coefficients $x_0, \ldots, x_n$. When the latter are drawn from a continuous distribution, a symmetry among the roots corresponds to a zero-probability instance.[1]

The name f for the polynomial is apt because when evaluated at $z = z_\omega = \exp(2\pi i\, \omega)$,

$$f(z_\omega) = \sum_{t=0}^{n} \exp(2\pi i\, \omega\, t)\, x_t \,,$$

we get the (unnormalized) discrete Fourier transform of the signal at frequency ω, where ω is an integer multiple of $1/(n+1)$. In phase retrieval we do not know the complex numbers $f(z_\omega)$, just the squared magnitudes (intensities)

$$|f(z_\omega)|^2 = f(z_\omega)f(z_\omega^*) = f(z_\omega)f(1/z_\omega) \,. \tag{3.7}$$

The last expression in (3.7) suggests we should also define the autocorrelation,

$$f(z)f(1/z) = \sum_{k=-n}^{n} a_k z^k$$

$$= a(z) \,,$$

with coefficients

$$a_k = \sum_{i=0}^{n}\sum_{j=0}^{n} \delta_{k\,(i-j)}\, x_i x_j \,.$$

Unlike the coefficients of $f(z)$, we *do* know the values of the $2n+1$ numbers $a_{-n}, \ldots, a_n$. Because of the negative powers, $a(z)$ is not a polynomial. To define a proper autocorrelation polynomial $\tilde{a}(z)$, we multiply by z^n. Given the known structure (3.6) of $f(z)$, the autocorrelation polynomial has the following structure

$$\begin{aligned}
\tilde{a}(z) &= z^n f(z)f(1/z) \\
&= F^2\,(z - \alpha_1)(1 - \alpha_1 z)\cdots(z - \beta_1)(1 - \beta_1 z)(z - \beta_1^*)(1 - \beta_1^* z)\cdots \,.
\end{aligned} \tag{3.8}$$

This unconventional way of writing the factorization is slightly nonunique because

$$(z - \alpha_i)(1 - \alpha_i z) = \alpha_i^2\,(z - 1/\alpha_i)(1 - z/\alpha_i) \tag{3.9}$$

leaves the structure unchanged while replacing α_i by its reciprocal. The same reciprocal-pair ambiguity applies to the complex roots. In both cases, the prefactor F^2 is changed between the two forms, by a factor α_i^2 when the root is real and $|\beta_i|^2$ when it is complex. By our no-symmetry assumption, these factors are never equal to 1, so we recover uniqueness by insisting the resulting prefactor F^2 is the smallest possible.

With this background, we are finally able to completely solve the phase retrieval problem. We start by forming the autocorrelation polynomial $\tilde{a}(z)$ from the known autocorrelation coefficients. Factoring this polynomial, we are struck by the fact that all the roots come in reciprocal pairs, and a unique factorization of the form (3.8) can be written down. Deciding which factors belong to $f(z)$ and which belong to $f(1/z)$ is largely up to us. Using (3.8) we can decide the factor with root α_i belongs to $f(z)$ and its partner belongs to $f(1/z)$, or vice versa. We have the same binary choice with the complex roots, where we must remember to keep conjugate pairs together (both in $f(z)$ or both in $f(1/z)$) in order that $f(z)$ is a real polynomial. Because the number of binary choices equals m, the sum of the number of real roots and conjugate root pairs, this gives 2^m phase retrieval solutions.

Our counting has left out a factor of two because the autocorrelation prefactor F^2 determines the prefactor of $f(z)$ only up to a sign ($\pm F$). This overall sign ambiguity of the constrast is not special to one dimension, and is usually resolved by appealing to nonnegativity of the contrast, when it applies. If it does apply in our one-dimensional problem, it will severely reduce the 2^m candidates to just those whose contrast coefficients all have the same sign.

In addition to the overall sign, one of the reciprocal-root ambiguities is also not special to one dimension. This is the refactorization of $\tilde{a}(z)$ that simply swaps $f(z)$ and $f(1/z)$. If $f(z)$ has contrast coefficients (3.5), then

$$z^n f(1/z) = x_n + x_{n-1}z + \cdots + x_0 z^n$$

is just the reflection of the contrast (reversing the time order of the signal). In any number of dimensions, reflecting in the origin,

$$x'_p = x_{-p},$$

has no effect on the autocorrelation. The only way to eliminate reflection ambiguity in phase retrieval is to use a noncentrosymmetric support constraint having the property that only one of the reflection-related contrasts fits inside the support.

The discrete, exponential-in-number nonuniqueness in one dimension is sometimes portrayed as a great failing of phase retrieval. This is unfair when we consider how easily uniqueness is restored. The marginal value $\Omega = 1$ applies only when the support is convex – an interval. If instead $\mathcal{S}$ is an interval, say of length a, from which an interior subinterval of length b has been excised, then

$$\Omega = a/(a - b) > 1$$

and the problem is overconstrained. Now there has to be a conspiracy in the autocorrelations (Fourier magnitudes) for a solution to exist at all. It is not even necessary to specify the positions of the subinterval pixels where the constrast vanishes. Simply the constraint that the contrast vanishes on *some* fraction of the pixels – positions unknown – is enough to confer uniqueness.

3.2 Crystallographic Phase Retrieval

The concluding remark of the previous section is the main insight for this one. But to do justice to this variant of phase retrieval, we start with some background.

By far the most important application of phase retrieval is *crystallography*: atomic structure determination from crystal-diffraction data. The computational grid for representing contrast – when the support is known – is periodic for computational convenience only. In this section, the contrast is truly periodic because the data is obtained from a periodic material. Retrieving phases for periodic contrasts has more of a discrete character because one is dealing with the terms of a Fourier *series*. In general, there is no continuity of the phase (or even magnitude) between adjacent series coefficients.

Phase retrieval with a known support is not such a hard problem. We even saw one case – positive contrast in a triangular support – where the computation of the unique solution is explicit. From what we know at present, crystallographic phase retrieval is a much harder problem. Circumstantial evidence in support of that claim is the fact that mathematicians and computer scientists have, apart from a few outliers [53, 17], completely avoided it. As of 2024 all work on the crystallographic problem has been of the heuristic variety, such as what is expounded in this book. Some of the pioneers of crystallographic phase retrieval – Hauptman, Sayre – had mathematical training and brought rigor to the subject. But that rigor never extended to the solution process itself, the retrieval of phases.

Every year about 50 000 crystal structures are uploaded to the Cambridge Structural Database [50], each one solved by a heuristic algorithm. These are typically inorganic compounds of interest to materials scientists. At a higher level of complexity are the biomolecule (protein, enzyme, virus, ...) structures deposited in the Protein Data Bank (PDB) [18]. The PDB is a vital resource for biomedical researchers and as of 2024 contains over 150 000 structures determined by heuristic phase retrieval algorithms. Credit for the growth of the PDB mostly goes to the "structural biologists" whose expertise is coaxing biomolecules to form crystals. Once crystallized and measured with X-rays (diffraction intensities), the phase retrieval is performed by software developed principally by George Sheldrick [98] beginning in the 1970's. A large part of what goes on in Sheldrick's code is imposing a sparsity constraint on the contrast.

3.2.1 Crystals and Symmetry

Most proteins have irregular shapes and form crystals whose periodicity has odd repeat distances and angles. In general, the periodicity is defined by three (space) vectors $\mathbf{p}_1$, $\mathbf{p}_2$, $\mathbf{p}_3$ of arbitrary length and subtending arbitrary angles. Similar to wave microscopy, contrast *voxels* are indexed with three integers: $p = (p_1, p_2, p_3)$. The position of a voxel in space is

$$\mathbf{p} = \left(\frac{p_1}{\ell_1}\right) \mathbf{p}_1 + \left(\frac{p_2}{\ell_2}\right) \mathbf{p}_2 + \left(\frac{p_3}{\ell_3}\right) \mathbf{p}_3 ,$$

where p_1 runs from 0 to $\ell_1 - 1$ and similarly for p_2 and p_3. The vectors $\mathbf{p}_1$, $\mathbf{p}_2$, $\mathbf{p}_3$ are the edges of a parallelepiped-motif, or *unit cell* of the crystal.

The diffraction data also has three repeat-vectors, now defining the set of possible *changes* in the wave vectors (momenta) of the scattered X-rays. Having units of inverse length, $\mathbf{q}_1$, $\mathbf{q}_2$, $\mathbf{q}_3$ are completely determined by the crystal repeats via the relations

$$\mathbf{p}_i \cdot \mathbf{q}_j = \delta_{ij} . \tag{3.10}$$

For example, to construct $\mathbf{q}_1$ we select the vector orthogonal to the plane defined by $\mathbf{p}_2$ and $\mathbf{p}_3$ having unit dot product with $\mathbf{p}_1$.

Now consider data at the wave vector (change), or *Bragg peak*

$$\mathbf{q} = q_1 \, \mathbf{q}_1 + q_2 \, \mathbf{q}_2 + q_3 \, \mathbf{q}_3 ,$$

where $q = (q_1, q_2, q_3)$ are integers. Let x_p be the contrast at voxel p of the unit cell, which is repeated at positions

$$\mathbf{p} + n_1 \, \mathbf{p}_1 + n_2 \, \mathbf{p}_2 + n_3 \, \mathbf{p}_3 ,$$

where the integers n_1, n_2, and n_3 have an enormous range that covers all the unit cells of the crystal. The wave interference at Bragg spot $\mathbf{q}$, from the scattering by all these repeating contrast elements, is given by the following sum:

$$x_p \sum_{n_1, n_2, n_3} \exp\Big(2\pi i (\mathbf{p} + n_1\,\mathbf{p}_1 + n_2\,\mathbf{p}_2 + n_3\,\mathbf{p}_3) \cdot \mathbf{q}\Big)$$

$$= x_p \exp(2\pi i\,\mathbf{p}\cdot\mathbf{q}) \sum_{n_1, n_2, n_3} 1\,.$$

This follows from the q_i and n_i being integers and relation (3.10). The remaining sum is independent of $\mathbf{p}$ and $\mathbf{q}$ and proportional to the number of unit cells in the crystal. The enormous magnitude of the latter is responsible for the concentration of the intensity data at "peaks."

Finishing up the wave interference calculation, at the same $\mathbf{q}$ but now including all the contrast voxels of the unit cell, we obtain, after normalization,

$$\hat{x}_q = \frac{1}{\sqrt{|\Lambda|}} \sum_{p \in \Lambda} \exp(2\pi i\,\mathbf{p}\cdot\mathbf{q})\, x_p \qquad (3.11)$$

$$= \frac{1}{\sqrt{|\Lambda|}} \sum_{p \in \Lambda} \exp\left(2\pi i \left(\frac{p_1 q_1}{\ell_1} + \frac{p_2 q_2}{\ell_2} + \frac{p_3 q_3}{\ell_3}\right)\right) x_p\,.$$

The sum is over the same kind of orthogonal grid Λ we had in wave microscopy and defines a set of diffraction amplitudes $\hat{x}_q$, again as the discrete Fourier transform of the contrast. The amplitudes $\hat{x}_q$ live on a similarly shaped grid $\widetilde{\Lambda}$. Because we only care that there is a proportionality between $|\hat{x}_q|^2$ and the Bragg-peak intensity I_q, the normalization of the sum is arbitrary. The normalization choice in (3.11) preserves norms (Parseval's theorem).

The fact that the data I_q live on a 3D grid $\widetilde{\Lambda}$ is at odds with X-ray detectors being 2D arrays of photon counters. Figure 3.9 tries to resolve the matter in the more easily visualized case where both dimensions are smaller by one (2D crystallography and 1D detector). Every incident photon has wave vector $\mathbf{k}$. The wave vectors $\mathbf{k}'$ of the scattered photons have the same magnitude, because energy is conserved, and lie on a sphere named after Paul Ewald (a circle in our cartoon). There is no actual scattering, or constructive interference, unless the wave vector change $\mathbf{k}' - \mathbf{k}$ matches one of the $\mathbf{q}$'s as derived earlier from the crystal's periodicity. We see that the $\mathbf{q}$'s form a 2D lattice $\widetilde{\Lambda}$ in the wave vector space and in general do not intersect Ewald's sphere at all. To target individual Bragg peak $\mathbf{q}$'s, the crystal must be suitably rotated. There is a whole branch of crystallography devoted to best practices in Bragg peak measurement that take into account the finite widths of the peaks. In any rotation of the crystal, relatively few peak are intersected by the sphere and are "seen" in the measurement. The

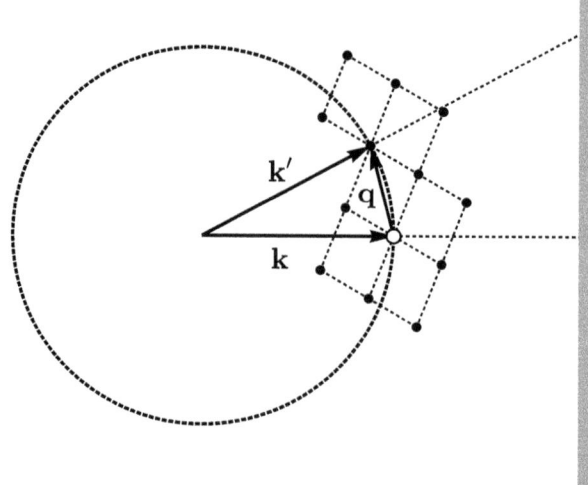

Figure 3.9 Geometry of crystallographic data in wave vector space, shown here in one lower dimension. The origin of the space is the white dot. There is strong scattering whenever there exist scattered-photon wave vectors $\mathbf{k}'$ whose difference $\mathbf{k}' - \mathbf{k}$ from the incident-photon wave vector matches one of the discrete set of wave vectors $\mathbf{q}$ determined by the periodicity of the crystal. Because the wave vectors $\mathbf{k}'$ lie on a circle (Ewald's sphere in three dimensions), the crystal must be rotated to map out the lattice of $\mathbf{q}$'s and measure the corresponding intensity of scattered photons.

2D position of a peak can be inferred from the known rotation of the crystal and the angle $\mathbf{k}'$ subtends with respect to the incident $\mathbf{k}$ (which is aimed at the center of the detector).

Very often the crystal-periodicity vectors $\mathbf{p}_1$, $\mathbf{p}_2$, and $\mathbf{p}_3$ have special symmetries (e.g., 60° and 90° angles) because the irregular molecules first combine into symmetrical pairs, triples, or even larger motifs. When this happens, the voxels of the unit cell fall into finite symmetry orbits $\{p_1, p_2, \ldots\}$, all sharing the same contrast values ($x_{p_1} = x_{p_2} = \cdots$). Though this reduces the number of unknowns, the virtual dimension is unchanged because the number of equations changes by the same amount. That's because the Bragg peak indices fall into symmetry orbits as well, with equal intensity values ($I_{q_1} = I_{q_2} = \cdots$). The premise that underlies the symmetry independence of the virtual dimension is that the contrast and crystal lattice have the same symmetry group. Though there is a technical sense in which this premise is always upheld, we next consider a situation where the two groups are different in a practical sense.

If one ignores the genetic material encased by the capsid shell, icosahedral

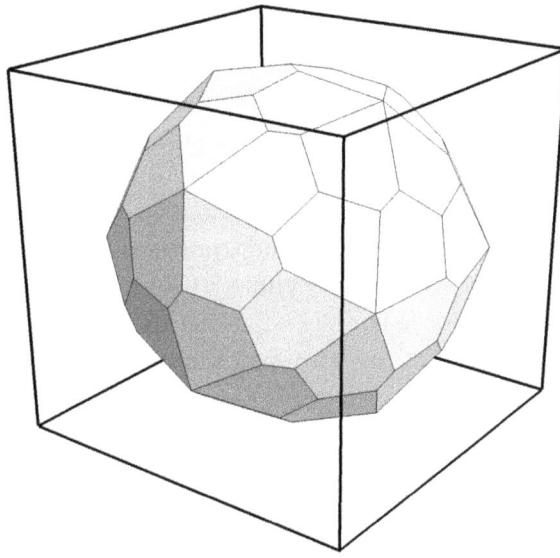

Figure 3.10 Cartoon of a virus crystal. Sixty copies of the irregular pentagon form an icosahedral polyhedron in much the same way that 60 identical capsid molecules assemble into an icosahedral virus particle. The surrounding cube is the unit cell of the crystal and its axes are aligned with three of the virus particle's 15 axes of 2-fold rotational symmetry.

virus particles have the second largest symmetry group in three dimensions: the 60-element icosahedral group of the 60-sided polyhedron shown in Figure 3.10. This figure is also meant to serve as a cartoon for a virus crystal. The unit cell has the shape of a cube and encloses one virus particle. Its symmetry group is just a 12-element subgroup of the virus symmetry group. This means that for the voxels in the capsid there are five times fewer variables than there are equations from the cubic-crystal diffraction data. In a strict sense this is not quite true, because the cubic packing of the virus particles in the crystal perturbs the perfect icosahedral symmetry. Pairs of viruses make contacts along just three of their 15 axes of 2-fold rotation symmetry, thereby reducing the symmetry to the 12-element subgroup. But if the deformation resulting from these contacts is small, the near-icosahedral symmetry of the capsid is a conspiracy that can and should be exploited [77, 58]!

The virus crystal is instructive in another way. Most crystals are periodic only in an average sense, as the contrast may vary from unit cell to unit cell. As long as its average exists and is nontrivial (nonconstant), the diffraction data comprises the same set of Bragg peaks as a perfectly periodic crystal. In general, the Bragg peak intensities provide only information about the average contrast, and it is only this that phase retrieval can hope to reconstruct. In a virus crystal, both the genetic material within the capsid and the solvent molecules outside (filling the vacuum between particles) differ greatly from unit cell to unit cell. From the Bragg-peak intensities one can only hope to infer their averaged contrast. Because the genetic material has an icosahedrally symmetric enclosure, its contrast average will also have icosahedral symmetry. On the other hand, the contrast average of the solvent on the outside will have the lower symmetry of the cubic crystal packing. Both averages may be very featureless, the limit of no features being a "flat" (constant) contrast described by a single number.

3.2.2 Nonnegativity and Sparsity

Though nonnegativity of the contrast in X-ray diffraction is a solid constraint, by itself it is not useful in crystallography. That's because most of the incident X-rays pass straight through the crystal, making the measurement of the $q = (0, 0, 0)$ intensity (forward scattering) infeasible. The corresponding amplitude is the constant term in the Fourier series. When unconstrained by a measurement, any contrast can be made nonnegative by a suitably large value of this constant term.

Suppose the unit cell of the crystal has N voxels. Its Fourier transform will then also have N amplitudes (Fourier series coefficients). For the purpose of counting variables and constraints, we care about only one property of these amplitudes: Are their phases continuous variables? Since this property only depends weakly on the number of dimensions d, we expand the analysis to cover any d (not just the case $d = 3$).

For most of the amplitude indices $q \in \widetilde{\Lambda}$, the following is true:

$$q \neq -q \quad \mod \widetilde{\Lambda} .$$

This is never true at the origin, where all components of q are zero. Also, if any of the sizes (periods) ℓ of $\widetilde{\Lambda}$ is even, then because $\ell/2 \equiv -\ell/2 \mod \ell$, there are some other q's that are equivalent to their reflections in the origin. Altogether there are 2^e such q, where e is the number of even sizes among $\ell_1, \ldots, \ell_d$. From the complex conjugation property $\hat{x}_{-q} = \hat{x}_q^*$ of real contrasts, all 2^e reflection-equivalent amplitudes are real and therefore do not have a continuous phase variable. Because the number of real coefficients, 2^e, is much smaller than N when the contrast is well resolved (all ℓ's are large), and because $\varphi_{-q} = -\varphi_q$ (inversion-related amplitudes share the same phase variable), we can always use the estimate $N/2$ for the number of continuous phase variables.

Nonnegativity is only a useful constraint when combined with sparsity. When we pick random values for the $N/2$ phases, one of the voxels $p \in \Lambda$ will have the smallest contrast, and by adjusting the (unmeasured) constant term in the Fourier series we get zero contrast on just that one voxel. If we now take full advantage of the $N/2$ free parameters (phases), it should be possible, again by adjusting the constant term, to have $N/2$ zero-contrast voxels. A significantly larger number of zero-contrast voxels, by setting just $N/2$ phases, can only be explained by a conspiracy!

Highly conspiratorial crystal contrasts are said to have the "atomicity" property. The averaged contrast in well-ordered crystals, where the cell-to-cell variability is small, can be very sparse if the data extends to large enough q that atoms are well resolved. Though atoms have a nonzero size, their combined footprint (support) is a small fraction of the volume of the unit cell. The number of voxels containing positive atom contrast is typically much smaller than $N/2$, leaving a number of zero-contrast voxels that is much larger than $N/2$. If an algorithm discovers settings of the $N/2$ phases for which the number of zero-contrast voxels is much larger, we can be very confident that the correct phase retrieval is at hand.

Just as in the case of the triangular fixed support of wave microscopy, in the high-sparsity limit of the crystallographic problem the retrieval of phases is essentially direct when analyzed using the contrast autocorrelation. We will describe the method graphically for a two-dimensional contrast in a unit cell of 9×9 pixels ($2^e = 1$). Suppose there are exactly M "atoms." In our model this means that M pixels have positive contrast, and all others are zero. The leftmost panel in Figure 3.11 shows an example contrast with $M = 4$.

By taking the Fourier transform of the intensity, we can construct the contrast-autocorrelation a, shown next to the contrast in Figure 3.11. The origin has been placed at the center of the 9×9 grid to aid in visualizing the $a_{-k} = a_k$ symmetry. Thanks to the high sparsity, for each of the $4 \times 3 = 12$ nonzero k's for which a_k is nonzero, there is exactly one atom pair that has this k as its separation.

Let's start with $k = (2, 1)$. Since $a_k > 0$, we know there must be a pair of atoms with that separation. Moreover, because the autocorrelation is unchanged when the contrast is uniformly translated, we are free to place a pair of atoms with this separation anywhere in the unit cell. The two gray pixels, p_1 and p_2 in the third panel of Figure 3.11, is such a placement. These pixels represent just the support of the contrast. The contrast values on the support will be reconstructed once we are finished reconstructing the support.

Let $\mathcal{A}$ be the support (set of pixels) of the autocorrelation. The third panel of Figure 3.11 also shows the pixels we get when $\mathcal{A}$ is translated so its origin

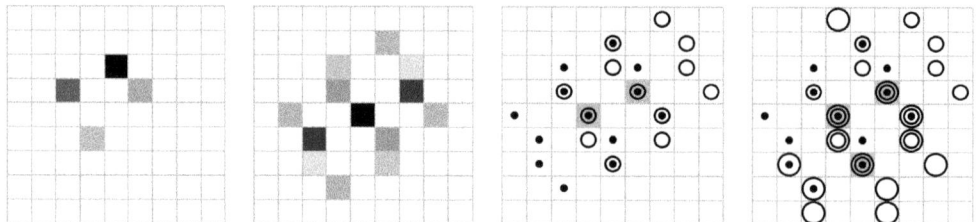

Figure 3.11 The first two panels show a contrast supported on four pixels and its autocorrelation. In the other two panels, the 4-pixel support is reconstructed, iteratively, from translations of the autocorrelation support.

first coincides with p_1 (dots), and then p_2 (circles). A pixel that belongs to both of these sets (dot inside circle) is a candidate for another pixel of the contrast support, because such a pixel should have a positive autocorrelation with both p_1 and p_2. We see there are four such candidates. The rightmost panel of Figure 3.11 shows what happens when we select one of these for the third contrast pixel p_3, and repeat the exercise of translating $\mathcal{A}$, now including p_3 to be the origin. We see that now there is just a single pixel in the three-way intersection of the $\mathcal{A}$-translations, and in fact this is the last remaining contrast pixel. Though we had a choice of four candidate pixels when selecting p_3, any of them would have led to the same 4-pixel contrast support or its inversion (which has the same autocorrelation).

After the contrast support is reconstructed, it is easy to reconstruct the contrast values when the support is very sparse. Suppose there are three pixels in the support whose separations k are unique, that is, not shared by any other pair of contrast-support pixels. For example, any three of the contrast pixels in Figure 3.11 has this property. The three independent autocorrelation equations for the corresponding k's take the form

$$
\begin{aligned}
x_{p_1} x_{p_2} &= |\Lambda|\, a_{p_1 - p_2}\,, \\
x_{p_2} x_{p_3} &= |\Lambda|\, a_{p_2 - p_3}\,, \\
x_{p_3} x_{p_1} &= |\Lambda|\, a_{p_3 - p_1}\,,
\end{aligned}
$$

because the autocorrelation sums just have a single term. Using the prior information that all the contrast values (on the support) are positive, the

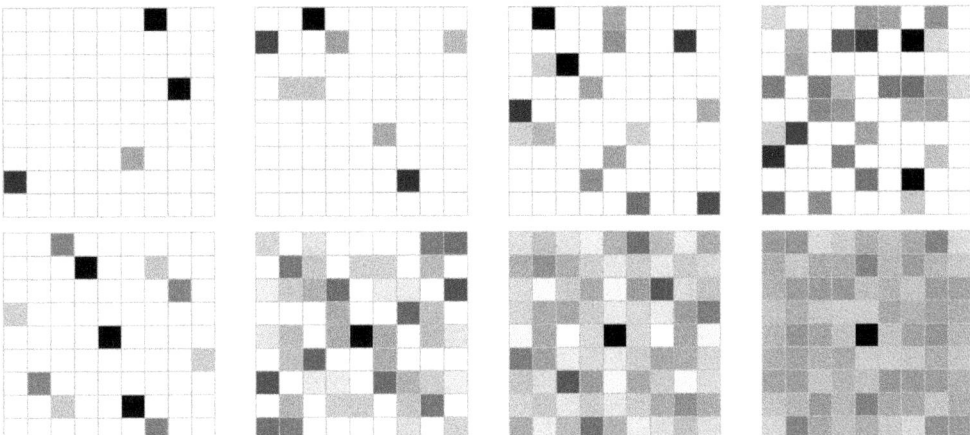

Figure 3.12 Examples of contrasts (top row) and corresponding autocorrelations (bottom row) in a unit cell of 9×9 pixels as the number of support pixels of the contrast is doubled.

solution is unique:

$$x_{p_1} = \sqrt{|\Lambda| \, \frac{a_{p_1-p_2} \, a_{p_3-p_1}}{a_{p_2-p_3}}} \; ,$$

$$x_{p_2} = \sqrt{|\Lambda| \, \frac{a_{p_2-p_3} \, a_{p_1-p_2}}{a_{p_3-p_1}}} \; ,$$

$$x_{p_3} = \sqrt{|\Lambda| \, \frac{a_{p_3-p_1} \, a_{p_2-p_3}}{a_{p_1-p_2}}} \; .$$

There may be many such unique-autocorrelation triangles, even when three times the number of triangles (the number of equations) greatly exceeds the number of contrast pixels. The contrast value of particular pixels will then be determined multiple times, and their equality – or near equality because of noise – firmly establishes the conspiracy in case we had any doubts.

The feasibility of contrast reconstruction by autocorrelation analysis drops precipitously as the number of support pixels M increases. Figure 3.12 compares contrasts and their autocorrelations for $M = 4, 8, 16, 32$, still for a unit cell of $N = 9^2 = 81$ pixels. The last of these has $32 \times 31 = 992$ (nonzero) atom-atom separations, bringing the average number of terms in the autocorrelation sums to $992/80 = 12.4$. To solve for the contrast, we would first have to confront the combinatorial explosion in the number of terms. Already for $M = 16$ the autocorrelation support covers almost the entire unit cell. The iterative construction shown in Figure 3.11, of intersecting translates of the autocorrelation support, is completely useless for reconstructing the support

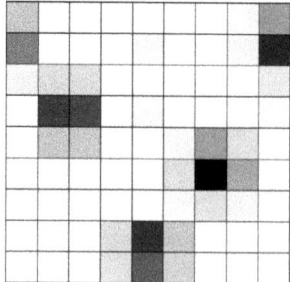 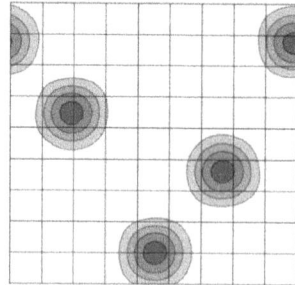

Figure 3.13 *Left*: Coarse sampling of the contrast where four atoms are given a combined support of 32 pixels. *Right*: Fourier interpolation of the coarse contrast defines atom centers at subpixel resolution.

of the contrast. On the other hand, conspiracy theory tells us all of these phase retrieval instances have unique solutions. In the $M = 32$ instance, the contrast has $81 - 32 = 49$ zeros, whereas the $80/2 = 40$ free phases combined with the single free constant term can account for only 41 zeros.

The "atomic" contrasts in Figure 3.12 are both good and bad in terms of realism. Real atoms possess size, so having them occupy an entire pixel (in the 2D model) is not unreasonable. What's bad is that all the atoms are precisely located at the pixel centers. In effect there is an unphysical integrality in the relative atomic positions. This shortcoming is easily fixed by allowing each atom to be supported on more than one pixel. The left panel of Figure 3.13 shows four Gaussian atoms, of standard-deviation-width 0.6 pixel-spacings, supported on a total of $4 \times 8 = 32$ pixels. Eight support pixels per atom is plenty to define the atom centers at subpixel resolution. To see this, we perform Fourier interpolation of the coarse 81-pixel contrast and render its contours in the right panel. Fourier interpolation pads the Fourier transform of the coarse contrast with zeros and then inverse transforms the result to produce a smoother contrast. The centers of the contours of the smooth contrast define continuously variable atom positions that have been freed of the integrality artifact. In practical terms, this means we can estimate the number of (3D) support voxels M knowing the number of atoms and their width in voxel units. Then, provided we can assert the number of zero voxels, $N - M$, is greater than $N/2$ by a healthy margin, we can be confident our phase retrieval solution is unique.

A contrast can be sparse in ways that does not involve well-resolved atoms. In crystals of complex biomolecules, there is much variability between unit cells. The average of the contrast, over unit cells, is a highly blurred image in which only larger-scale molecular features are discernible (chains, rings). Luckily, the space between biomolecules is often occupied by small solvent molecules. Because the latter are disordered, their contrast average is "flat" and characterized by a single constant whose value is much smaller than the biomolecule contrast. We therefore are presented once again with a sparse positive contrast. The main difference is that unlike the case of atoms, the size of the support is harder to estimate.

The Protein Data Bank could not have been created without several clever extensions of crystallographic phase retrieval that go beyond the scope of this book. These extensions exploit specific characteristics of the molecules of interest as well as the physics of the X-ray scattering process. Like cryptographers that break codes by finding weaknesses, crystallographic phase retrievers have been similarly resourceful in unlocking a wealth of biological secrets.

3.3 Quantum State Tomography

To verify the operation of a new chip, a computer engineer will try combinations of input voltages and measure the corresponding output voltages. A *quantum computer* engineer faces a much harder task. Measurement outcomes are intrinsically probabilistic. Also, it is not immediately obvious what combination of measurements suffices to determine the "state" of the output, a quantum mechanical entity. Engineering undertakings aside, there is a line of quantum mechanics research where the challenge is to create, in the lab, increasingly "entangled" states of some number of qubits – the quantum analogs of bits. To achieve success in that endeavor, one also needs to carry out measurement protocols. The design of protocols, and the algorithmic reconstruction of quantum states from the outcome of measurements, is called *quantum state tomography* (QST). The term tomography suggests that multiple measurements are required, each "slice" revealing something new about the state. The crystal rotations in crystallography designed to slice through different Bragg peaks could also be called a form of tomography. In fact, we will see at the end of this section that phase retrieval is a special case of QST.

We will introduce the QST problem in mathematical terms. No prior background in quantum mechanics is expected. In the course of finding a bipartisan formulation of QST, we will see that once again we are confronted by a system of quadratic equations. Unlike phase retrieval, where translational symmetry (via the Fourier transform) confers considerable structure to the system of equations, the QST problem is posed with almost no structure at all.

3.3.1 Quantum Formalism

We forgo Dirac's notation for quantum states and operators and stay with the notational conventions of this book. A vector ψ is understood to be a row of numbers while ψ^{T} is the same set of numbers arranged in a column. The symbol † combines transposing and taking the complex conjugate. The only thing unconventional about our notation is the use of lower case for matrices when these are unknowns. We do this to emphasize the fact that matrices are unknown about as much as vectors. To help navigate the world of lower-case symbols, we use the standard symbol and naming conventions of quantum mechanics. The names will mean something to readers with background in quantum mechanics but that meaning plays no role in QST as a mathematical exercise. Upper-case symbols are reserved for known (constant) entities, such as data.

Consider an apparatus, such as a quantum computer, whose *state* is modeled by normalized *wave functions* $\psi \in \mathbb{C}^\ell$, $\|\psi\|^2 = \psi\psi^\dagger = 1$. Information about the state is gained through measurements associated with $\ell \times \ell$ Hermitian matrices H. The eigenvalues of each H are the possible (discrete) measurement outcomes and the projections of ψ onto the H-eigenvectors give the probabilities, after squaring, of the measurement outcomes. Say

$$H = U^\dagger \operatorname{diag}(E_1, \ldots, E_\ell)\, U \ ,$$

then

$$\psi\, U^\dagger U = \sum_{i=1}^{\ell} (\psi U_i^\dagger)\, U_i$$

$$= \sum_{i=1}^{\ell} \psi_i\, U_i$$

is the expansion of the state in the eigenbasis of H and $|\psi_i|^2$ is the probability of measurement outcome (H-eigenvalue) E_i. When the apparatus is prepared in state ψ multiple times, and each is subjected to measurement, then the expectation value of the measurements is

$$\langle H \rangle = \sum_i E_i |\psi_i|^2$$

$$= \sum_i E_i \left(\sum_j \psi_j U_{ij}^* \right) \left(\sum_k \psi_k^* U_{ik} \right)$$

$$= \sum_{ijk} \psi_k^* \psi_j \, U_{ij}^* E_i U_{ik}$$

$$= \mathrm{Tr}\left(\psi^\dagger \psi H \right) .$$

The rank-1 matrix $\rho = \psi^\dagger \psi$ is the *density matrix* associated with the state ψ.

In general, even a quantum apparatus interacts weakly with an environment whose state cannot also be interrogated by measurement. The apparatus is then modeled as being in any of m states $\psi_1 \ldots, \psi_m$ with probabilities $p_1, \ldots, p_m$. Measurement expectation values are then subject also to these probabilities. This is conveniently covered by replacing the rank-1 *pure state* density matrix with the rank-m *mixed state* density matrix

$$\rho = \sum_{k=1}^m p_k \, \psi_k^\dagger \psi_k . \tag{3.12}$$

The problem of reconstructing this density matrix from measurement expectation values is the mathematical statement of quantum state tomography.[2] Since ρ has many parameters, and each measurement expectation value is a single real number, the reconstruction must use data from multiple Hermitian measurement matrices:

$$\langle H_i \rangle = \mathrm{Tr}\left(\rho H_i \right) , \quad i = 1, \ldots, n . \tag{3.13}$$

The matrices H_i can be assumed to be linearly independent. That's because, for example, if $\langle H_1 \rangle$ is the measurement result of H_1, and $\langle H_2 \rangle$ is the measurement result of H_2, then we already know that the result of measuring $H_1 + H_2$ would be $\langle H_1 \rangle + \langle H_2 \rangle$.

Given the size ℓ of the quantum states, a bound m on the number of states contributing to the density matrix mixture, what number n of expectation value data is required for a unique reconstruction of ρ?

3.3.2 Counting Variables and Constraints

The equations (3.13), like the equations of phase retrieval, are quadratic in the unknowns, if we take these to be the complex elements of the m wave functions. But unlike phase retrieval, where the structure of the equations only depends on the geometry of the support $\mathcal{S}$, the quadratic equations of quantum state tomography depend on all details of the n measurement matrices H_i. We should cast aside hopes of a miracle transformation, like the Fourier transform, that manages to convert all these equations into diagonal form, like the $|\hat{x}_q|^2 = I_q$ of phase retrieval.

For quantum state tomography we use an approach where the quadratic entity itself, ρ, assumes the role of the unknown. We break with our convention of always using x for the unknown, and continue to use the symbol ρ. This also avoids confusion later, when we revisit phase retrieval (with its own definition of x) as a special case of QST. To get started, we ask what constraints need to be imposed on ρ. From the definition (3.12) we see that ρ is Hermitian, has rank bounded by m, and is positive semi-definite because the probabilities p_k are nonnegative. Since the wave functions are normalized and the probabilities sum to one, we also have the constraint

$$
\begin{aligned}
\operatorname{Tr} \rho &= \sum_{k=1}^{m} p_k \left(\sum_{j=1}^{\ell} \psi_{kj}^{*} \psi_{kj} \right) \\
&= \sum_{k=1}^{m} p_k \\
&= 1
\end{aligned}
$$

on the trace of ρ. All of the above properties combined define constraint set A:

$$
A = \left\{ \rho \in \mathbb{H}^{\ell} \colon \operatorname{rank}(\rho) \leq m, \ \rho \succcurlyeq 0, \ \operatorname{Tr} \rho = 1 \right\}.
$$

In the declaration, $\mathbb{H}^{\ell}$ denotes the space of $\ell \times \ell$ Hermitian matrices. Of course A is only a suitable constraint set if there is an efficient projection that takes arbitrary $\rho \in \mathbb{H}^{\ell}$ and maps them to the nearest point in A. This turns out to be a special case of the *spectral projections* covered in Section 4.4.2.

The formerly quadratic equations are now linear equations in ρ and define constraint set B:

$$
B = \left\{ \rho \in \mathbb{H}^{\ell} \colon \operatorname{Tr}(\rho H_i) = \langle H_i \rangle, \ i = 1, \ldots, n \right\}. \tag{3.14}
$$

Recall that the real numbers $\langle H_i \rangle$ are the data for the reconstruction, analogous to the intensity measurements in phase retrieval. The projection to the hyperplane defined by an arbitrary system of linear equations is covered in Section 4.2.1.

To answer the data sufficiency question for unique density matrix reconstruction, we start by counting the number of continuous dimensions d_A of set A. This is the number of variables in the reconstruction problem. Then, because each of the linearly independent equations that define set B reduces the dimension of the set intersection by one, the virtual dimension is

$$d^* = d_A - d_B ,$$

where $d_B = n$ is the codimension of B. To have confidence in a reconstruction, d^* should not just be negative, but negative by a healthy margin. We can convey this with the same ratio used in known-support phase retrieval, of the number of equations per unknown:

$$\Omega = \frac{d_B}{d_A} .$$

Each point ρ in A is a Hermitian matrix with certain properties. Consider the eigendecomposition of ρ. There will be m nonnegative eigenvalues that sum to one and m orthogonal basis vectors. The basis vectors are normalized and can be multiplied by a phase with no effect on ρ. The first basis vector is therefore completely specified by $2\ell - 2$ real parameters. In order for the second basis vector to be orthogonal to the first, a complex inner product must vanish, reducing its number of parameters by 2. The third basis vector has 4 fewer parameters, and so on, until we arrive at the mth basis vector with $2\ell - 2m$ parameters. The total number of continuous parameters associated with the eigenbasis is therefore

$$\sum_{k=1}^{m} (2\ell - 2k) = 2\ell m - m(m+1) .$$

The m eigenvalues that sum to 1 add another $m - 1$ real parameters giving a grand total of

$$d_A = 2\ell m - m^2 - 1$$

continuous dimensions for set A. From this we get an equation-per-variable ratio

$$\Omega = \frac{n}{2\ell m - m^2 - 1} . \tag{3.15}$$

As a point of reference, in wave microscopy Ω is usually at least equal to 2.

If we succeed at reconstructing a unique density matrix, can we then proceed to reconstruct m unique states ψ and their associated probabilities [56]? A similar question came up in the kissing spheres problem, where the unknown was the Gram matrix of the sphere-center vectors, not the vectors themselves. There we found a continuous nonuniqueness in reconstructing the vectors, implicit in the definition of the Gram matrix. Not surprisingly, the reconstruction of quantum states, when there is more than one $(m > 1)$, is also beset by nonuniqueness, though unfortunately not as comprehensible as a rotation of space. The eigendecomposition of the density matrix can serve as a canonical state reconstruction, one where the m states are orthogonal, but there is nothing in the quantum model that singles out this reconstruction from others. Further exploration of this question is taken up in one of the exercises.

3.3.3 Phase Retrieval as QST

To help us interpret phase retrieval as an example of QST, we will start to refer to the contrast x as the *exit wave*. To maintain our conventions for phase retrieval, will not go as far as replacing x with the wave function symbol ψ, though a quantum wave is absolutely the right model for the coherent, laser-like radiation that exits the specimen. Up to now we have only considered the effect of opacity on the exit wave. If we include the full effects of the specimen's index of refraction, the incident wave also experiences a position-dependent phase rotation by the time it has exited the specimen. The combined effect of opacity and phase rotation is a complex-valued exit wave x.

For computations we sample the exit wave exactly as we did the contrast in phase retrieval: on an $\ell \times \ell$ grid Λ. We will assume the specimen is contained in the aperture of an opaque screen, so x is supported on only a subset $\mathcal{S} \subset \Lambda$ of the pixels. The number of support pixels, $|\mathcal{S}|$, takes the place of the length ℓ of the complex vector (wave function) of QST. We will continue to use integer coordinates $p \in \Lambda$ to index the components of x.

The density matrix for phase retrieval is the rank-1, $|\mathcal{S}| \times |\mathcal{S}|$ Hermitian matrix

$$\rho = x^\dagger x \ .$$

More precisely, this is the density matrix when the radiation that illuminates the specimen is perfectly coherent. The case of partially coherent illumination is modeled exactly as in QST, by a set of exit waves $x_1, \ldots, x_m$ that arise with probabilities $p_1, \ldots, p_m$. The corresponding density matrix is again the rank-m matrix

$$\rho = \sum_{k=1}^{m} p_k\, x_k^\dagger x_k \ .$$

Besides incoherent illumination, mixed-state density matrices also arise when the illumination is coherent, but the state of the specimen is not static [106]. Atomic vibrations are a well-known example of this, but not an interesting one because the corresponding m (atomic configurations) is far too large. Much more feasible is the case of a biomolecule that can be in two conformations. Unlike crystallographic phase retrieval, which can only reconstruct the average of the two conformations, a reconstructed rank-2 density matrix would deliver two images, one for each conformation.

The free-space modes (states) of radiation propagation are characterized by a three-dimensional wave vector $\mathbf{q}$. We index these by the same pairs of integers $q \in \Lambda$ we used in wave microscopy to label the pixels of the detector. This works because the position on the detector is a scale factor applied to the two components of $\mathbf{q}$ in the plane of the specimen and the third component of $\mathbf{q}$ is determined from the other two because $\|\mathbf{q}\|$ is a constant. A scale factor relates the integer-coordinate point samples q to the in-plane components of $\mathbf{q}$.

The exit wave is a quantum superposition of waves of all $\mathbf{q}$, and detector pixel q measures just the waves closest to that particular point sample. The projection of the exit wave onto point sample q is the wave

$$\hat{x}_q = (x\, F_q^\dagger)F_q \ , \tag{3.16}$$

where

$$(F_q)_p = \frac{1}{\sqrt{|\mathcal{S}|}} \exp(2\pi i\, p \cdot q/\ell)$$

is the pth component of the normalized wave (function) with in-plane wave vector q. The probability the wave is detected, as a photon, in detector pixel q is equal to

$$\begin{aligned} \|\hat{x}_q\|^2 &= \hat{x}_q \hat{x}_q^\dagger \\ &= x\, F_q^\dagger F_q\, x^\dagger \\ &= \mathrm{Tr}(\rho H_q) \ , \end{aligned} \tag{3.17}$$

where

$$H_q = F_q^\dagger F_q$$

is the $|\mathcal{S}| \times |\mathcal{S}|$ Hermitian measurement matrix for detector pixel q. As H_q is a rank-1 projection matrix, there is one measurement value (eigenvalue) of 1 (photon detected), and all others are 0 (no photon detected).

Previously, when the symbol $\hat{x}_q$ was used for phase retrieval, it was just a number: the value of the Fourier transform of the contrast at wave vector q. Up to normalization, this is exactly the inner product $x\,F_q^\dagger$ in (3.16). Since F_q is a unit vector, the quantum expectation value of QST (3.17) is therefore nothing other than the intensity of phase retrieval, again up to normalization:

$$\mathrm{Tr}(\rho H_q) = \langle H_q \rangle = I_q \ . \tag{3.18}$$

To give phase retrieval the full QST interpretation, we would say each incident photon from the radiation source is a preparation of the same quantum state (the exit wave) or a relaxation of that when the state preparation is mixed ($m > 1$).

All that remains in our QST interpretation of phase retrieval is the question of the linear independence of the measurement matrices H_q. To analyze this we consider, for all $k \in \Lambda$, the sums

$$\sum_{q\in\Lambda} \exp(2\pi i\, k \cdot q/\ell)\, (H_q)_{pp'} = \frac{1}{|\mathcal{S}|} \sum_{q\in\Lambda} \exp\left(2\pi i(k - p + p') \cdot q/\ell\right)$$

$$= |\Lambda|\, (A_k)_{pp'} \ , \tag{3.19}$$

where

$$(A_k)_{pp'} = \frac{1}{|\mathcal{S}|}\, \delta_{p\,(p'+k)}$$

are suitably normalized autocorrelation matrices. Applying the inverse Fourier transform to (3.19),

$$H_q = \sum_{k\in\Lambda} \exp(-2\pi i\, k \cdot q/\ell)\, A_k \ ,$$

we see that H_q is expressed in terms of explicitly linearly independent matrices. Moreover, since $A_k = 0$ (the zero matrix) whenever $k \notin \mathcal{A}$ (falls outside the autocorrelation support), the number of independent measurement matrices, what we called n in QST, is equal to $|\mathcal{A}| = |\mathcal{S} - \mathcal{S}|$.

When we use the general QST equation-per-variable ratio (3.15) for phase retrieval, we obtain

$$\Omega = \frac{|\mathcal{S} - \mathcal{S}|}{2|\mathcal{S}|m} \tag{3.20}$$

in the limit where $|\mathcal{S}| \gg m$. Aside from the factor of m in the denominator, this result matches the ratio (3.3) we obtained earlier (which assumed perfect coherence). Note that here the $2|\mathcal{S}|$ in the denominator counts the two real variables per complex exit-wave pixel. In real-contrast phase retrieval, the number of real equations in the numerator was $|\mathcal{S} - \mathcal{S}|/2$, because $I_{-q} = I_q$. When the exit wave is complex, the intensities at $\pm q$ are independent, thereby also doubling the number of equations.

The factor m in the denominator of (3.20) is prohibitive for standard support shapes. Even for $m = 2$ and a convex, centro-symmetric support, more measurement constraints (in the numerator) are required than the standard diffraction apparatus delivers. A clever extension that meets this challenge, called *ptychography* [90, 106], is based on two physical principles. First, exit waves are a pointwise multiplicative combination of two functions (of the pixel-position p). Second, multiple diffraction experiments can be made where the only thing being changed is the relative translation of these functions.[3] According to the model, the kth exit wave in the rth experiment has the structure

$$(x_{k\,r})_p = (y_k)_p \, (z)_{p+r} \,,$$

where y_k is the kth mode of the illuminating radiation and z is the specimen's *transmission function*. The set of translations $r \in \mathcal{R} \subset \Lambda$ is chosen so the regions of specimen (after translation) that fall into the support of the y_k have significant overlap. In order to reconstruct the m illumination modes y_k (in addition to the specimen function z), each part of the specimen needs to be measured in about m experiments (elements of $\mathcal{R}$). Though the design of the sets $\mathcal{R}$ and the equations-per-variable analysis for ptychography is beyond the scope of this book, it must be said that of all the conspiracies described so far, this is one of the most benevolent.

In the simple coherent case ($m = 1$), the nonconvexity of the Fourier magnitude constraints on $\hat{x}$ in phase retrieval is replaced by the nonconvexity of the rank-1 constraint on ρ in the QST reformulation. But while the relative challenges posed by the two kinds of nonconvexity are up for debate, the demands on computational resources put the QST approach at an enormous disadvantage. That's because the rank-1 constraint must be lifted when imposing the linear constraints (3.18) on ρ. But a general Hermitian ρ requires $|\mathcal{S}|^2$ real numbers of memory compared to the $2|\mathcal{S}|$ real numbers of the complex exit wave x.

3.4 Complex Lines

There is perhaps no greater conspiracy than that exhibited by complex equiangular lines, also known as SIC-POVMs.[4] The standard version of the problem asks for n normalized complex vectors (states) ψ_i in d dimensions such that all pairs have the same inner product magnitude, or "angle" :

$$|\psi_i \psi_j^\dagger|^2 = \alpha , \quad i \neq j .$$

In this section, we consider the generalization

$$|\psi_i \psi_j^\dagger|^2 = \alpha_{ij} , \quad i \neq j , \tag{3.21}$$

where the data $\alpha_{ij} = \alpha_{ji}$ are analogous to the intensity data in phase retrieval. Instead of asking for equiangular configurations that achieve the
maximum n, the question is now to determine the minimum n for which
the data determine the complex lines uniquely, up to symmetry. From the
lines, of course, the phases of $\psi_i \psi_j^\dagger$ can be retrieved.

If we include the normalization of the vectors in the data as $\alpha_{ii} = 1$,
the generalized complex line problem is very similar to the kissing spheres
problem. Instead of the inequalities in set A, we have magnitude constraints
on the elements of the Hermitian Gram matrix $x_{ij} = \psi_i \psi_j^\dagger$:

$$A = \left\{ x \in \mathbb{H}^n : |x_{ij}|^2 = \alpha_{ij} \right\} .$$

Set B is the direct complex generalization of the set B for kissing spheres:

$$B = \left\{ x \in \mathbb{H}^n : x \succeq 0, \text{ rank}(x) \leq d \right\} .$$

The inequality just means the lines may occupy fewer dimensions, if that is
possible.

Determining the minimum $n_0(d)$ for which a configuration of complex lines
in d dimensions is uniquely determined by angle data is a counting problem
with three parts. As in phase retrieval and quantum state tomography, we
need to determine the number of real unknowns d_A and the number of real
constraint equations d_B. But the uniqueness transition is not found by setting
these equal, because a continuous symmetry group G with d_G parameters
can act on the unknowns with no effect on the constraint equations. The
virtual dimension should therefore be defined as

$$d^* = d_A - d_G - d_B .$$

A positive d^* means solutions have continuous parameters, even beyond those
attributable to symmetry. Negative d^* implies solutions should exist only
when the data are designed.

Because each normalized complex line has $2d - 1$ real paremeters, and
there are n lines,

$$d_A = (2d - 1)n .$$

And since there is a real-valued constraint (3.21) for every pair of lines,

$$d_B = n(n-1)/2 \ .$$

There are three symmetries that have no effect on the inner product magnitudes. The first is a basis transformation,

$$\psi'_i = \psi_i u \ ,$$

where u is an arbitrary $d \times d$ unitary matrix. This adds d^2 parameters to d_G. Second, each line may be multiplied by an independent phase:

$$\psi'_i = e^{i\varphi_i} \, \psi_i \ .$$

This only adds $n - 1$ parameters to d_G because applying the same phase to all the lines is possible through u. Finally, complex conjugation

$$\psi'_i = \psi_i^* $$

just conjugates the inner products (when applied to all of them). But this is not a continuous symmetry and does not change d_G. Altogether the symmetry group has

$$d_G = d^2 + (n-1)$$

continuous parameters.

Combining the three counts we arrive at the following formula for the virtual dimension:

$$d^* = (2d-1)n - d^2 - (n-1) - n(n-1)/2 \ .$$

For each d there is a minimum $n_0(d)$ such that $d^* < 0$ for all $n > n_0(d)$. This is the minimum number of lines for which a unique reconstruction from designed data is possible. For large d this criterion implies

$$n_0(d) \sim (2 + \sqrt{2})d \ .$$

Had we forgotten to include symmetry (d_G), this would have been replaced by $n_0(d) \sim 4d$.

To appreciate the depth of the *equiangular* complex line conspiracy, we recall Zauner's conjecture [113] that such configurations exist for n as large as d^2! One of the exercises will help you discover why the equal angle must have the special value $\alpha = 1/(d+1)$. On the other hand, a design principle for equiangular configurations in general d remains elusive. Configurations with the special angle and fewer lines, but $n \geq n_0(d)$ for unique reconstruction, are probably always subsets of the highly symmetric configurations having all d^2 lines.

3.5 Boolean Satisfiability

When the sets A and B are discrete, we cannot count variables and equations to get a handle on the existence of solutions or to assess the degree to which a designed instance is a conspiracy and can be counted on to have a unique solution. We will use the Boolean satisfiability problem to introduce a different approach based on counting.

In Boolean satisfiability, one seeks to find an assignment of variables that makes a particular Boolean formula evaluate to TRUE. Formulas have the *conjunctive normal form*, comprising a conjunction of clauses, where each clause is the disjunction of some number of variables or their negations.

Since Boolean satisfiability is NP-complete, we cannot hope to define an ensemble that does justice to the diversity of hard problems within its scope. Narrowing the focus to 3-SAT, the subset of formulas where each clause contains exactly three variables, does not help. Though 3-SAT is also NP-complete [61], clearly what makes any instance easy or hard is strongly dependent on the structure of the formula. On the other hand, the simple fact that more clauses make a formula harder to satisfy, all else being the same, is nicely captured by the *random k-SAT* ensemble. Here k is the internationally agreed upon symbol for the number of variables per clause.[5] The case $k = 3$ already includes all the hardest problems – when cleverness is invested in their design. Random k-SAT formulas may not be all that hard, but serve to model the transition from satisfiable, SAT, to unsatisfiable, UNSAT, as the number of clauses is increased.

The random k-SAT ensemble is defined by three integers: k, m, and n. In each instance, there are n clauses formed by uniformly sampling k of the m variables without repetition. Negation is applied to the variables with 50% probability. There are 2^m possible TRUE/FALSE assignments to the m variables, and we use the symbol # for the number of assignments that evaluate to SAT. We should think of # as a random variable defined on the random k-SAT ensemble. Expectation values are denoted with angled-brackets, so that $\langle \# \rangle$ is the expected number of SAT assignments (solutions) in an ensemble specified by k, m and n.

The expected number of solutions might be a biased indicator for the probability that a typical instance is SAT. For example, if almost all instances are UNSAT, but there is a tiny minority of SAT instances, each with an enormous number of solutions, then $\langle \# \rangle$ could still be large. To avoid this, we would study the random variable $\theta(\#)$, where

$$\theta(\#) = \begin{cases} 1, & \text{if } \# > 0, \\ 0, & \text{otherwise.} \end{cases}$$

Unfortunately, while the analytical calculation of $\langle \# \rangle$ is easy, the calculation of $\langle \theta(\#) \rangle$ is much harder. Note that $\theta(\#) \leq \#$ implies the inequality

$$\langle \theta(\#) \rangle \leq \langle \# \rangle . \tag{3.22}$$

To calculate $\langle \# \rangle$, consider one of the 2^m variable assignments, denoted x, and define indicator variables

$$y_1(x), \ldots, y_n(x) ,$$

one for each clause of the formula, with the rule that $y_i(x)$ equals 1 if clause i evaluates to TRUE for x and is 0 otherwise. Because the formula is TRUE only when the product of all the y's equals 1 and is FALSE otherwise, the formula is TRUE for assignment x with probability

$$\text{prob}(x) = \langle y_1(x)\, y_2(x) \cdots y_n(x) \rangle$$

in the random k-SAT ensemble. The y's are independent random variables because the Boolean variables and negations are sampled independently for each clause. No matter the constitution of the clause in variables (there are no repetitions), there will be exactly one negation state, out of 2^k, which makes the clause FALSE and its y equal to zero. From this it follows that

$$\text{prob}(x) = \left(1 - 2^{-k}\right)^n .$$

Because this is independent of x, the expected number of assignments for which the formula evaluates to TRUE is

$$\langle \# \rangle = 2^m \left(1 - 2^{-k}\right)^n . \tag{3.23}$$

The expected-solution-number growth rate per variable is defined by taking the logarithm of (3.23) and dividing by m :

$$\begin{aligned} \gamma(\alpha) &= \frac{1}{m} \log_2 \langle \# \rangle \\ &= 1 - (n/m) \log_2 \left(1/(1 - 2^{-k})\right) \\ &= 1 - \alpha/\alpha_0 . \end{aligned}$$

This is linear in the clause-to-variable ratio $\alpha = n/m$, crossing zero at $\alpha = \alpha_0$:

$$\alpha_0 = 1/\log_2 \left(1/(1 - 2^{-k})\right) .$$

The difference of terms in γ is analogous to the difference between the numbers of real variables and equations in a problem with continuous sets A and B. Instead of the solution set having several continuous dimensions,

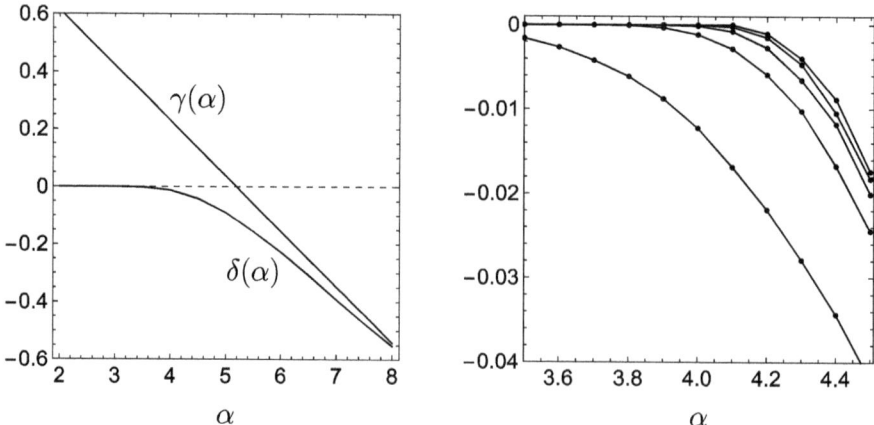

Figure 3.14 *Left*: Growth rate γ of the expected number of solutions, and the decay rate δ of the solution probability, of 3-SAT as a function of the clause-to-variable ratio, $\alpha = n/m$. *Right*: Detail of δ in the transition region, showing the development of a nonanalytic function with increasing m. The lowest curve, for $m = 20$, is the same as the δ on the left, and is followed by $m = 100, 200, 300, 400$.

$\gamma > 0$ implies an exponential growth, $2^{\gamma m}$, in the number of solutions with the problem size. The case $\gamma < 0$ roughly corresponds to there being more equations than continuous variables, or no solutions at all in the continuous case. By inequality (3.22), the exponentially decreasing expectation number, $2^{\gamma m}$, is an upper bound on the probability of solutions.

To study the solution probability in more detail, we define

$$\delta(\alpha) = \frac{1}{m} \log_2 \langle \theta(\#) \rangle$$

in analogy with γ. This notation is misleading because unlike γ, δ is not just a function of the ratio $\alpha = n/m$. Figure 3.14 compares the two functions for $k = 3$ and also shows the evolution of the function δ with m. It is believed that $\delta(\alpha)$ has a singular limiting form, just like the functions that describe phase transitions in statistical mechanics [79]. The limiting function is zero for $\alpha < \alpha_1$, where solutions have probability 1, and turns negative for $\alpha > \alpha_1$. In the latter case, the solution probability $2^{\delta m}$ decays exponentially with m, becoming a step function in the limit $m \to \infty$. Inequality (3.22) implies $\alpha_1 < \alpha_0$. The trend with m shown in Figure 3.14 is consistent with nonanalytic behavior at $\alpha_1 \approx 4.267$, the current best estimate of this number for $k = 3$ [75].

3.6 Edge-Matching Tiles

The expressive power of square tiles with four colored edges to encode problems is enormous. In fact, there is a proof that just the edge-matching constraint has the capacity to express problems that are NP-complete [27]. In other words, a suitable set of tiles can encode any instance of Boolean satisfiability. We leave to the reader to decide which is the more human-friendly medium for serving up this highest level of difficulty.

As we saw earlier (Figure 3.5), edge-matching tile puzzles are ideal for problem design. Taken to the extreme of many colors, where every tile edge has only one like-colored mate, the puzzle is reduced to a jigsaw puzzle. To keep the puzzle interesting, we want fewer colors and unforced choices at many stages of the assembly. At the same time, the color diversity should be above the threshold that ensures the creator's wonderful graphic design is the *only* solution. Knowing where this threshold lies, in the random tile ensemble of the same number of colors, is very useful information and the subject of this section.

A case where it was important to create a puzzle on the other side of the transition, where even a random instance has solutions, was the puzzle called *Pauling's Revenge* [36]. To commemorate the 25th anniversary of the discovery of quasicrystals, a 5×5 puzzle with five colors was created with the expectation rather than the guarantee it would have a solution.

Since the 5-fold quasicrystal symmetry was at odds with the square shape of the tiles, it was decided to realize this symmetry through the colors. There are five ways of leaving out one of the five colors so the edges can be colored using all of the remaining colors. Each of these combinations of four colors can be arranged in $4!/4 = 6$ ways, if we don't count tile rotations, for a total of $5 \times 6 = 30$ tiles. This has five too many tiles and is unacceptable because the coloring scheme is completely permutation symmetric. To reduce the number to 25 and the color symmetry to cyclic 5-fold symmetry, one can do the following. Define the 5-fold symmetry by a particular arrangement of the five colors around a circle. Next, review the six permutation-symmetric colorings from earlier (that leave out one color), and eliminate the one that has the same ordering around the symmetry-defining color circle.

Nothing about this elaborate coloring scheme builds confidence the puzzle will have a solution! The best we can do is treat it as any other instance of a random ensemble for which we can work out the expected number of solutions. Below we calculate this number in the simplest random ensemble, where the colors on the tile edges are independently sampled from the uniform distribution. This ensemble is defined by two integers, the number of

Figure 3.15 *Left*: The *Pauling's Revenge* tile set designed via a color-symmetry principle and no guarantee of a solution. *Right*: One of the 4 457 600 solutions. Because of the periodic boundary conditions, the tiling extends to an infinite periodic crystal.

colors k and the size of the assembled square, n. We will see later, that puzzles in the $(k, n) = (5, 5)$ ensemble have on average 196 628 solutions, making the prospect of a solution for the *Revenge* tile-set quite good. This number could have been many orders of magnitude larger and made the puzzle much less challenging than it is. The actual number of solutions, 891 520, after the conspiratorial 5-fold symmetry is divided out, is about 4.5 times the average for random puzzles.[6]

The "revenge" is manifested by the use of periodic boundary conditions. In a solution, like the one shown in Figure 3.15, the tile edges on the left side of the puzzle match the colors on the right side, and similarly for the top and bottom edges. Because the tiling can then be extended indefinitely as a periodic crystal, Pauling was right all along [88]! Periodic boundary conditions greatly simplify the expectation number calculation because all the tiles are subject to the same constraints, independent of position.

To emphasize the similarity with random k-SAT, we keep the notation the same as much as possible. Instead of x being an assignment of Boolean variables, x is now a particular arrangement of the tiles of the tile set to form the assembled puzzle. Since n^2 tiles get matched with n^2 positions, and each tile may be rotated in four ways, there are $(n^2)! \times 4^{n^2}$ choices for x.

Corresponding to the indicator variables y for the truth of clauses, we now have an indicator y_i at each edge i of the assembled puzzle. Thanks to the periodic boundary conditions, there are two such edges per tile, or $2n^2$. Indicator $y_i(x)$ equals 1 when the two tiles adjacent to edge i have matching colors on that edge in arrangement x; it is 0 otherwise. Fixing x and understanding that the angled brackets now denote an average over random tile sets, the probability that x is a solution is given by

$$\text{prob}(x) = \langle y_1(x)\, y_2(x) \cdots y_{2n^2}(x) \rangle \ .$$

The indicators are independent random variables in our ensemble because the two colored edges of the tile set that $y_i(x)$ depends on affect the value of only this indicator when x is fixed. Since the colors on those two edges are sampled independently and uniformly from the k colors, $\langle y_i(x) \rangle = 1/k$ for all i and independent of x. The probability that arrangement x is a solution is therefore

$$\text{prob}(x) = k^{-2n^2} \ .$$

Multiplying this x-independent probability by the number of different x, we obtain the formula

$$\langle \# \rangle = (n^2)!\, 4^{n^2} k^{-2n^2}$$

for the expected number of solutions in the random tile ensemble. This is the source of the numbers quoted earlier.

Taking the logarithm, dividing by n^2, and using Stirling's approximation for the factorial, we get the growth rate per tile:

$$\gamma(\alpha) = \frac{1}{n^2} \log \langle \# \rangle$$
$$\sim \log \left(\frac{4}{e} \left(\frac{n}{k} \right)^2 \right), \quad (n,k) \to \infty$$
$$= 2 \log (\alpha/\alpha_0) \ .$$

As in Boolean satisfiability, this is a function of a ratio, in this case $\alpha = n/k$. Typical instances are underconstrained when the argument of the logarithm is greater than 1, overconstrained otherwise. For large puzzles, the transition occurs at the ratio

$$\alpha_0 = \sqrt{e/4} = 0.824361 \ .$$

Exercises

3.1   The square numbers $A = \{1, 4, 9, \ldots\}$ and triangular numbers $B = \{1, 3, 6, \ldots\}$ are easy to generate and not exceptionally exotic. Show[7] that the set $A \cap B = \{1, \ldots\}$ is also infinite but much, much smaller than either A or B.

3.2   Wave microscopy works even with perfectly transparent specimens, where the only action of the specimen is to introduce a position-dependent phase rotation of the incident wave (via the position dependence of the index of refraction). In the simplest design, perfectly transparent material (e.g., amoeba in water) is placed in the aperture of an opaque screen. The contrast x can then be modeled as a position-dependent pure phase inside a support $\mathcal{S}$ defined by the aperture and zero everywhere else:

$$x_p = \begin{cases} e^{i\varphi_p}, & p \in \mathcal{S}, \\ 0, & \text{otherwise} . \end{cases}$$

(a) How is the constant magnitude of the complex-valued x within $\mathcal{S}$ related to the measured Fourier magnitudes $|\hat{x}|$? This constant magnitude will be used as a constraint when reconstructing x.

(b) Keeping in mind the constraint above, repeat the analysis of Section 3.1.3 to derive the analog of the ratio Ω for pure-phase-contrast phase retrieval.

3.3   Supply the details in the direct algebraic reconstruction of contrast from autocorrelation when the contrast is strictly positive in an isoceles-right-triangle support. The idea is to order the contrast pixels as shown in Figure 3.8 for a triangle of side $s = 7$. Use the symbol $x(i, j)$ for the contrast at pixel (i, j), where the three sides of the triangle correspond to $i = 0$, $j = 0$, and $i + j = s - 1$. Use the same notation for the autocorrelation data, $a(i-i', j-j')$, where the index pair is the difference of two contrast-pixel coordinates.

(a) Write down three equations involving just the three pixels marked 0 in Figure 3.8 and solve them.

(b) For each of the pixels marked 1, write down one equation in which the autocorrelation sum has just 1 term and solve each of these.

(c) For each of the pixels marked 2, write down one equation in which the autocorrelation sum has just 2 terms, only one of which involves (linearly) a pixel marked 2, and solve each of these.

(d) Explain how this continues. What is the depth of the hierarchy in terms of s?

(e) How many autocorrelation data were used to solve for the contrast by this method?

3.4 In crystallography two structures are said to be *homometric* if they possess the same set of interatomic separation vectors and yet are not related by rotations, translations, and reflections. When represented by contrast voxels, the two contrasts are not related by a symmetry while their autocorrelations have exactly the same support. In the extreme case of the phenomenon, even the autocorrelation values are identical.

In this exercise you construct a pair of homometric structures where all the nonzero contrast values are equal to 1. For simplicity you will work in one dimension, but the construction generalizes to crystals in any number of dimensions.

As in known-support phase retrieval in one dimension, we use a polynomial with real coefficients

$$p(z) = x_0 + x_1 z + \cdots + x_{n-1} z^{n-1}$$

to represent the contrast at n equally spaced points in the unit cell. The difference is that the crystal-periodicity means the powers of z are integers mod n. For example, the crystal with reflected contrast has polynomial $p(1/z)$ and the autocorrelation polynomial is simply

$$a(z) = p(z)p(1/z)$$

because there is no need to multiply by $z^n = z^0 = 1$.

(a) Suppose the contrast polynomial can be factored as $p(z) = f(z)g(z)$ where f and g also have real coefficients. Show that $p(z)$ and the crystal $q(z) = f(z)g(1/z)$ have exactly the same autocorrelation.

Now consider polynomials where the coefficients (contrasts) are all either 0 or 1. When there are k 1's we say the polynomial is k-sparse, or that there are k "atoms" in the unit cell.

(b) For $n = 13$ and the 3-sparse factors

$$f(z) = 1 + z^2 + z^7,$$
$$g(z) = 1 + z^3 + z^4$$

work out the contrasts $p(z)$ and $q(z)$ and show that they are both 9-sparse and not related by a symmetry (reflection and translation). Since $p(z)$ and $q(z)$ have the same autocorrelation, they are homometric.

(c) Consider the $N = \binom{n}{k}$ possible crystals with k atoms (when we choose an origin for the unit cell). What is the probability, p_h, there is a homometric structure among them? Start by showing this is zero unless k is a composite integer.

(d) Next show that the number of homometric structures (again with a choice of origin) is upper bounded by[8]

$$N_h = \sum_{(k_1,k_2)} \frac{1}{n} \binom{n}{k_1} \binom{n}{k_2},$$

where the sum is over all proper factorizations $k_1 k_2 = k$ and $k_1 \leq k_2$.

(e) Suppose k is an even square number. For the case of a sparse crystal, $k \ll n$, compare the sizes of the extreme terms in the sum, $k_1 = 2$ and $k_1 = \sqrt{k}$.

(f) Bounding p_h by N_h/N, and estimating the sum for N_h by just the dominant term, work out how the bound on p_h decays with k, again for a sparse crystal.

3.5 Can a virus be reconstructed from crystallographic data, even when the resolution is too coarse to use atomicity as a constraint? To answer this question, model the virus particles as spheres that pack with six contacts in a cubic unit cell as shown in Figure 3.10. The interiors of the spheres have the 60-element icosahedral symmetry (and the contacts ensure that all virus capsids have the same orientation). Model the contrast in the solvent region outside the spheres/viruses as "flat" (constant). What is the corresponding equation-per-variable ratio Ω? A successful reconstruction would of course only reveal the averaged contrast of the genetic material inside the capsid.

3.6 Recall that in generalized phase retrieval (Exercise 2.5), you are given the magnitudes $|y|$ of n complex numbers with the promise that y was generated by multiplying m complex numbers x by a known complex $m \times n$ sensing matrix G. On the assumption that G has no special structure (like the Fourier matrix), and the only symmetry is the single phase that may be applied to x (with no effect on $|y|$), compute the virtual dimension for generalized phase retrieval. What should be the relationship between m and n for reconstructions of y to be unique, up to an overall phase?

3.7 If we replace all the elements of a unitary matrix u by their squared magnitudes, the result is a doubly stochastic matrix w. Now suppose we are given a w constructed in this way. What are the prospects of reconstructing u? A 3×3 real-world example is the reconstruction (from particle physics data) of the Cabibbo–Kobayashi–Maskawa matrix.

Before addressing this mathematically appealing phase retrieval problem, we make a few observations. First, there certainly exist doubly stochastic matrices that are *not* derived from unitary matrices. Consider the one parameter example,

$$w = \begin{bmatrix} 0 & p & 1-p \\ p & 1-p & 0 \\ 1-p & 0 & p \end{bmatrix},$$

where $0 < p < 1$.

(a) Prove that there is no unitary u such that $|u|^2 = w$. This shows that some degree of design is involved for this version of phase retrieval to have solutions.

Second, we are quite familiar with instances of this problem where solutions are very far from unique: w's with all elements equal, for which any complex Hadamard matrix is a solution. When the order n of the matrix is additionally divisible by four, there are real solutions whose number grows rapidly with n. We attribute the conspiracy to the highly symmetrical nature of that particular problem instance.

(b) Because the unitary matrices form a continuous group, you can get a handle on the number of variables by looking at matrices near the identity: $u = I + ih$, where $|h| \ll 1$ and the square root of -1 is thrown in for convention. What is the most general h such that $uu^\dagger = I + O\left(h^2\right)$? Count the number of real parameters d_A in h required to specify an arbitrary $n \times n$ unitary matrix.

(c) There is at least one kind of symmetry orbit that has no effect on the magnitudes $|u|$. Any row and any column of u may be multiplied by a phasor without changing $|u|$ while keeping u unitary. Count the number of independent parameters d_G for this continuous symmetry group.

(d) Finally, count the number of independent real-valued constraint equations, d_B. Approach this by first asking how many independent data are provided by w.

If you didn't make any mistakes, you should end up with $d^* = 0$. Phase retrieval of unitary matrices is a marginal case, where the solutions of random instances form discrete sets.

3.8 In Exercise 2.7 you learned that complex equiangular lines have a nice bipartisan formulation. In this follow-up, you determine the maximum number of such lines and the value of their common angle (which came up in Section 3.4).

Since this involves matrices, we use matrix notation, even for the lines. A row vector $\psi \in \mathbb{C}^d$, with normalization convention $\psi\psi^\dagger = 1$, defines a complex line. Associated with a line ψ is the rank-1 Hermitian matrix $\rho = \psi^\dagger\psi$. In addition to these quantum entities (states and density matrices), we also define a purely geometrical one:

$$\tilde{\rho} = \mathrm{Re}(\rho) + \mathrm{Im}(\rho) . \tag{3.24}$$

Note that if ρ is a general $d \times d$ Hermitian matrix, not necessarily rank-1, then the two terms in (3.24) are general symmetric and skew-symmetric real matrices, that is, $\tilde{\rho} \in \mathbb{R}^{d \times d}$.

(a) For any $\tilde{\rho}$ derived from a complex line, show that

$$\mathrm{Tr}\,\tilde{\rho} = 1 .$$

Since this is a linear constraint, the map $\psi \mapsto \rho \mapsto \tilde{\rho}$ takes complex lines into codimension-1 hyperplanes $\Lambda_1 \subset \mathbb{R}^{d \times d}$.

(b) Now consider lines ψ_i and ψ_j (which might be the same), with associated real matrices $\tilde{\rho}_i$ and $\tilde{\rho}_j$, and show that

$$|\psi_i\psi_j^\dagger|^2 = \mathrm{Tr}\,\tilde{\rho}_i\tilde{\rho}_j^\mathsf{T} . \tag{3.25}$$

Note that the last expression is just the usual inner product formed by the matrix elements flattened into a vector.

(c) Next suppose there are n distinct lines and all pairs have the same relative "angle":

$$|\psi_i\psi_j^\dagger|^2 = \alpha, \quad i \neq j .$$

Using (3.25), show that this property implies the associated real matrices $\tilde{\rho}_i$ form an equidistant set. That is, the n points in the hyperplane Λ_1 have all pairwise distances equal.

From the geometrical fact that in $d^2 - 1$ dimensions there can be at most d^2 equidistant points, we infer that there can be at most $n = d^2$ equiangular lines. Because this maximal set of equidistant points always forms a regular simplex, we can use the geometry of the simplex to determine the constant α. The only fact we need is that the circumcenter of the regular simplex coincides with its centroid.

(d) Suppose $v_1, \ldots, v_{m+1}$ are the vectors to the points of a regular simplex in m dimensions, relative to its circumcenter/centroid. Use the centroid equation

$$0 = v_1 + \cdots + v_{m+1}$$

to show that

$$v_i \cdot v_j = -\frac{1}{m}\|v\|^2, \quad i \neq j . \tag{3.26}$$

(e) In order to use (3.26), you first have to find the circumcenter c of the points $\tilde{\rho}_i$ in Λ_1. Since the norms (case $i = j$ of (3.25)) of the $\tilde{\rho}_i$ are equal, and all lie in Λ_1, argue that the circumcenter c of the simplex is obtained by projecting the origin into Λ_1.

(f) Compute c (a real matrix), construct the (unflattened) vectors $v_i = \tilde{\rho}_i - c$, and use (3.26) (with the appropriate m) to find the common α.

3.9 The kissing spheres problem was reformulated as finding the Gram matrix of the sphere-center inner products. When recovering the sphere centers, an arbitrary orthogonal transformation may be applied to the Cholesky decomposition of the Gram matrix solution. In this exercise, you tackle the counterpart of this kind of nonuniqueness in quantum state tomography.

Suppose you have obtained enough measurements so that criterion $\Omega > 1$ for (3.15) is comfortably satisfied and your reconstruction of a rank-m density matrix ρ is unique. What can you say about the states and probabilities in the decomposition (3.12)?

Begin by finding the eigendecomposition of ρ:

$$\rho = \sum_{k=1}^{m} q_k \, \chi_k^\dagger \chi_k \ .$$

The χ_k correspond to orthogonal and normalized states, and the q_k can be interpreted as probabilities because they are nonnegative and sum to one. Assume the latter are distinct, so the decomposition is unique, and are ordered as

$$q_1 < \cdots < q_m \ .$$

To make your work easier, define unnormalized orthogonal states by absorbing the probabilities:

$$\tilde{\chi}_k = \sqrt{q_k} \, \chi_k \ .$$

(a) Show that unnormalized states defined by

$$\psi_k = \sum_{k'=1}^{m} u_{kk'} \, \tilde{\chi}_{k'} \ , \tag{3.27}$$

where u is an arbitrary $m \times m$ unitary matrix, are consistent with your reconstructed density matrix:

$$\rho = \sum_{k=1}^{m} \psi_k^\dagger \psi_k \ .$$

(b) Next show that the probabilities $p_k = \psi_k \psi_k^\dagger$ are related to the probabilities of the eigendecomposition by a relation of the form

$$p_k = \sum_{k'=1}^{m} w_{kk'} q_{k'} \, ,$$

where w is doubly stochastic, that is, nonnegative with all row and column sums equal to one.

(c) Use this fact to prove that all the probabilities satisfy

$$q_1 \leq p_k \leq q_m \, .$$

3.10 Configurations of equidistant points are always regular simplices (equilateral triangles, regular tetrahedra, ...), and that's about all to be said on that problem. The problem of equidistant lines, on the other hand, is wide open.

    The distance between two lines, in any number of dimensions, is defined to be the minimum distance between all pairs of points on the two lines. Let $d > 2$ be the number of dimensions and consider configurations of $n > d$ lines. We are free to set the common distance between all lines at 1.

(a) In a bipartisan formulation of this problem, set A comprises all configurations of n lines in d dimensions (not necessarily equidistant). Count the number of unknowns, or the number of dimensions d_A of set A.

(b) Membership in set B imposes the distance constraint between the lines. Each of these is like a single real-valued equation and counting them will give you d_B.

(c) What are the continuous symmetry transformations G that preserve the distances between all the lines? Count the number of parameters of G to obtain d_G, and combine this with d_A and d_B to obtain the virtual dimension d^*.

(d) What is the largest n for which an equidistant line configuration in $d = 3$ is not ruled out by the virtual dimension? How many parameters would such configurations have, not counting the continuous symmetries? In answering these and the following questions, you should assume there are no conspiracies at work, even in these special (equidistant) instances.

(e) Find a pair (d, n) for which $d^* = 0$, so that the equidistant configurations, if they exist, form a discrete set (up to symmetry).

(f) Find the relationship between d and n, in the limit of large d, at the under/overconstrained transition.

3.11 As explained in Section 3.6, the *Pauling's Revenge* tile set is invariant under a cyclic permutation of the five colors (shades of gray). Find this permutation using the data in Figure 3.15.

3.12 Eternity II is an edge matching tile puzzle that appeared on game store shelves in 2007. A \$2 million prize in a solving competition that ended in 2010 was never claimed, and there are no (publicly) known solutions.

Like the puzzles of Section 3.6, Eternity II comprises n^2 square tiles that are arranged, with rotations, into a $n \times n$ square with edge-matched colors. Instead of periodicity (no boundary), Eternity II has an all-white boundary with three kinds of tiles: 4 corner tiles, $4(n - 2)$ boundary tiles, and $(n - 2)^2$ interior tiles.

Corner tiles have two adjacent white edges and boundary tiles have one white edge. These tiles are forced to occupy the boundary positions of the $n \times n$ square with their white edges facing outward. The two non-white edges of the corner tiles have ℓ possible colors, and those same colors are used in the boundary tiles, on the two edges adjacent to the white edge. The matching of the ℓ "boundary colors" is therefore a constraint among just the corner and boundary tiles.

The remaining edge of each boundary tile and all the edges of the interior tiles have k possible "interior colors," distinct from the boundary colors. The ensemble of random Eternity-II puzzles, therefore, has three integers, (k, ℓ, n), in its specification.

(a) Using the method of Section 3.6, obtain a formula for the expected number of solutions $\langle \#(k, \ell, n) \rangle$. Do not apply Stirling's approximation to the factorials.

(b) Eternity II has $n = 16$. Using your formula for $\langle \#(k, \ell, 16) \rangle$, find the k and ℓ that give the smallest expected number of solutions greater than 1. These match the numbers in Eternity II.

(c) The puzzle instructions include, as a clue, the position of one of the tiles. Assuming a solution exists, what does this tell you about its uniqueness?

3.13 A generalization of the Fermat equation is

$$x^p + y^q = z^r, \tag{3.28}$$

where x, y, z, p, q, r are all positive integers, and the three exponents need not be equal. Without calling on any background in number theory, in this exercise you learn how to estimate the growth rate of the number of solutions.

Imagine the task of checking a solution when the arithmetic is done in base 2. Suppose z^r has n bits, where n is large. When there are many solutions, x^p and y^q will also have very close to n bits. After computing the strings of bits for x^p and y^q, and writing one below the other, the check that the sum equals z^r amounts to confirming, in n cases, that the sum of two bits matches a third bit, not forgetting to include the carry-bit.

(a) Estimate the number of possible triples of n-bit strings, one for each of x^p, y^q, and z^r, for large n. Treating these as random strings, what is the probability that a triple of strings is a solution? Multiply this probability by the number of triples to get the expected number of solutions, $\langle\#\rangle$. Finally, take the base-2 logarithm of $\langle\#\rangle$ and divide by n to get the growth rate per bit, γ. Your answer for γ should be a simple expression involving only the exponents p, q, and r.

(b) Is the sign you get for γ when $p = q = r = 2$ consistent with what you know about Pythagorean triples?

(c) By treating the three bit-strings as random, your analysis is undercut by a number-theoretical conspiracy! Suppose $p = 2$ and the other two exponents are odd, so that $qr + 1 = 2s$ is even. Check that $x = c^s$, $y = ac^r$, and $z = bc^q$ is a solution for arbitrary pairs of positive integers a and b, provided $c = a^q + b^r$. In all of these solutions x, y, and z have $c > 1$ as a common factor. We can eliminate such conspiracies with the qualification that x, y, and z are coprime.

(d) Since γ is positive for $p = 2$, $q = 3$, $r = 5$, the equation $x^2 + y^3 = z^5$ should have "many" coprime solutions. Using a computer, find four with $x < 10^6$.

3.14 One way to decide whether we should be surprised by the existence of Hadamard matrices is to design a problem ensemble that includes Hadamard matrices as one problem instance. By designing the ensemble in a way that lets us calculate the expected number of solutions, we will at least know whether Hadamard matrices are a typical or atypical instance.

Our ensemble applies to matrices $u \in \{-1, +1\}^{n \times n}$, where n is even. An instance is specified by a set of signs $S_{(i,j)k} \in \{-1, +1\}$, where (i, j) ranges over all $\binom{n}{2}$ pairs of matrix indices and k is a matrix index.

Solutions u satisfy the following constraints:

$$\forall (i,j):\ \sum_{k=1}^{n} S_{(i,j)k}\, u_{ik} u_{jk} = 0 \ .$$

Setting $S_{(i,j)k} = 1$ for all (i,j) and k gives the Hadamard instance.

(a) Work in the ensemble where the signs $S_{(i,j)k}$ are uniformly distributed. Calculate the expected number of solutions $\langle\#\rangle$ by first calculating the probability that a particular u is a solution.

(b) What is the smallest n for which $\langle\#\rangle$ is less than 1?

(c) Use the large n approximation

$$\binom{n}{n/2} \sim \sqrt{\frac{2}{\pi n}}\, 2^n$$

to determine the growth rate γ of the expected number of solutions, $\langle\#\rangle \sim 2^{\gamma n^2}$. You will find that γ grows logarithmically more negative with n, a clear sign that the robust empirical growth in the number of Hadamard matrices is atypical in the ensemble (a conspiracy).

3.15 Consider the following combinatorial problem with integer parameters m and n. You are given a $2m \times n$ matrix M, of ± 1 elements, and the freedom to flip all the signs on any subset of its $2m$ rows. Can this be done in such a way to make all n column sums equal to zero?

The obvious bipartisan formulation uses the sets

$$A = \left\{ x \in \mathbb{R}^{2m} : x \in \{-1,1\}^{2m} \right\}\ ,$$
$$B = \left\{ x \in \mathbb{R}^{2m} : xM = 0 \right\}\ .$$

Though B is continuous, to allow for an efficient projection, for locating the under/overconstrained transition, we use the counting approach in this discrete problem.

(a) Fix a solution x (a point of A) and for this x find the probability that a uniformly sampled M has x as one of its solutions. Your answer will not depend on x, so the expected number of solutions $\langle\#\rangle$ is the product of the probability and the number of all possible x.

(b) As in k-SAT and edge-matching tiles, locate the transition by the condition $\langle\#\rangle = 1$. Show that the expected number of solutions behaves exponentially with m, but the location of the transition is not at a special ratio of n and m, but something close to that.[9]

4

Projections

No doubt some readers had their first encounter with projections in George Gamow's wonderful book *One two three ... infinity* [43]. A sketch, reproduced in Figure 4.1, was meant to show the wrong and correct way of making something that exists in a higher dimension – a horse – accessible to inhabitants of a lower-dimensional world. Gamow's "correct" notion was to make all points differing in some of their coordinates equivalent, which just means eliminating those coordinates. In the case of points in three dimensions projected into two, the projection transformation would be

$$(x_1, x_2, x_3) \mapsto (x_1, x_2) \,.$$

In one respect, Gamow's "wrong" notion is closer to how we think about projections in this book. The objects being projected, usually just individual points, always live in the same space as their projections. In this same space, there are sets called A and B and the projections will always lie in these sets. Continuing the earlier example, if A is the plane defined by the equation $x_3 = 0$, then the transformation would be replaced by

$$(x_1, x_2, x_3) \mapsto (x_1, x_2, 0)$$

and written as

$$P_A(x_1, x_2, x_3) = (x_1, x_2, 0) \,.$$

This is the most technically demanding chapter of the book. That statement may come as a surprise, since the actual algorithm doesn't make an appearance until Chapter 5. However, it is the constraint projections that perform the work of the algorithm and the feasibility of implementing the algorithm rests entirely on being able to compute the projections efficiently. You are encouraged to study all the constraint projections, even those that have no obvious relevance to the problem that led you to this book. Though

Figure 4.1 Wrong (left) and correct (right) way to squeeze a three-dimensional horse into two dimensions, according to George Gamow [43]. Reprinted from Gamow (2003) with permission from Dover Publications Inc.

a bipartisan problem formulation calls for only two projections, each responsible for solving roughly half of the problem, familiarity with the doable projections – those with efficient algorithms – can be a great source of inspiration for dividing a problem creatively into "halves."

Our survey of projections will also give you a renewed respect for the Euclidean distance. There are other ways of defining distance – a whole continuum of norms – but it is only with the 2-norm that projections become a versatile and intuitive tool for building algorithms.

4.1 General Principles

In general, the projection $P_A(x)$ of a point x to the set A is the set of all points $y \in A$ that minimize the distance to x. In math notation,

$$P_A(x) = \arg\min_{y \in A} \|y - x\| .$$

If we do not also qualify how projections are used by our algorithm, this would not be a practical definition. The problem is that the sets $P_A(x)$

can sometimes be very large. Suppose $A = \{-1, 1\}^n$ is the hypercube in n dimensions, then $P_A(0, \ldots, 0) = A$ comprises 2^n points. Even worse, if A is the origin-centered n-sphere, then again $P_A(0, \ldots, 0) = A$ and the number of points is infinite. We could avoid large projection sets $P_A(x)$ by declaring problematic x as off-limits. But there is a better approach based on how projections will be used.

The idea is to ensure that $P_A(x)$ is always a single point whenever x is close to A. But before we analyze this case, let's examine one reason why convex sets are so popular with mathematicians.

4.1.1 Convexity

Convex sets are nice because of this:

Theorem 4.1 (Convex sets have unique projections) *If A is convex, and x is any point, then $P_A(x)$ is a single point.*

Proof Suppose that $P_A(x) = \{a_1, a_2, \ldots\} \subset A$ had two (or more) points, all with distance d to x. Since A is convex and contains a_1 and a_2, it also contains the points

$$\alpha a_1 + (1 - \alpha) a_2$$

for any $\alpha \in [0, 1]$. In particular, consider the point $a = (a_1 + a_2)/2$ for which the distance to x is

$$
\begin{aligned}
\|a - x\| &= \|(a_1 - x) + (a_2 - x)\|/2 \\
&= \|u + v\|/2 \\
&\leq \|u\|/2 + \|v\|/2
\end{aligned}
$$

by the triangle inequality. The equality case corresponds to the vectors u and v being related by a nonnegative multiplier, but because $\|u\| = \|v\| = d$, that multiplier would be 1 and therefore $a_1 - x = a_2 - x$. Because the equality case implies $a_1 = a_2$ and contradicts the hypothesis, the triangle inequality is strict and implies $\|a - x\| < d$, or that a_1 and a_2 are not elements of $P_A(x)$, contradicting the first hypothesis.   □

In this and future formal mathematical statements, we omit details considered to be understood. For example, all sets may be considered subsets of $\mathbb{R}^m$ for some m (after flattening the "shape" of x). There is no need to always state this explicitly, especially when the dimension m never comes up.

But convex sets for both A and B in a bipartisan formulation is too limiting. Fortunately, we will only need the mildest form of nonconvexity, where we can still count on A and B to be *locally affine*.[1] Affine subsets of Euclidean space are simply the hyperplanes in the space. This includes single points – affine sets of dimension zero. Affine sets are even better behaved than general convex sets because the entire line passing through a pair of distinct points also lies in the set. *Locally* affine sets "look" affine when examined on a small enough scale. This property limits how the sets A and B can intersect and greatly simplifies the analysis of the algorithm we will use to find points in their intersection.

4.1.2 Locally Affine Sets

Informally, a set A is locally affine if for any point x close to A there is a unique projection $P_A(x) = a(0)$ and A is well approximated near $a(0)$ by an affine set $\bar{A}$ that passes through $a(0)$. Because we often work with sets A defined by systems of equations, the affine approximation at $a(0)$ is the same as the tangent space at that point.

By the implicit function theorem, a set A defined by equations can locally be represented by a *function graph*. Suppose A is m-dimensional and lives in a space of $m + n$ dimensions. By appropriate choice of coordinates, points $a(y) \in A$ parameterized by $y \in \mathbb{R}^m$ have the following function graph representation,

$$a(y) = (\, y_1, \, \ldots, y_m, \, z_1(y), \, \ldots, z_n(y)\,) \,, \tag{4.1}$$

where $z(y)$ is a real-analytic function from $\mathbb{R}^m$ to $\mathbb{R}^n$. We may further refine our parameterization so that the first m coordinates coincide with the tangent space at $a(0)$. To satisfy this, the gradient of $z(y)$ must vanish at $y = 0$:

$$\forall i \in \{1, \ldots, m\} : \left.\frac{\partial z}{\partial y_i}\right|_{y=0} = 0 \,. \tag{4.2}$$

Using the local representation of A defined by (4.1) and (4.2), we can prove a simple result that will be useful later:

Lemma 4.2 (Tangent-space orthogonality of projections) *If set A is locally affine at $a = P_A(x)$, then $a - x$ is orthogonal to the tangent space of A at a.*

Proof Use the parameterization above for points $a(y) \in A$ near the projection $a(0) = P_A(x)$. Since projections are locally distance minimizing,

$$\forall i \in \{1, \ldots, m\} : \quad 0 = \left(\frac{\partial}{\partial y_i} \|a(y) - x\|^2 \right) \Big|_{y=0}$$

$$= \left(2(y_i - x_i) + \sum_{j=1}^{n} 2(z_j - x_{m+j}) \frac{\partial z_j}{\partial y_i} \right) \Big|_{y=0}$$

$$= -2x_i .$$

This shows the first m components of x are zero, just as the first m components of $a(0)$ (by choice of parameterization). Since that m-dimensional subspace is the tangent space of A at $a(0)$, the vector $a(0) - x$ is orthogonal to that subspace. □

To further explore the projection locally, we choose the coordinate origin of our representation so that $z(0) = 0$ (and therefore $a(0) = 0$). By suitably rotating the last n coordinates, we may further assume that only the first of these is nonzero and equal to d :

$$x = (0 , \ldots , 0 , d , 0 , \ldots , 0) .$$

The squared distance between x and a general point $a(y) \in A$ is then

$$\|a(y) - x\|^2 = \sum_{i=1}^{m} y_i^2 + (z_1(y) - d)^2 + \sum_{j=2}^{n} z_j^2(y)$$

$$= d^2 + \sum_{i=1}^{m} y_i^2 - 2dz_1(y) + O(z^2) , \tag{4.3}$$

and $d = \|a(0) - x\|$ is the distance of the projection. The squared distance has an extremum at $y = 0$ (as it should) because the gradient of $z_1(y)$ is zero at that point. The Taylor series of (4.3) therefore begins with the quadratic terms from the Taylor series of z_1. Rotating the y coordinates to put the quadratic expression into a diagonal form with (signed) principal curvatures $\kappa_1, \ldots, \kappa_m$, we obtain

$$\|a(y) - x\|^2 = d^2 + \sum_{i=1}^{m} (1 - d\kappa_i) y_i^2 + O(y^3) .$$

For $a(0)$ to be the unique projection in our local representation of A, it is sufficient that

$$d < \frac{1}{\kappa_i} \tag{4.4}$$

holds for all the positive curvatures.

If the curvatures of A were bounded above, we could use (4.4) to bound the distance d of x from A such that any projection (within that distance of A) would be unique. But this fails even for the much used rank constraint. For example, let A be the set of rank-1 2×2 symmetric matrices. In the parameterization

$$A = \left\{ \begin{pmatrix} z - x & y \\ y & z + x \end{pmatrix} : (x, y, z) \in \mathbb{R}^3 \setminus (0, 0, 0), \ x^2 + y^2 = z^2 \right\}$$

we see that A is geometrically a double cone (minus its apex). Since one of the curvatures diverges as the projection moves toward the apex (where the rank is zero), there is no uniform safe-distance bound to make the projection unique.

A more direct way to disqualify the rank-1 constraint is to observe that this set fails to be locally affine at the apex point (there is no approximating affine set). On the other hand, it seems a shame to throw away an extremely useful constraint just because of the trouble caused by one rogue point! The problematic rank-0 point should be thought of as the *boundary* of the rank-1 constraint. More generally, the boundary of the set of rank-k symmetric matrices is the set of matrices with rank-$(k - 1)$.

Another very useful constraint that is in violation of the locally affine property is any set defined by inequalities (in addition to equations). This comes up in the kissing spheres problem, where the off-diagonal Gram matrix elements are subject to an upper bound constraint. Any set that includes inequalities in its definition looks locally like a half-space and fails to be locally affine on its boundary.

A uniform way of avoiding potential problems posed by sets that are not locally affine is to insist that A and B are locally affine at least near some solutions, that is, near some points in the intersection $A \cap B$. We formalize this with a definition:

Definition 4.3 A bipartisan formulation is *regular* when there exist solutions $x \in A \cap B$ such that both A and B are locally affine near x.

It's easy to keep bipartisan formulations regular, that is, to design sets A and B so that solutions are not on any boundaries that these sets may have. Consider the kissing spheres problem. In three dimensions, and for even the maximum number of 12 kissing spheres, there is wiggle room in the solution configurations. No two spheres have to be tangent in a solution, or on the boundary of the A constraint. This changes in four dimensions, in the optimal case of 24 spheres [80], where many sphere pairs saturate the inequality bound. To restore regularity, all we need to do is replace the inner-product upper bound of $1/2$ by $1/2 + \epsilon$ for some small positive ϵ (making the spheres slightly smaller than the central one). In this regularized formulation, the constraints are locally affine near almost all solutions. Of course, when we examine any of these 24-sphere solutions, we will be struck by the many sphere pairs having inner product just slightly greater than $1/2$. We can then rerun the search, but with the inequality constraints on the near-tangent spheres replaced by equality constraints. This is again a regular formulation because the solution points are not on the boundary of a set.

4.1.3 Ubiquity of Projection Uniqueness

It is only when x is close to a point of $A \cap B$ that we insist the projections $P_A(x)$ and $P_B(x)$ are unique and the two sets can be approximated as affine near their respective projections. As we will see in Chapter 5, this guarantees local convergence to solutions. On the other hand, over most of the search one cannot count on these nice properties. Could nonunique projections still pose a problem? Here is a general fact that will help us answer this question:

Proposition 4.1 (Positive codimension of nonunique projections) *Let A be a smooth set, possibly with boundary. If X^* is the set of points $x \in X^*$ such that $P_A(x)$ is not a unique point, then X^* has a positive codimension.*

Proof Consider any $x \in X^*$. Then $P_A(x)$ contains at least two points, a_1 and a_2. If a_1 is an interior point of A, let y_1 be any vector of the tangent space at a_1. If a_1 lies on the boundary, then restrict y_1 to lie in the interior of the boundary cone at a_1. In either case, $a_1 + y_1$ approximates all possible points of A near a_1 when y_1 is small. Define vectors y_2 analogously for the point a_2. Since a_1 and a_2 locally minimize the distance to x, by Lemma 4.2

$$\begin{aligned}(a_1 - x) \cdot y_1 &= 0 \,, \\ (a_2 - x) \cdot y_2 &= 0 \,.\end{aligned} \tag{4.5}$$

Next consider points $x + z$, where z is small and $x + z \in X^*$. The projection $P_A(x + z)$ will still have at least two points, and two of them, a_1' and a_2' will respectively be near a_1 and a_2. By the function-graph representation (4.1), the coordinate differences orthogonal to the tangent space are $O(y^2)$, so we can write

$$\begin{aligned}a_1' &= a_1 + y_1 + O(y_1^2) \,, \\ a_2' &= a_2 + y_2 + O(y_2^2) \,.\end{aligned}$$

Since $x + z \in X^*$, the two projection distances (squared) must be equal:

$$0 = \|a_1' - x - z\|^2 - \|a_2' - x - z\|^2$$
$$= \|a_1 + y_1 - x - z\|^2 - \|a_2 + y_2 - x - z\|^2 + O(y_1^2, y_2^2) . \qquad (4.6)$$

The $O(1)$ terms of (4.6) are consistent because $\|a_1 - x\| = \|a_2 - x\|$. The $O(y_1, y_2, z)$ terms give us the following constraint:

$$0 = (a_1 - x) \cdot (y_1 - z) - (a_2 - x) \cdot (y_2 - z) .$$

Using (4.5) this simplifies to

$$(a_1 - a_2) \cdot z = 0$$

and shows that the codimension of X^* is at least 1. $\qquad\square$

Though the study of the generalized Douglas–Rachford iteration rule (2.11) will begin in earnest not until Chapter 5, it may already be plausible that as a map[2] it is full rank. Then, if we think of the initial point as a distribution with a full-dimensional support, the iterates will continue to be distributions with full-dimensional support. This means there is zero probability of iterates x landing in the codimension-1 (or higher) set X^*. The phenomenon of nonunique projections is something that simply never comes up during the course of a search!

The case of sets X^* with codimension 1 is quite common and responsible for *branching*. For $x \in X^*$ there are two equidistant points in set A, say a_1 and a_2. But on either side of X^* there is just a single point, and this point switches discontinuously from a_1 to a_2 (or vice versa) as X^* is crossed. Each crossing corresponds to taking a different branch in the search. In Chapter 5 we will even see examples where codimension-1 surfaces are crisscrossed repeatedly, apparently keeping both options open when neither is clearly favored.

Nonchalance about nonunique projections comes naturally in computations. Consider the k-sparsity constraint on vectors of length m, or the rank-k constraint on $m \times m$ symmetric matrices. When computing the projection to these constraints, a key step is ranking the magnitudes of m numbers (the vector's components or the eigenvalues of the matrix). The projection is distance minimizing when the k highest-magnitude numbers are preserved and the rest are set to zero. There is nonuniqueness only when there is not a unique subset of k highest-magnitude numbers. But the probability of this is zero when the m numbers are sampled from a full-dimensional, continuous distribution.

4.1.4 Unions

That even the hardest problems are within the scope of bipartisan formulations is expressed more simply by the statement that finding elements of $A \cap B$ can be very hard even when finding elements of A and B individually, or projecting to these sets, is easy. By contrast, projecting to the union is very easy, requiring only as much work as projecting to A and B individually.

Consider any $y \in P_{A \cup B}(x)$. Then $y \in A$, or $y \in B$, with the possibility that both of these are true. If $y \in A$, then to be distance minimizing with respect to x, $y \in P_A(x)$. Otherwise, when $y \notin A$, we know that $y \in B$ and to be distance minimizing, $y \in P_B(x)$. The candidates for the projection to the union are therefore contained in the union of the projections, and we have the formula

$$P_{A \cup B}(x) = \underset{y \in P_A(x) \cup P_B(x)}{\arg\min} \|y - x\| .$$

This generalizes to the union of any number of sets.

Often one of the constraint sets in a bipartisan formulation is the union of a small number of sets, say $A = A_1 \cup \cdots \cup A_k$. By the ubiquity of projection uniqueness, the individual projections $P_{A_1}(x), \ldots, P_{A_k}(x)$ each comprise a single point. Projecting to A is then a simple matter of selecting among these k points the one that is closest to x. Because a wealth of shapes can be constructed as the union of simple geometric shapes (planes, spheres), unionization is a great principle for designing constraints!

4.1.5 Sparsity and Nonnegativity

The support constraint of wave microscopy is the statement that some elements of a vector are known to be zero. In spite of its simplicity, in a bipartisan context, the support constraint is a powerful and nontrivial source of information. This continues to be true when knowledge of the location of zeroes is replaced by knowledge of the number of zeroes. The property conferred by this less specific constraint on the support is called *sparsity*.

Sparsity is a natural constraint in crystallography because crystallographers have solid information about the existence and numbers of atoms, but no information on their positions. The nonzero contrast (in the X-ray or electron diffraction process) contributed by one atom occupies a small fraction of the image (unit cell of the crystal), leaving most of the image with zero contrast. The fewer the number of pixels allowed to have nonzero contrast, the stronger the sparsity constraint. When it is safe to assume that no more than k pixels (voxels) are nonzero, we say the image is k-sparse.

The sparsity constraint is an example of a nonconvex set expressible as the union of simple convex sets. Consider an image of three pixels (x, y, z), only one of which – identity unknown – is allowed to be nonzero. Geometrically the image is restricted to the union of three lines, the x-axis, y-axis, and z-axis of a three-dimensional space. Because the contrast in X-ray crystallography (electron density) is nonnegative, the three lines would be replaced by half-lines. In either case, the sparsity constraint is a nonconvex union of convex sets.

The k-sparsity property of a vector $x \in \mathbb{R}^m$ is often written using functions, $\|x\|_0 \leq k$, or $|\text{supp}(x)| \leq k$. But the 0-norm is not a true norm, and these functions – whatever we call them – are not very nice functions! In this book we avoid functions to express sparsity and introduce instead the symbol $\mathbb{R}_k^m$ for the set of k-sparse vectors in $\mathbb{R}^m$. If x is a k-sparse $m \times n$ matrix, we write $x \in \mathbb{R}_k^{m \times n}$.

It would be a mistake to follow the general prescription for projecting to unions (Section 4.1.4) when projecting to the sparsity constraint. The set $\mathbb{R}_k^m$ is the union of $\binom{m}{k}$ Cartesian subspaces. Fortunately there is a much more efficient projection to this set than trying out all these subspaces.

Proposition 4.2 (Sparsity projection) *To project $x \in \mathbb{R}^m$ to the k-sparsity constraint, the k components of x with the largest absolute value are preserved and all others are set to zero.*

Proof Each component i that is set to zero contributes x_i^2 to the squared projection distance. The minimum distance is achieved when the $m - k$ components with the smallest contribution are selected to be zero. Regarding the remaining k components, the projection distance is minimized when these are not changed at all. □

When nonnegativity is included, there is a small modification:

Proposition 4.3 (Nonnegative-sparsity projection) *To project $x \in \mathbb{R}^m$ to the nonnegative k-sparsity constraint, all negative components are set to zero and the k-sparsity projection is applied to the positive components.*

Proof If a negative component was *not* set to zero, but a positive value (to comply with the constraint), the contribution to the squared projection distance would be greater and the number of nonzero components would be increased relative to setting the value to zero. This leaves no ambiguity on what should be done with the negative components. To the positive components we can apply the k-sparsity projection because this preserves nonnegativity. □

Sparsity and nonnegative sparsity are examples of constraints that make no reference to how the m variables are packaged – as a vector, matrix, and so on. To compute the projection in the worst case, when k is not significantly smaller than m, one needs to rank m numbers. Like sorting, this can be done with $O(m \log m)$ operations [63].

4.1.6 Product Spaces

The Fourier magnitude constraint of wave microscopy, the element-wise constraints on Hadamard and kissing-sphere-Gram matrices, and also the constraints acting on the two sides of the bipartite graph of set-cover, are all examples of constraints on *product spaces*. Formally, we can think of constraints acting independently on disjoint sets of variables as defining an orthogonal decomposition

$$X = X_1 \times \cdots \times X_n ,$$

where X_1 is the subspace of all variables participating only in constraint 1, and so on. If constraint set A has this product structure, then the projection of a general point $x = (x_1, \dots, x_n) \in X$ has the form

$$P_A(x) = P_1(x_1) \times \cdots \times P_n(x_n) , \qquad (4.7)$$

where P_1 is a projection acting on the subspace X_1, and similarly for the other subspaces.

Though the product space concept is almost too obvious to deserve comment, as an abstraction it will prove useful when projecting to "concurring" variables that also satisfy a side constraint (Section 4.1.7). Also, by drawing attention to this structure when it exists, there is a smaller chance of its advantages being overlooked! Consider a set A having the shape of a square in the space of two variables. Computing P_A looks like it might involve some messy case analysis (corners, edges, interior) until we realize the square has a product structure in the right coordinate system:

$$A = \left\{ (x_1, x_2) \in \mathbb{R}^2 \colon |x_1| \le 1, \ |x_2| \le 1 \right\} .$$

Computing $P_A(x)$ now reduces to two independent projections to the interval $[-1, 1]$, the first of which is

$$P_1(x_1) = \begin{cases} 1, & x_1 > 1 , \\ -1, & x_1 < -1 , \\ x, & \text{otherwise.} \end{cases}$$

4.1.7 Side Constraints

A less obvious instance of product structure comes up when the concur in divide-and-concur is itself subject to a constraint, called a *side constraint*. For example, consider variables that live on the edges of a bipartite graph, as we used for set-cover. Suppose there are n broadcast nodes with edge degrees $m_1, \ldots, m_n$. The m_1 variables on the edges incident to node 1 are *replicas* of a single variable and concur (are equal) in a solution. These m_1 variables live in the space X_1, and similarly for the replicas at nodes $2, \ldots, n$. So far our variables have this product structure:

$$X = X_1 \times \cdots \times X_n . \tag{4.8}$$

In this section our focus is primarily on the product structure of each of the factors in (4.8). Each of these, like the first, has two factors:

$$X_1 = X_1^c \times X_1^\perp .$$

The subspace X_1^c corresponds to equality of all m_1 replicas and no restrictions on the equality-value itself, while $X_1^\perp$ is the orthogonal complement of that subspace (ways of being unequal). If the constraint on the 1-replicas was a simple concur (no side constraint), the constraint set for that factor would be

$$A_1 = X_1^c \times \{0\} ,$$

and similarly for the other factors. Points in this set are m_1-tuples of the form $(\bar{x}_1, \ldots, \bar{x}_1)$, where $\bar{x}_1$ is the concur value of the variable associated with node 1. As we saw already in Section 2.4, the distance-minimizing concur value is the mean of the m_1 replicas:

$$\bar{x}_1 = \frac{1}{m_1}(x_{1\,1} + \cdots + x_{1\,m_1}) .$$

When there are side constraints, the corresponding constraint set S lives in just the concur factors:

$$S \subset X_1^c \times \cdots \times X_n^c .$$

Thanks to the fact that S is independent of the variables in the complements $X_1^\perp, \ldots, X_n^\perp$, there is a simple rule for projecting to concur with a side constraint.

Proposition 4.4 (Concur with side constraint) *Let $\bar{x}_1, \ldots, \bar{x}_n$ be the concur values (equality projections) for independent sets of respectively $m_1, \ldots, m_n$*

*variable-replicas. If the concur values are additionally subject to the side con-
straint S, then the concur values are replaced by*

$$\arg\min_{(\bar{y}_1,\dots,\bar{y}_n)\in S}\left(\sum_{i=1}^{n} m_i \left\|\bar{y}_i - \bar{x}_i\right\|^2\right). \tag{4.9}$$

Proof The formula follows directly from the fact that a whole m_i-tuple of
identical concur values is changed for replica-set i. □

The numbers $m_1,\dots,m_n$ in (4.9) are irrelevant if constraint S also has the
product structure (4.8). When this is not the case, one can think of the concur
values of variables with many replicas being less compliant distance-wise than
variables with few replicas. In the set-cover problem, the side constraint S is
an inequality on the sum $\bar{y}_1 + \cdots + \bar{y}_n$ and the distance multipliers $m_1,\dots,m_n$
do matter. Computing projections to linear constraints with this wrinkle is
covered in Section 4.1.8.

4.1.8 Diagonally Generalized Metric

Constraint B in the set-cover problem is a system of linear equations: n
equality constraints and a simple sum constraint on the equality values. As
we saw in Section 4.1.7, to project to this constraint we first project to the
equality values (concur) and then project these to satisfy the side constraint
on the sum. In set-cover the side constraint is an inequality. If the concurred
values already satisfy the inequality, nothing else needs to be done; otherwise
we project to the equality case of the inequality constraint. Assuming the
latter, then according to (4.9), the side constraint projection has the form

$$P_S(x) = \arg\min_{y\in S} \sum_{i=1}^{n} g_i \left\|y_i - x_i\right\|^2, \tag{4.10}$$

where

$$S = \left\{y \in \mathbb{R}^{n\times k}: \sum_{i=1}^{n} y_i = c\right\}, \tag{4.11}$$

and c, like the x_i and y_i, are vectors in $\mathbb{R}^k$ ($k = 1$ for the set-cover problem).
The positive numbers g_i are interpreted as diagonal metric coefficients. In
the set-cover problem $g_i = m_i$, the number of replicas of variable i. Quite
apart from the case of unequal numbers of concurring variables, a diagonally generalized metric offers advantages over the isotropic metric even in
highly symmetric problems, such as the Turán numbers (Section 2.4). In

Section 6.3.2 we describe a heuristic that uses gap information to automatically adjust the metric coefficients. In this section, the numbers g_i are given constants.

A routine method for working out projections when constraints are defined by equations is to use Lagrange multipliers. Here we have k constraint equations and need a vector of multipliers $\lambda \in \mathbb{R}^k$. The Lagrangian

$$\mathcal{L}(y) = \sum_{i=1}^{n} g_i \left\| y_i - x_i \right\|^2 + \lambda \cdot \left(\sum_{i=1}^{n} y_i - c \right)$$

is the sum of the function being optimized and terms comprising all the constraint equations, each multiplied by its own Lagrange multiplier. Using the method of Lagrange multipliers, we can prove the following.

Proposition 4.5 (Sum-constraint projection with diagonally generalized metric) *The projection* (4.10) *to sum constraint* (4.11) *has the following form:*

$$P_S(x_1, \ldots, x_n) = (y_1, \ldots, y_n) \,,$$

where

$$y_i = x_i - \frac{\Delta}{g_i} \tag{4.12}$$

are metric-dependent shifts and

$$\Delta = \frac{\sum_{i=1}^{n} x_i - c}{\sum_{i=1}^{n} 1/g_i}$$

is the normalized constraint discrepancy.

Proof Because the Lagrangian is stationary at the optimizing y,

$$\forall i \colon 0 = \frac{\partial \mathcal{L}}{\partial y_i} = 2g_i(y_i - x_i) + \lambda \,.$$

Solving for y_i in terms of λ and applying constraint (4.11) to that expression to determine λ, we arrive at the stated result. The optimizing point is unique because the constraint is convex (and does not have to be tested to be minimizing). $\qquad\square$

Formula (4.12) is consistent with the intuition that the least metric-compliant (large g_i) variables change least.

Just about as common as the simple sum constraint (4.11) is the equality (concur) constraint:

$$C = \left\{ y \in \mathbb{R}^{n \times k} \colon y_1 = \cdots = y_n \right\} \,.$$

The projection to C with the diagonally generalized metric takes the following intuitively appealing form:

Proposition 4.6 (Concur projection with diagonally generalized metric)

$$P_C(x_1, \ldots, x_n) = (\bar{x}, \ldots, \bar{x}) \,,$$

where

$$\bar{x} = \frac{\sum_{i=1}^{n} g_i \, x_i}{\sum_{i=1}^{n} g_i} \tag{4.13}$$

is the metric-weighted concur value.

Proof   The minimizing concur value $\bar{x}$ minimizes the squared distance:

$$d^2 = \sum_{i=1}^{n} g_i \, \|\bar{x} - x_i\|^2 \,.$$

The result follows upon completing the square,

$$d^2 = \left(\textstyle\sum_{i=1}^{n} g_i\right) \left\|\bar{x} - \frac{\sum_{i=1}^{n} g_i \, x_i}{\sum_{i=1}^{n} g_i}\right\|^2 + \cdots \,,$$

where $\cdots$ are terms independent of $\bar{x}$.   □

Formulas (4.12) and (4.13) are standard repertoire for anyone using constraint projections to solve problems. They both owe their simplicity to the Euclidean norm's definition of distance!

4.1.9 Selector Variables

Indicator variables are constrained to be either 0 or 1. Very often these are packaged as *selectors*: vectors where exactly one element is 1 and the rest are 0. Because these are so common, we add the following to the standard repertoire:

Proposition 4.7 (Selector projection) *Let $x \in \mathbb{R}^m$ be a selector subject to the squared distance*

$$\|x' - x\|_g^2 = \sum_{i=1}^{m} g_i \, (x_i' - x_i)^2 \,.$$

The distance-minimizing selection is the element i that maximizes $g_i(2x_i - 1)$.

Proof Let d_i be the projection distance when element i is selected and d_0 be the distance when x' is set to the all-0 vector. Since

$$d_i^2 - d_0^2 = g_i \left((1 - x_i)^2 - (0 - x_i)^2 \right) \tag{4.14}$$
$$= g_i - 2g_i x_i \ ,$$

the stated result follows. $\square$

We take this simple result for granted until we ask what would be changed had we used the 1-norm instead of the Euclidean 2-norm. In the case where the analogous 1-norm coefficients g_i are equal, (4.14) would be replaced by

$$d_i - d_0 = |1 - x_i| - |0 - x_i|$$
$$= \sigma(x_i) \ ,$$

where σ is the step function

$$\sigma(x) = \begin{cases} +1, & x < 0 \ , \\ 1 - 2x, & 0 \le x \le 1 \ , \\ -1, & 1 < x \ . \end{cases}$$

Now if multiple elements have $x_i > 1$, any of them would be distance-minimizing selections!

4.1.10 Quasi-Projections

It sometimes happens that projections to sets defined by relatively simple equations have extraordinarily complex formulas. For such projections, one should consider an alternative to evaluating the explicit formula. There is no better example of this technical question than the simple product constraint,

$$A = \left\{ (x, y) \in \mathbb{R}^2 : xy = c \right\} \ ,$$

where the constant c is assumed to be positive. Minimizing the distance to A from an arbitrary point results in a quartic equation whose solutions can be written explicitly in terms of radicals. However, the complexity of the formula makes it very unpleasant to implement. One should also bear in mind that there are hidden computational costs when expressions include radicals.

A better way to project to A is to use *quasi-projections*. In general, the quasi-projection of x is an element of the constraint set that might not be distance minimizing to x. Often the set A has a combinatorial character that the quasi-projection can take advantage of.

The product constraint set is symmetric with respect to reflections in the

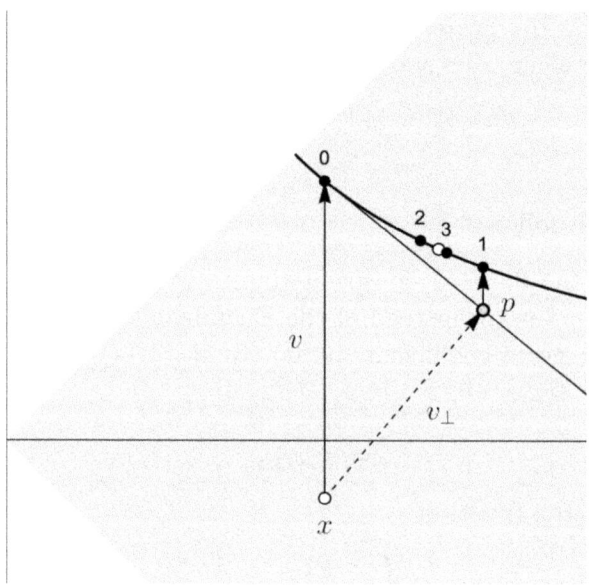

Figure 4.2 The piece A_{++} of the product constraint that lies in the rotated quadrant X_{++} (gray region). The first step in the computation of the projection of x is the quasi-projection a_0 (vertical arrow). This is followed by tangent-space refinement (a_1, a_2, ...), the first iteration of which is the dashed arrow followed by another quasi-projection (short vertical arrow). The next two tangent space refinements are marked 2 and 3, and the true projection is the white dot.

pair of orthogonal lines $x + y = 0$ and $x - y = 0$. From this property, we know the signs of $x + y$ and $x - y$ match the signs of $x_A + y_A$ and $x_A - y_A$ in the true projection. This breaks the projection down to four cases, and without loss of generality we consider just the case $x + y > 0$, $x - y > 0$, or $(x, y) \in X_{++}$, and denote the corresponding constraint set A_{++} .

Here is a possible quasi-projection to A_{++} , singled out by its simplicity:

$$Q_{A_{++}}(x, y) = \begin{cases} (\sqrt{c}, \sqrt{c}) , & x \le \sqrt{c} , \\ (x, c/x) , & x > \sqrt{c} . \end{cases} \qquad (4.15)$$

Figure 4.2 shows a quasi-projection $Q_{A_{++}}$ along with some refinements, to which we turn next.

Because the distance-minimizing property of true projections is key to the fixed-point property of the Douglas–Rachford algorithm, quasi-projections have a serious shortcoming. Fortunately, it is computationally inexpensive to systematically improve the distance-minimizing quality of quasi-projections by the method of *tangent-space refinement*.

Diagram 4.2 serves as a guide for the general case. Let

$$a_0 = Q_A(x)$$

be a quasi-projection to set A of a point x. We orthogonally decompose the vector

$$v = a_0 - x$$
$$= v_\parallel + v_\perp \,,$$

where

$$v_\perp = P_{\perp(a_0)}(v)$$

is the projection of v into the space orthogonal to the tangent space of A at a_0. Computing $v_\perp$ is inexpensive when A is defined by equations because one has available the gradient of the corresponding level-set function at the point a_0.

Next, construct the point

$$p = a_0 - \lambda\, v_\parallel$$

in the tangent space, where $\lambda > 0$ is a parameter. Diagram 4.2 shows the construction of p for the case $\lambda = 1$. Quasi-projecting p gives us another point on A, a refinement of a_0 :

$$a_1 = Q_A(p) \,.$$

We can prove this is an improvement (a decrease in the projection distance) if the quasi-projection satisfies

$$\|Q_A(a_0 + u) - (a_0 + u)\| = O(\|u\|^2) \,, \tag{4.16}$$

where a_0 is in the interior of A and u is any short vector in the tangent space to A at a_0. We verify this property for the quasi-projection (4.15) later. For the general case, property (4.16) ensures

$$\|a_1 - p\| = O(\lambda^2) \,. \tag{4.17}$$

We now compare the distance to the refinement with the distance to the original quasi-projection, for small λ. Starting with

$$\|p - x\| = \|v - \lambda\,v_\parallel\|$$
$$= \sqrt{\|v_\perp\|^2 + (1 - \lambda)^2\|v_\parallel\|^2}$$
$$= \sqrt{\|v\|^2 - 2\lambda\,\|v_\parallel\|^2 + O(\lambda^2)}$$
$$= \|v\| - \lambda\,\frac{\|v_\parallel\|^2}{\|v\|} + O(\lambda^2)\,,$$

and the triangle inequality,

$$\|a_1 - x\| \le \|a_1 - p\| + \|p - x\|\,,$$

we obtain

$$\|a_1 - x\| \le \|a_0 - x\| - \lambda\,\frac{\|v_\parallel\|^2}{\|v\|} + O(\lambda^2)$$

after making use of (4.17). This shows that for suitably small λ, the distance of the quasi-projection can be decreased. A good practical strategy is to first try $\lambda = 1$, since we know $v_\parallel = 0$ in the true projection. If that aggressive choice turns out to increase the distance, we try $\lambda = 1/2, 1/4$ and so on until the distance is decreased. For the point x shown in Figure 4.2, it was possible to use $\lambda = 1$ for all the refinements.

To check (4.16), for any $y_0 \in (0, \sqrt{c})$ consider the point

$$a_0 = (c/y_0\,,\,y_0) \in A_{++}\,.$$

A general vector in the tangent space at a_0 has the form

$$u = \nu(c/y_0\,,\,-y_0),\ \nu \in \mathbb{R}\,,$$

but we only care about small ν. From

$$a_0 + u = \big((c/y_0)(1 + \nu)\,,\,y_0(1 - \nu)\big)\,,$$
$$Q_{A_{++}}(a_0 + u) = \big((c/y_0)(1 + \nu)\,,\,y_0(1 + \nu)^{-1}\big)\,,$$

we obtain

$$Q_{A_{++}}(a_0 + u) - (a_0 + u) = \left(0\,,\,y_0\left(\frac{\nu^2}{1 + \nu}\right)\right)$$

and find

$$\|Q_{A_{++}}(a_0 + u) - (a_0 + u)\| = O(\nu^2)$$
$$= O\big(\|u\|^2\big)\,.$$

4.2 Affine Constraints

While the simple-sum and equality constraints (with and without a diagonally generalized metric) are by far the most used affine constraints, it is good to know how to project to the general case.

There are two ways of specifying affine constraints: *primal* and *dual*. The first of these sounds uncivilized, but the name is explained by the second. Linear programs, or the optimization of a linear function subject to linear constraints, have "dual" versions that in effect interchange variables and constraints [74]. "Primal" was invented in order to be able to refer to the original problem. For us the primal case is where the constraint set is defined by linear equations. In the dual case, the set is specified as a linear function graph.

Included in this section are two special cases that are fairly common and for which the projections can be computed more simply and efficiently than using the general-case formulas.

4.2.1 Primal Case

The B constraint (3.14) of quantum state tomography is a general system of linear equations, the kind of set we consider in this section. All the variables of the m-dimensional space, $x_1, \ldots, x_m$, participate directly, or primally, in a system of n linear equations:

$$B = \left\{ x \in \mathbb{R}^m : xC = b \right\}.$$

Here $C \in \mathbb{R}^{m \times n}$ and $b \in \mathbb{R}^n$ are the data that define the constraint. We may assume that the columns of C are linearly independent and $1 \le n < m$. When $n = 0$ there is no constraint, and B is a single point when $m = n$ that can be precomputed (and the projection simply replaces x by that point).

The Gram–Schmidt factorization $C = QR$ [46], where $Q \in \mathbb{R}^{m \times n}$ has n normalized and orthogonal columns and R is upper triangular, defines an orthogonal decomposition of $\mathbb{R}^m$, or

$$\begin{aligned} x &= x_{\parallel} + x_{\perp} \\ &= xP_{\parallel} + xP_{\perp}, \end{aligned}$$

where

$$P_{\parallel} = QQ^{\mathsf{T}}$$

is the projection to the space spanned by the columns of Q and $P_{\perp} = I - P_{\parallel}$. Because many projections P_B get performed in the course of solving a problem, the matrices Q, R and matrices/vectors derived from them are precomputed and held in memory. Here is the explicit formula for the projection:

Proposition 4.8 (Projection to primal affine constraint)

$$P_B(x) = b_\| + x P_\perp \, ,$$
$$b_\| = b R^{-1} Q^\mathsf{T} \, .$$

Proof Since

$$b = xC = (x_\| + x_\perp)C = x_\| C \tag{4.18}$$

is just a constraint on the $x_\|$ part of x, for P_B to be distance minimizing only this part of x will be changed:

$$P_B(x_\| + x_\perp) = x'_\| + x_\perp \, .$$

Every $x'_\|$ is uniquely parameterized by a $y \in \mathbb{R}^n$ via the columns of Q as

$$x'_\| = y Q^\mathsf{T} \, . \tag{4.19}$$

Imposing constraint (4.18),

$$\begin{aligned} b &= (y Q^\mathsf{T})C \\ &= (y Q^\mathsf{T})(QR) \\ &= yR \, . \end{aligned}$$

Multiplying on the right by R^{-1} to obtain y, equation (4.19) evaluates to

$$x'_\| = b R^{-1} Q^\mathsf{T} \, .$$

$\square$

4.2.2 Dual Case

Not to play favorites, the dual type of affine constraint will be called A:

$$A = \left\{ y C^\mathsf{T} + a : y \in \mathbb{R}^n \right\} \, .$$

With C having the same shape as in the primal case and $a \in \mathbb{R}^m$, this defines a hyperplane of dimension n in a space of m dimensions. Exactly this type of constraint, with $a = 0$, was one of the constraints in generalized phase retrieval (Exercise 2.5) with sensing matrix $G = C^\mathsf{T}$.

In the dual case, the constraint collapses to a single point, $\{a\}$, when $n = 0$, and there is no constraint when $m = n$. Not surprisingly, the projection formula also has a complementary relationship to the primal case:

Proposition 4.9 (Projection to dual affine constraint)

$$P_A(x) = x P_\| + a_\perp \,,$$
$$a_\perp = a P_\perp \,.$$

Proof Taking advantage of the same orthogonal decomposition defined by C as in the primal case, a general element x of A has the form

$$x_\| + x_\perp = y\, C^\mathsf{T} + a_\| + a_\perp \,, \tag{4.20}$$

where $y\, C^\mathsf{T}$ realizes any point in the $\|$-subspace for suitable y. There is therefore no constraint on $x_\|$ and P_A will be distance minimizing by preserving $x_\|$ (the corresponding y will be such that $y\, C^\mathsf{T} = x_\| - a_\|$). On the other hand, only the value $x_\perp = a_\perp$ satisfies the constraint so the projection reduces to making this replacement. □

4.2.3 Single Linear Constraint

A very common linear constraint for which it would be a mistake to project with a general formula is the case of a single linear equation or inequality. We saw the inequality case in the A constraint (2.17) of the spin glass problem. Note that the single inequality in that constraint involves only the off-diagonal elements of x and is therefore decoupled from the other constraints of set A.

In this section we consider the constraint

$$A = \left\{ x \in \mathbb{R}^m : b \cdot x = c \right\} , \tag{4.21}$$

where $b \in \mathbb{R}^m$ and $c \in \mathbb{R}$ parameterize the set. If the constraint is instead an inequality, the computation of $P_A(x)$ begins with a check of the inequality. If it is satisfied, then x is unchanged and the computation terminates. If the inequality is not satisfied, the distance minimizing x is the one that realizes the equality case, that is, it reduces to constraint (4.21).

The following projection formula deserves to be included in the standard repertoire:

Proposition 4.10 (Projection to single linear equation)

$$P_A(x) = x + \left(\frac{c - b \cdot x}{\|b\|^2} \right) b \,. \tag{4.22}$$

Proof Let y be the change in x to satisfy the constraint, and decompose as $y = y_\| + y_\perp$, where these are respectively parallel and perpendicular to *b.* Since only $y_\|$ changes the value of the constraint equation, in a distance-minimizing change $y_\perp = 0$. The distance-minimizing y is therefore a multiple of b for the unique multiplier that satisfies the constraint. ☐

Though we have seen the following special case several times already, this concise way of stating it is key to the projection we consider in Section 4.2.4:

Corollary 4.1 (Simple sum projection) *When projecting a vector so its components have a given sum, all components are changed by the same amount.*

The amount of change is contained in formula (4.22) when we take b to be the all-1 vector.

4.2.4 Linear Equations with Symmetry

The following affine constraint, on a rectangular table of numbers, arises in many applications:

$$A = \left\{ x \in \mathbb{R}^{m \times n} : \sum_{i=1}^{m} x_{ij} = c_j, \ \sum_{j=1}^{n} x_{ij} = r_i \right\} . \qquad (4.23)$$

The columns of the table are required to have column-sums $c_1, \ldots, c_n$ and similarly for the rows, with given row-sums $r_1, \ldots, r_m$. This set of $n + m$ numbers is the data for the constraint. It's important to realize that data-consistency requires

$$\sum_{j=1}^{n} c_j = t = \sum_{i=1}^{m} r_i , \qquad (4.24)$$

where t is the grand-total constraint on all the matrix elements.

    An example of constraint A is the 9×9 table of indicators for, say, the digit 4 in sudoku. Since there should be exactly one 4 in each row and column of a solution, all the column and row sums of constraint A are equal to 1. The rules of sudoku have a very high degree of symmetry. This symmetry is reflected in the equality of all the x-coefficients.

    The matrix Q upon which the primal-general-case projection is based would have mn rows and $m+n$ columns. If $m = n$, then computing anything with Q involves at least $O(m^3)$ operations (just to access every matrix element). But by taking advantage of the high symmetry of set A, the projection can be computed with only $O(mn)$ operations.

Claim 4.1 (Column and row sums projection) *To project to set A (4.23), the simple-sum projection is applied first to all the columns of x, and then to all its rows.*

Before giving the proof of this claim, we generalize constraint A in a way that highlights the properties responsible for the efficient projection.

Let U be the index set of a set of variables x, where U has k partitions into disjoint sets:

$$U = V_1^1 \sqcup \cdots \sqcup V_{m_1}^1$$
$$= V_1^2 \sqcup \cdots \sqcup V_{m_2}^2$$
$$\vdots$$
$$= V_1^k \sqcup \cdots \sqcup V_{m_k}^k .$$

Moreover, these partitions are *pairwise-regular* in that all pairs of parts from the same pair of partitions have equal-sized intersections:

$$\forall i \neq j, \ p \in \{1, \ldots, m_i\}, \ q \in \{1, \ldots, m_j\}: \quad |V_p^i \cap V_q^j| = c^{ij} . \tag{4.25}$$

Here is the generalization of set A:

$$A = \left\{ x \in \mathbb{R}^{|U|} : \forall i \in \{1, \ldots, k\}, \ p \in \{1, \ldots, m_i\}, \ \sum_{v \in V_p^i} x_v = s_p^i \right\} . \tag{4.26}$$

Equality of all the sums-of-sums generalizes the data consistency condition (4.24):

$$\forall i \in \{1, \ldots, k\}: \quad \sum_{p=1}^{m_i} s_p^i = \sum_{u \in U} x_u = t . \tag{4.27}$$

Theorem 4.4 (Pairwise-regular simple sum projections) *To project to set A (4.26), the simple-sum projection is applied, in turn, to the sums defined by partitions $V^1, \ldots, V^k$.*

Proof The projection P_A is performed in k rounds. In round 1, simple-sum projection is applied to the m_1 independent sets of variables with index sets in $V_1^1, \ldots, V_{m_1}^1$. Doing the same thing for the partition $V_1^2, \ldots, V_{m_2}^2$ in round 2 would be distance minimizing for those constraints but might modify the sums in round 1. To see that this doesn't happen, we use property (4.25).

Starting with the variables in V_1^2, these are all shifted by the same δx to satisfy the sum constraint. By (4.25), exactly c^{12} variables in each of $V_1^1, \ldots, V_{m_1}^1$ is shifted by the same amount, δx_1 (they are, after all, the same variables with different names). This has the effect of shifting the sums in round 1 from the values s_p^1 to $s_p^1 + c^{12}\delta x_1$.

Continuing with the variables in V_2^2, we again get equal distance-minimizing shifts δx_2 and a combined shift in the sums in round 1 to $s_p^1 + c^{12}(\delta x_1 + \delta x_2)$. After projecting to the remaining simple-sum constraints of partition V^2, we satisfy all the constraints for that partition but shift the sums from round 1 to

$$s_p^1 + c^{12}(\delta x_1 + \cdots + \delta x_{m_2}) = s_p^1 + c^{12}\delta x .$$

This has the effect of shifting the sum-of-sums (4.27) for $i = 1$ from the value t to

$$\sum_{p=1}^{m_1}(s_p^1 + c^{12}\delta x) = t + m_1 c^{12}\delta x .$$

But after projecting to the sums in the V^2 partition, we know that the sum of all the elements of x has the correct grand total t. This implies that $\delta x = 0$ and the sums in round 1 were in fact not shifted at all by the projections in round 2.

By exactly the same line of argument, we can show that the simple-sum projections for partition V^3 do not shift the sums in rounds 1 and 2, and so on, for all the remaining partitions. $\square$

A natural setting for pairwise-regular simple sum constraints is tensor variables. We write

$$x \in \mathbb{R}^{m_1 \times \cdots \times m_k}$$

when x is an order-k tensor of the given dimensions. The partitions $V^1, \ldots, V^k$ correspond to the k ways of "slicing" such a tensor. In partition/slicing V^i, the ith tensor index is fixed at respectively $p \in \{1, \ldots, m_i\}$, with s_p^i being the sum of those elements of x. The constant size of the intersection of any slice of index i with any slice of index $j \neq i$ is

$$c^{ij} = \frac{m_1 \times \cdots \times m_k}{m_i \times m_j} .$$

Claim 4.1 is the simplest, or order-2 case, where $c^{12} = 1$.

It is not often that treating interdependent constraints (e.g., on rows and columns) as independent gives the correct result, but simple sums on partitions with the regular intersection property is one of those cases. Most of the credit goes to the high symmetry of the sum constraints. Symmetry plays an even larger role in Section 4.3.

4.3 Distribution Constraints

In a distribution constraint, we care only about the values of a set of variables. Which variables get which values has no relevance beyond what projections always do, which is to minimize the distance to the constraint.

We will use the term *permutation-symmetric set* to describe the geometry of a distribution constraint set. If only the set of values taken by the variables matters, then any permutation applied to the variables is also an element of the constraint set. We have already seen two examples of constraints that are permutation-symmetric: k-sparsity (Section 4.1.5) and selectors (Section 4.1.9). A k-sparse vector has to only satisfy the simple property that at most k of its elements are nonzero. Any permutation of the elements of a k-sparse vector is also k-sparse.

A vector $x \in \mathbb{R}^m$ that satisfies the selector constraint has a single 1-element and the rest 0. This time there are just m distinct permutations, but they all belong to the constraint set. This set of m points, a regular m-simplex, is the most symmetric possible, geometrically. The metric symmetry is appropriate when the m choices have a symmetry in the application. In sudoku, for example, it would be a mistake to use just a single variable x and $A = \{1, 2, \ldots, 9\}$ as the constraint on what may appear in one of the cells. The extremes $x = 1$ and $x = 9$ would be metrically inequivalent to the middle choice, $x = 5$. In the 9-simplex all nine choices are metrically equivalent.

We first consider distribution constraints where the set of values taken by the variables is specified explicitly. The set of allowed radii in the disk packing puzzle of Section 1.5 is an example of this kind of constraint.

We then consider distributions where the values correspond to probabilities: nonnegative numbers that sum to 1. Vectors of numbers with these properties occupy the interior of a regular simplex.

Because distribution constraints apply to any shape of variable, in this section we let them be vectors for simplicity.

4.3.1 Value-List Constraints

The data for this constraint is just the set of m values

$$V = \{v_1, \ldots, v_m\} \tag{4.28}$$

for the m variables. Without loss of generality, we will assume these are given as an ordered list:

$$v_1 \le v_2 \le \cdots \le v_m . \tag{4.29}$$

The algorithm for projecting to the value-list constraint makes use of the following fact:

Lemma 4.5 (Order-preserving matching) *Of the two variable assignments (1) $x_i = v_p$, $x_j = v_q$ and (2) $x_i = v_q$, $x_j = v_p$, the projection distance is minimized when x_i and x_j have the same ordering as v_p and v_q.*

Proof From the difference of the squared distances,

$$d_2^2 - d_1^2 = \left((v_q - x_i)^2 + (v_p - x_j)^2\right) - \left((v_p - x_i)^2 + (v_q - x_j)^2\right)$$
$$= 2(v_p - v_q)(x_i - x_j),$$

we see that assignment (2) can only increase the distance when the product is nonnegative, or when x_i and x_j have the same ordering as v_p and v_q. $\square$

The projection to the value-list constraint follows directly from the lemma:

Proposition 4.11 (Value-list projection) *In the projection of $x \in \mathbb{R}^m$ to the values in the list V (4.28), the elements of x are matched in an order-preserving manner with the values in V .*

Proof We may assume the elements of x have a strict ordering, as the case of even two equal elements has codimension-1 and arises with probability zero. By the lemma, if x_i and x_j are consecutive increasing elements in the ordering, then these must be assigned values $x_i = v_p$, $x_j = v_q$ where $v_p \leq v_q$. A distance-minimizing matching is therefore achieved when, starting with the smallest, the elements of x in increasing order are matched with the non-decreasing sequence of values (4.29) starting with v_1. $\square$

In an implementation, the nondecreasing list (4.29) would be precomputed. The projection of x starts by ranking its elements so they form an increasing sequence. This ordering is then used when replacing each element with the corresponding element of V. The work to compute the projection is dominated by the ranking of the x-elements, and requires $O(m \log m)$ operations.

Publishing houses could use this projection to imprint their brand on the grayscale images in their publications. For example, the *Fair Contrast Press* can deliver on its promise of all grayscale values being equally represented by projecting m-pixel images to the set $V = \{1, 2, \ldots, m\}$, where 1 is black and m is white. Because projections are distance minimizing, the transformed image is the best possible likeness of the original for the chosen contrast distribution. Figure 4.3 shows an example of a projection to the fair-contrast distribution.

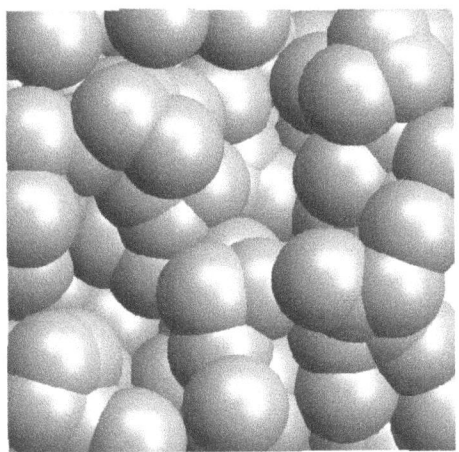 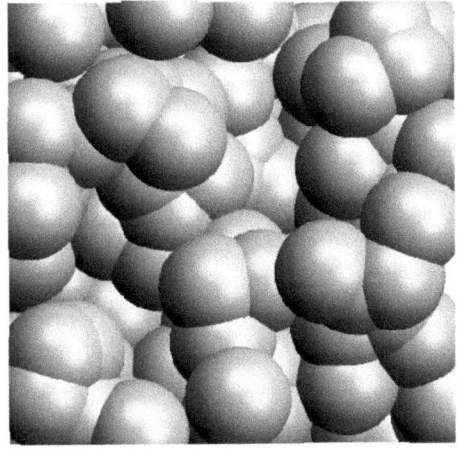

Figure 4.3 Original image (left) projected to the fair-contrast distribution (right), where all grayscales, white to black, are equally represented.

4.3.2 The Probability Simplex

This constraint generalizes the selector constraint. Instead of just the vertices of the regular m-simplex, the constraint set now includes all of the simplex interior:

$$S = \left\{ x \in \mathbb{R}^m : x \geq 0 , \sum_{i=1}^{m} x_i = 1 \right\} .$$

Projecting to this constraint combines what we have learned about the simple sum projection, permutation symmetry, and makes use of the order-preserving matching lemma:

Theorem 4.6 (Probability simplex projection) *When projecting x, let y be the elements of x arranged in decreasing order:*

$$y_1 > y_2 > \cdots > y_m .$$

If y' are the corresponding elements of the projection, then

$$y_1' = y_1 + \delta_\ell$$
$$y_2' = y_2 + \delta_\ell$$
$$\vdots$$
$$y_\ell' = y_\ell + \delta_\ell \qquad (4.30)$$
$$y_{\ell+1}' = 0$$
$$\vdots$$
$$y_m' = 0 \, ,$$

where the shift δ_ℓ is chosen so that

$$\sum_{i=1}^{\ell} y_i' = 1 \qquad (4.31)$$

and $\ell > 0$ (the number of positive probabilities) is the largest integer for which $y_\ell' > 0$.

Proof We are free to rearrange the elements of x into the decreasing sequence y because the constraint is permutation symmetric. If y' are the corresponding elements after projection, then we know that all permutations of these are contained in the constraint set. Viewing this as an instance of a value-list constraint, by Proposition 4.11 we know that

$$y_1' \geq y_2' \geq \cdots \geq y_m' \, .$$

Since the y' are nonnegative, there are m cases to consider, corresponding to the number of positive probabilities. By the simple-sum projection rule, the positive y' are given by the same shift applied to the corresponding y. The value of the sum, constraint (4.31), ensures the positive y' are proper probabilities (do not exceed 1). If $\ell > 0$ is the number of positive probabilities, then the projection must have the form (4.30).

The projection also has the property that the ℓ shown in (4.30) is the maximum possible. That is, either $\ell = m$ or shifting $\ell + 1$ elements produces a negative $y_{\ell+1}'$. To check this claim, consider the possibility of a projection $y \mapsto y''$ with only $\ell^* < \ell$ positive probabilities. If we consider the first ℓ elements of both maps, we notice that they both satisfy the simple-sum constraint, with y'' differing from y' in having $\ell - \ell^*$ zero elements. But this shows y'' would not be distance minimizing, since the unique projection, obtained by applying the same shift δ to the elements of y, is the projection y'. $\square$

An efficient algorithm for projecting to the probability simplex closely follows the proof above. After ranking the elements of x, there is a loop that increments the number of nonzero x', calculating the shift for each case, terminating when the next case would produce a negative x'. Because the shifts can be computed recursively, the work is dominated by the initial sort of x.

It's also easy to project to the probability simplex if we additionally want the probabilities to be k-sparse. Since now at least $m - k$ of the ordered x' are zero, by the order preserving matching lemma, this same number of the smallest of the ordered x will get replaced by zero in the projection. The projection computation for the remaining k elements of x is identical to the one described above, with k replacing m.

4.4 Matrix Constraints

When variables x are packaged as matrices, the distance invariance unique to matrices is exploited to lay bare the constraints on x. Most generally these take the form of orthogonal or unitary transformations applied on the left and right. In the complex case, for $x \in \mathbb{C}^{m \times n}$,

$$
\begin{aligned}
\|x\|^2 &= \mathrm{Tr}\left(x x^\dagger\right) \\
&= \mathrm{Tr}\left(x\, u u^\dagger x^\dagger v v^\dagger\right) \\
&= \mathrm{Tr}\left(v^\dagger x\, u u^\dagger x^\dagger v\right) \\
&= \mathrm{Tr}\left(v^\dagger x\, u\, (v^\dagger x\, u)^\dagger\right) \\
&= \|v^\dagger x\, u\|^2 ,
\end{aligned}
$$

where u and v are arbitrary $n \times n$ and $m \times m$ unitary matrices. The first projection we consider, to the nearest matrix of rank ℓ, is one of the most direct applications of this invariance, where u and v are chosen to reveal the singular values of x.

4.4.1 Low Rank

The singular value decomposition (SVD) of a rank k matrix $x \in \mathbb{C}^{m \times n}$ has the form

$$
x = s\, \mathrm{diag}(\sigma_1, \ldots, \sigma_k)\, v ,
$$

where the columns of $s \in \mathbb{C}^{m \times k}$ and the rows of $v \in \mathbb{C}^{k \times n}$, called singular bases, are orthogonal and have norm 1. Up to phase rotations of the singular bases, the SVD is unique when the positive singular values are distinct because we can decide to order them as

$$\sigma_1 > \sigma_2 > \cdots > \sigma_k \ .$$

When projecting x to have rank $\ell < k$, we will not worry about codimension-1 exceptions where two singular values are equal.

One of the cornerstones of data science is the following theorem:

Theorem 4.7 (Eckart–Young [33]) *If $y \in \mathbb{C}^{m \times n}$ is any matrix of rank $\ell < k$, then*

$$\|x_\ell - x\|^2 \le \|y - x\|^2 \ , \tag{4.32}$$

where

$$x_\ell = s \ \mathrm{diag}(\sigma_1, \ldots, \sigma_\ell, 0, \ldots, 0) \, v \ . \tag{4.33}$$

Because the x_ℓ defined by (4.33) has rank ℓ, inequality (4.32) identifies this matrix as the rank ℓ projection of x. Had we instead been projecting a real matrix $x \in \mathbb{R}^{m \times n}$, the only thing that changes is that the singular bases s and v are real.

4.4.2 Spectral Projection

In several of the problems encountered in previous chapters, the variables were packaged as real-symmetric or Hermitian matrices. The Gram matrix of sphere-center vectors was used in the kissing spheres problem, and Hermitian matrices proved to be the natural variables for quantum state tomography and configurations of complex lines. The symmetry of these matrices give them a real spectrum, and at least one of the constraints in these problems was purely a constraint on the spectrum.

Constraints on the real spectrum of a real-symmetric or Hermitian matrix are, automatically, distribution constraints. That's because spectra are inherently unordered. The best way to understand that statement is to express the eigen-decomposition as a sum of rank-1 matrices. If $u_1, \ldots, u_m$ are the normalized eigenvectors with corresponding eigenvalues $\sigma_1, \ldots, \sigma_m$ of some $x \in \mathbb{S}^m$, then

$$x = \sum_{i=1}^{m} \sigma_i u_i u_i^{\mathsf{T}} \ .$$

Any constraint on the distribution of eigenvalues is permutation symmetric because the sum has permutation symmetry.

The A constraint on the Hermitian density matrix ρ of quantum state tomography is a nice example of a distribution constraint on the eigenvalues of ρ. These must be nonnegative, k-sparse, and have sum equal to 1. This is the same as the constraint that defines the k-sparse probability simplex and whose projection we worked out in Section 4.3.2. What we will learn in this section is that we can use exactly the same projection in quantum state tomography, now applied to the eigenvalues of ρ. Remarkably, the projection leaves the eigenvectors of ρ unchanged!

Our goal in this section is the *spectral projection theorem* for Hermitian and real-symmetric matrices. The first step is a lemma that establishes an optimizing property of the eigenvectors. Though our notation addresses the Hermitian case, all that changes for real-symmetric matrices is that the eigenvectors u are real and $u^\dagger$ can be replaced by u^T.

The lemma refers to two diagonal matrices:

$$d = \operatorname{diag}(\sigma_1, \ldots, \sigma_m) \,,$$
$$\tilde{d} = \operatorname{diag}(\tilde{\sigma}_1, \ldots, \tilde{\sigma}_m) \,.$$

These will later be related, respectively, to the input and output points of the projection. We may therefore assume the σ's are distinct with probability 1. On the other hand, we do not want to assume anything like this for the $\tilde{\sigma}$'s. For example, when the spectrum is k-sparse multiple $\tilde{\sigma}$'s are zero.

The lemma concerns the critical points of a function on the $m \times m$ unitary matrices, $\mathcal{U}^m$:

Lemma 4.8 (Diagonal matrix-pair criticality) *If u is a critical point of the function*

$$f(u) = \operatorname{Tr}\left(d\,u\,\tilde{d}\,u^\dagger \right) \tag{4.34}$$

from $\mathcal{U}^m$ to $\mathbb{R}$, then $c = u\,\tilde{d}\,u^\dagger$ is a diagonal matrix (a permutation of the elements of $\tilde{d}$).

Proof To study the critical points of f, we consider the linear variation with v of $f(u + v)$, where v is constrained to keep $u + v$ unitary. From

$$I = \left(u + v + O\left(v^2\right)\right)\left(u + v + O\left(v^2\right)\right)^\dagger$$
$$= I + vu^\dagger + \left(vu^\dagger\right)^\dagger + O\left(v^2\right) \,,$$

we see that this constraint is that $vu^\dagger$ is anti-Hermitian. Let a be any $m \times m$ anti-Hermitian matrix, then $v = au$. Using this parameterization of the perturbation v, we obtain

$$f(u + au) = \text{Tr}\left(d\left(u + au + O\left(a^2\right)\right)\tilde{d}\left(u + au + O\left(a^2\right)\right)^\dagger\right)$$
$$= f(u) + \text{Tr}\left(a(cd - dc)\right) + O\left(a^2\right) ,$$

where we used $a^\dagger = -a$. The linear term has the form $\text{Tr}(ab)$, where b is also anti-Hermitian. At a critical point the linear term vanishes, and $\text{Tr}(ab) = 0$ implies $b = 0$, since a was any anti-Hermitian matrix. Expressing the (i,j) element more explicitly for $i \neq j$, we obtain

$$b_{ij} = (d_{jj} - d_{ii})c_{ij}.$$

Since these are all zero, and $d_{ii} = \sigma_i \neq \sigma_j = d_{jj}$, all the off-diagonal elements of c must be zero. $\qquad\square$

We are now ready to state and prove the spectral projection theorem. The symbol $\Sigma \subset \mathbb{R}^m$ denotes the set of spectra that define the constraint: a distribution constraint on the m-tuple of eigenvalues. For any spectrum $\sigma \in \mathbb{R}^m$,

$$P_\Sigma(\sigma) = \arg\min_{\tilde{\sigma} \in \Sigma} \|\tilde{\sigma} - \sigma\|^2$$

is the projection to a point in Σ. The constraint set, now to be imposed on $m \times m$ Hermitian matrices, is

$$A = \left\{x \in \mathbb{H}^m : \text{spec}(x) \in \Sigma\right\}.$$

Theorem 4.9 (Spectral projection) *The projection of*

$$x = u\,\text{diag}(\sigma)u^\dagger = u\,d\,u^\dagger$$

to constraint A is given by

$$P_A(x) = u\,\text{diag}\left(P_\Sigma(\sigma)\right)u^\dagger.$$

Proof The projection is a point

$$x_A = v\,\text{diag}(\tilde{\sigma})v^\dagger = v\,\tilde{d}\,v^\dagger \tag{4.35}$$

for some $v \in \mathcal{U}^m$ and spectrum $\tilde{\sigma} \in \Sigma$ that minimizes the squared distance

$$\|x - x_A\|^2 = \|d - w\,\tilde{d}\,w^\dagger\|^2 . \tag{4.36}$$

In (4.36) we used the invariance of the distance under left/right multiplication by u and defined $w = u^\dagger v$. Expressing the squared distance as a trace,

$$\|x - x_A\|^2 = \mathrm{Tr}\left(d - w\tilde{d}w^\dagger\right)\left(d - w\tilde{d}w^\dagger\right)^\dagger$$
$$= \mathrm{Tr}\left(d^2 + \tilde{d}^2\right) - 2\,\mathrm{Tr}\left(dw\tilde{d}w^\dagger\right), \qquad (4.37)$$

we see that its dependence on w, and up to a sign and a factor of 2, is the function (4.34). Since x_A is a minimizer of the squared distance, the corresponding v is a critical point and $w = u^\dagger v$ is a critical point of $f(w)$. By Lemma 4.8,

$$w\tilde{d}w^\dagger = \mathrm{diag}\left(\tilde{\sigma}'\right),$$

where $\tilde{\sigma}'$ is any permutation of $\tilde{\sigma}$. Since $\tilde{\sigma}' \in \Sigma$ for any $\tilde{\sigma} \in \Sigma$, and the trace of $\tilde{d}^2$ is unchanged by a permutation, we are free to replace $\tilde{\sigma}$ and $\tilde{d}$ by their permuted counterparts. Expression (4.37) for the squared distance simplifies to

$$\|x - x_A\|^2 = \mathrm{Tr}\left(d - \tilde{d}\right)^2$$
$$= \|\sigma - \tilde{\sigma}\|^2,$$

so the distance-minimizing spectrum in (4.35) is given by

$$\tilde{\sigma} = P_\Sigma(\sigma).$$

Finally, since

$$x_A = u\left(w\tilde{d}w^\dagger\right)u^\dagger$$
$$= u\,\mathrm{diag}(\tilde{\sigma})u^\dagger,$$

both parts of the projection formula are confirmed.   □

4.4.3 Orthogonal Matrices

The set of orthogonal matrices made an appearance in the bipartisan formulation of Hadamard matrices. And as we will see in Section 4.5.1, these matrices also play an important role when one is working with fixed-shape geometrical objects that are free to rotate in space. In that setting, the rows of the orthogonal matrix are interpreted as the body axes of the object.

Two sets of orthogonal matrices come up in practice. The larger set is defined by

$$O^m = \left\{u \in \mathbb{R}^{m \times m} : uu^\mathsf{T} = I\right\}.$$

In some applications, only the subset corresponding to proper or orientation-preserving rotations is admissible:

$$\mathcal{O}^m = \{u \in \mathbb{R}^{m \times m} : uu^\mathsf{T} = I, \det u = 1\} .$$

The *handedness* of the symbols for the two sets is meant to convey whether their rotations preserve this property, that is, when the object being rotated has a handedness.

The projections, in particular the projection to $\mathcal{O}^m$, rely on a lemma involving diagonal matrices. One of these is the diagonal matrix in the SVD of the projection input:

$$x = s\,d\,v ,$$

$$d = \mathrm{diag}(\sigma_1, \ldots, \sigma_m) .$$

The singular values σ are positive and distinct with probability 1. The lemma uses critical points to identify another diagonal matrix:

Lemma 4.10 (Diagonal matrix criticality) *If w is a critical point of the function*

$$f(w) = \mathrm{Tr}(wd)$$

from $\mathcal{O}^m$ to $\mathbb{R}$, then $w = \mathrm{diag}(s)$ where $s \in \{-1, 1\}^m$.

Proof As in the proof of Lemma 4.8, we parameterize the perturbation of w as aw, where a is any antisymmetric $m \times m$ matrix:

$$f(w + aw + O(a^2)) = \mathrm{Tr}\left(\left(w + aw + O\left(a^2\right)\right)d\right) .$$

At a critical w, the term linear in a vanishes. This implies that $b = wd$ is a symmetric matrix, so that $\mathrm{Tr}(ab) = 0$ for any antisymmetric a. From the definition of the symmetric b, we have

$$d^2 = b^\mathsf{T}b = b\,b^\mathsf{T} = w\,d^2\,w^\mathsf{T} .$$

The equality of the left and right expressions identifies w as the set of normalized eigenvectors of the diagonal matrix d^2. Since the elements of d are distinct, the eigenvectors are unique up to sign: a ± 1 in position i for the ith eigenvector, zeroes elsewhere. □

Using this lemma, the projections to the two sets of orthogonal matrices differ in only a small detail:

Theorem 4.11 (Orthogonal matrix projection) *If* $x \in \mathbb{R}^{m \times m}$ *has SVD* $x = s \, d \, v$, *then*

$$P_O(x) = s \, v \,,$$

$$P_{\mathcal{O}}(x) = \begin{cases} s \, v \,, & \text{if } \det(s \, v) = 1 \,, \\ s \, r(d) \, v \,, & \text{otherwise} \,, \end{cases}$$

where $r(d)$ is a diagonal reflection matrix comprising a -1 at the position of the smallest element of d and $+1$'s elsewhere.

Proof Start with the set of general orthogonal matrices:

$$
\begin{aligned}
P_O(x) &= \arg\min_{u \in O^m} \|u - x\|^2 \\
&= \arg\min_{u \in O^m} \text{Tr}(u - x)(u - x)^\mathsf{T} \\
&= \arg\max_{u \in O^m} \text{Tr}\,(u \, x^\mathsf{T}) \\
&= \arg\max_{u \in O^m} \text{Tr}\,((s^\mathsf{T} \, u \, v^\mathsf{T}) \, d) \\
&= s \left(\arg\max_{w \in O^m} \text{Tr}\,(wd) \right) v \,.
\end{aligned}
$$

The maximizing w is also critical for the function $\text{Tr}(wd)$ of Lemma 4.10. Clearly the choice where w is the identity matrix (all elements $+1$) gives the maximum and the projection formula for the set O^m. This choice is inadmissible for $\mathcal{O}^m$ in the case that $\det(s \, v) = -1$. To fix the sign of the determinant, only one sign needs to be flipped. Flipping the sign for which the corresponding element of d is smallest maximizes $\text{Tr}(wd)$. $\square$

4.4.4 A Fiber Perspective for Spectral Constraints

In all the matrix projections we have seen, all having to do with constraints on the spectrum (or singular values), the eigenvectors (or singular bases) of the projection input x always "pointed" to the projection output, $P_A(x)$. Is this a coincidence, or is there a deeper reason? In this section we present a *fiber perspective* that unifies these results and shows how they generalize.

As a warm-up, we take a break from matrices and consider the simple (and much used) norm constraint. Here it is for vectors:

$$A(r) = \{x \in \mathbb{R}^m \colon \|x\| = r\} \,. \tag{4.38}$$

The projection is so obvious it hardly deserves a formal proof:

$$P_{A(r)}(x) = \left(\frac{r}{\|x\|} \right) x .$$

(4.39)

The same formula applies to any other shape of variable, provided the 2-norm is a sum of squares with equal coefficients on all the terms (an isotropic metric). In phase retrieval $m = 2$, where the two components correspond to the real and imaginary parts of a complex number. When the constraint is an inequality on the norm, a rescaling to the equality case of the inequality is performed only when the inequality is not satisfied. In the upper bound case, the constraint set is a ball, a convex set.

But now let's prove all these simple things in a way that generalizes to matrices. We start by defining *singular fibers*. For the space $\mathbb{R}^m$ and any point $x \neq 0$, the singular fiber $F(x)$ is the unique line that passes through x and the origin:

$$F(x) = \left\{ \lambda \hat{x} \ : \ \hat{x} = \frac{x}{\|x\|} , \ \lambda \in \mathbb{R} \right\} .$$

(4.40)

The odd way to parameterize the line, with a unit vector, is not unlike how we like to decompose matrices:

Definition 4.12 Let $x \in \mathbb{R}^{m \times n}$ have rank k and singular value decomposition

$$x = \sum_{i=1}^{k} \sigma_i u_i^{\mathsf{T}} v_i ,$$

where the positive singular values $\sigma_1, \ldots, \sigma_k$ are distinct. Then the normalized rank-1 matrices $u_1^{\mathsf{T}} v_1, \ldots$ are unique and

$$F(x) = \left\{ \sum_{i=1}^{k} \lambda_i u_i^{\mathsf{T}} v_i \ : \ (\lambda_1, \ldots, \lambda_k) \in \mathbb{R}^k \right\}$$

is the singular fiber of x.

The argument of F is the *generator* of the fiber. Notice that not all x generate fibers. In the vector example, we have to avoid the x with norm zero; in the matrix example, we avoid x where one of the singular values is zero, or two singular values are equal. In the computational realm, these are not relevant restrictions, because we model x as having a continuous, full-dimensional distribution where the exceptional cases have zero probability.

Fibers are vector spaces. If y_1 and y_2 are two elements of $F(x)$, then so is

$c_1 y_1 + c_2 y_2$ for arbitrary real numbers c_1 and c_2. The parameters

$$(\lambda_1, \ldots, \lambda_k)$$

are coordinates in the fiber, and thanks to the unit-vector parameterization in the vector case, and the orthogonality of the rank-1 bases in the matrix case, the distance between points y and y' of the same fiber $F(x)$ is the distance for orthogonal coordinates:

$$\|y' - y\|^2 = \sum_{i=1}^{k} (\lambda_i' - \lambda_i)^2 .$$

A property of fibers that follows directly from their definition is so important we record it formally:

Lemma 4.13 (Fiber elements are generators) *If $y \in F(x)$, then with probability 1, $F(y) = F(x)$. Moreover, since $x \in F(x)$, $x \in F(y)$.*

Proof From definition (4.12) we see that unless any of the λ's are zero or equal, the singular values of y are nonzero and distinct, so that y has a unique representation as a sum of rank-1 matrices. But these are also the basis matrices of x. Because the fiber element corresponding to

$$(\lambda_1, \ldots, \lambda_k) = (\sigma_1, \ldots, \sigma_k)$$

is just x itself (the σ's are the singular values of x), $x \in F(x)$. □

The power of fibers comes from the fact that projections to constraints on the matrix spectrum are computed entirely within a single fiber! As we will see, this follows from the general geometrical property of projections given in Lemma 4.2. We restate this lemma in a form that is useful when the constraint set is specified as the level-set of a system of equations.

Lemma 4.14 (Projections are spanned by level-set gradients) *Define set A as the level-set of real-analytic functions $f_1, \ldots, f_k$ from $\mathbb{R}^m$ to $\mathbb{R}$:*

$$A(r_1, \ldots, r_k) = \{x \in \mathbb{R}^m \; : \; f_1(x) = r_1, \ldots, f_k(x) = r_k\} .$$

If $a = P_{A(r_1,\ldots,r_k)}(x)$ is the projection of x, then the vector $a - x$ lies in the span of the gradients $\nabla f_1, \ldots, \nabla f_k$, all evaluated at point a.

Proof Since the level-set functions are constant in the tangent space $f_\|$,

$$f_\| = (g_1)_\perp \cap \cdots \cap (g_k)_\perp ,$$

where g_i is the one-dimensional subspace generated by $\nabla f_i(x)|_{x=a}$ and $(g_i)_\perp$

is its orthogonal complement. By the general rule for the orthogonal complement of an intersection,

$$f_\perp = g_1 + \cdots + g_k ,$$

the orthogonal complement of the tangent space is the span of the gradients. But by Lemma 4.2 we know $a - x$ always lies in $f_\perp$.   □

Continuing with our warm-up exercise, let's apply this lemma to the norm constraint (4.38). For this constraint, there is just a single function,

$$f(x) = \|x\| ,$$

and by the lemma, if $a = P_{A(r)}(x)$, then $a - x$ is a multiple of

$$\nabla f(x)|_{x=a} = \hat{a} .$$

But "$a - x$ is a multiple of $\hat{a}$" is the same as the statement

$$a - x \in F(a) \qquad\qquad (4.41)$$

by our definition (4.40) of the singular fibers.

In the analogous calculation for matrices, we have the k functions

$$\sigma_1(x) , \ \ldots , \ \sigma_k(x) ,$$

which we know as the singular values of the rank-k matrix x (with some convention for their order). To calculate their gradients, we study the linear variation of the defining equation for the singular bases,

$$u_i\, x = \sigma_i(x)\, v_i ,$$

when x is replaced by $x + dx$, with all $m \times n$ elements of dx independent. Here is the equation relating the linear terms:

$$du_i\, x + u_i\, dx = d\sigma_i\, v_i + \sigma_i(x)\, dv_i , \qquad\qquad (4.42)$$

where

$$d\sigma_i = \mathrm{Tr}\left(dx^\mathsf{T}\, \nabla\sigma_i(x)|_{x=a}\right) \qquad\qquad (4.43)$$

and the components of the gradient are arranged in a $m \times n$ matrix, just like x. To simplify our expression, we notice that the normalization constraints

$$u_i\, u_i^\mathsf{T} = 1 ,$$
$$v_i\, v_i^\mathsf{T} = 1$$

imply

$$du_i \, u_i^\mathsf{T} = 0 \, ,$$
$$dv_i \, v_i^\mathsf{T} = 0 \, .$$

Multiplying (4.42) on the right by v_i^T (forming the inner product) and using the identities for v_i, we obtain

$$du_i \, x \, v_i^\mathsf{T} + u_i \, dx \, v_i^\mathsf{T} = d\sigma_i \, . \tag{4.44}$$

The first term can be seen to be zero when the companion singular bases relation

$$x \, v_i^\mathsf{T} = \sigma_i(x) \, u_i^\mathsf{T}$$

is multiplied on the left by du_i and the other orthogonality identity is used. Rewriting the surviving term of (4.44) on the left as

$$u_i \, dx \, v_i^\mathsf{T} = \mathrm{Tr} \left(dx^\mathsf{T} \, u_i^\mathsf{T} \, v_i \right) \, ,$$

and comparing with $d\sigma_i$ as written in (4.43), we obtain

$$\nabla \sigma_i(x)|_{x=a} = u_i^\mathsf{T} \, v_i \, . \tag{4.45}$$

Finishing the application of Lemma 4.14 to the matrix case, we notice that $a - x$ being in the span of the gradients (4.45) is the same as the statement

$$a - x \in F(a) \, ,$$

after we recall Definition 4.12 of the matrix fiber. This is the same as the relationship (4.41) we obtained for vectors subject to a norm constraint!

Now since $a \in F(a)$ and fibers are vector spaces, from (4.41) we infer

$$x \in F(a) \, .$$

By Lemma 4.13 we then conclude

$$a \in F(x) \, .$$

We have delivered on the promise that a local computation at the projection input x gives us the much smaller space, $F(x)$, where the projection a can be found.

While often the constraint on the norm of a vector will simply be that this norm equals a given number r, more generally we allow for constraints where r lies in some set R. For example, an upper bound r_0 on the norm corresponds to the set

$$R = \left\{ r \in \mathbb{R} : 0 \le r \le r_0 \right\} \, .$$

The fiber perspective makes it easy to know how to project to such generalizations. First, we realize that the generalized norm constraint

$$A = \{x \in \mathbb{R}^m : \|x\| = r\, , \, r \in R\} \tag{4.46}$$

is the union

$$A = \bigcup_{r \in R} A(r)$$

of given-norm constraints for which we already know the projection. Then, by the general rule for projecting to unions in Section 4.1.4,

$$P_A(x) = \operatorname*{arg\,min}_{y \in \bigcup_{r \in R} P_{A(r)}(x)} \|y - x\|\, , \tag{4.47}$$

we have a projection formula for our particular case. Implementing formula (4.47) is easy because all of the projections $P_{A(r)}(x)$ lie on a particular fiber $F(x)$, a subspace, with the result that the projection is an optimization over a fiber.

Proposition 4.12 (Projection to generalized norm constraint) *The projection of a vector x to the constraint (4.46) specified by a set R of admissible norms is*

$$P_A(x) = \operatorname*{arg\,min}_{a \in \{r\hat{x}\, : \, r \in R\}} \|a - x\|\, ,$$

where $\hat{x} = x/\|x\|$.

Proof Since all the projections in the union that appears in (4.47) lie on the fiber

$$F(x) = \{\lambda \hat{x}\, :\, \lambda \in \mathbb{R}\}\, ,$$

we can write the projection as $a = \lambda \hat{x}$ for some fiber coordinate λ. Since $\|x\|$ is the fiber coordinate of x,

$$\|a - x\|^2 = (\lambda - \|x\|)^2\, .$$

Interpreting constraint (4.46) as the constraint $\lambda \in R$ on the fiber coordinate of the projection, we arrive at

$$P_A(x) = \left(\operatorname*{arg\,min}_{\lambda \in R} (\lambda - \|x\|)^2 \right) \hat{x}\, ,$$

which is just a restatement of the proposition.   □

By analogy with norm constraints, the generalized singular value constraint for $m \times n$ real matrices is the set

$$A = \left\{ x \in \mathbb{R}^{m \times n} : \sigma(x) \in \Sigma \right\}, \tag{4.48}$$

where $\sigma(x)$ is the set of singular values of x, and Σ, most generally, is a distribution on $k = \min(m, n)$ real numbers. We depart from convention by allowing the singular values to have either sign. To restore the standard positive convention to a negative singular value σ_i, one only needs to apply a minus sign to either u_i or v_i of the corresponding singular basis. As a result of the sign irrelevance of the singular values, the distribution Σ is symmetric under independent sign flips of the σ_i. The simple rank $\ell < k$ constraint corresponds to the special case where $\Sigma = \mathbb{R}^k_\ell$, that is, where $\sigma(x)$ is constrained to be ℓ-sparse. We didn't give a proof of the Eckart–Young low-rank projection (Theorem 4.7) because the fiber approach covers that case too.

Theorem 4.15 (Projection to singular value distribution) *The projection to set (4.48), for some distribution Σ of singular values, has formula*

$$P_A(x) = \underset{a \in \left\{ \sum_{i=1}^k \lambda_i u_i^\mathsf{T} v_i \, : \, \lambda \in \Sigma \right\}}{\arg\min} \|a - x\|,$$

where $k = \min(m, n)$ and $(u_1, v_1), \dots, (u_k, v_k)$ is a set of singular bases for x.

Proof Let

$$x = \sum_{k=1}^k \sigma_i(x) \, u_i^\mathsf{T} v_i$$

be a singular value decomposition of x, where the singular basis vectors u_i and v_i are normalized (and $\sigma_i(x)$ are signed real numbers). Because the projection lies on the same fiber,

$$a = \left\{ \sum_{i=1}^k \lambda_i \, u_i^\mathsf{T} v_i \, : \, \lambda \in \Sigma \right\}.$$

Minimizing the distance between x and the projection a means minimizing

$$\|a - x\|^2 = \sum_{i=1}^k \left(\lambda_i - \sigma_i(x) \right)^2 \tag{4.49}$$

subject to $\lambda \in \Sigma$, because the rank-1 bases $u_i^\mathsf{T} v_i$ are orthogonal and normalized. But this is just a restatement of the theorem. $\qquad\square$

To minimize (4.49) for the case $\lambda \in \mathbb{R}_\ell^k$, the ℓ-largest $\sigma_i(x)$ (in absolute value) are reserved for the ℓ-allowed nonzero λ's. Setting the latter equal to those largest $\sigma_i(x)$ gives the Eckart–Young rank ℓ projection. The squared projection distance is the sum-of-squares of the $k - \ell$ smallest singular values, for which the corresponding λ's are zero.

Natural generalizations of low rank, when k is large, are singular-value distributions that are well approximated by a set of k samples. Distributions without singular features, such as the uniform distribution, fall in this class. The computation of the projection is exactly as we saw for value-list constraints in Section 4.3.1. Instead of replacing the ranked elements of a vector (e.g., pixel grayscales) with an ordered set of numbers, we apply this operation to the singular values of a matrix.

The set of full-rank $m \times m$ matrices with $\Sigma = \{-1, 1\}^m$ corresponds to the orthogonal matrices O^m. Minimization on the fiber with distribution Σ corresponds to the replacement

$$\sigma_i(x) \to \mathrm{sgn}\left(\sigma_i(x)\right) .$$

This is exactly what we found in Section 4.4.3. An advantage of using signed singular values is that the projection to proper orthogonal matrices, $\mathcal{O}^m$, just involves the modification

$$\sigma_i(x) \to -\,\mathrm{sgn}\left(\sigma_i(x)\right)$$

on the singular value with smallest absolute value, when $\det(x) < 0$, as this incurs the smallest extra distance to a projection with positive determinant.

The fiber approach to spectral projections also works when matrices are complex or symmetric. Because the modifications are minor, we leave the details to the reader. The projection formulas for the Hermitian case we already saw in Section 4.4.2. In the real-symmetric and Hermitian case, the level sets continue to be defined by real-valued functions, and the distribution Σ has no sign-flip symmetry, because the rank-1 bases $u_i^\mathsf{T} u_i$ (and $u_i^\dagger u_i$) are completely determined by normalization. For nonsymmetric complex matrices, the level set functions are complex-valued and the gradients are likewise complex. Because $u_i^\dagger v_i$ has an arbitrary phase, the distribution Σ is now invariant under independent phase rotations.

Fibers defined by a vector-norm singularity, or the singular values of a matrix, are powerful tools because they exploit a global structure of the whole space. To see that this is rather special, consider the projection to the union of two spheres. Each sphere, individually, defines a fiber decomposition of the space via gradients of a norm function. But there is no longer a unique fiber at any point when we consider the fibers generated by both spheres.

To compute a projection, we would have to try out two fibers, one for each sphere. But this is no different from the general rule for projecting to the union of two sets (even when these are not facilitated by fibers).

4.5 Geometrical Constraints

The geometrical foundation of the bipartisan approach makes problems with geometrical constraints an obvious target of the technique. This section examines four such constraints and their projections. The rigid-body constraint comes up when packing nonspherical objects. Coplanarity, for "planes" of any dimension, is a generalization of concur from points to affine sets. This constraint can be generalized to impose (zero and nonzero) "principal components" on sets of points, and is called a constraint on the covariance spectrum. Finally, distance constraints arise as inequalities when packing spheres, but also as equalities, say when reconstructing an object in space from knowledge of the distances between pairs of its points.

4.5.1 Rigid Bodies

Proteins are puzzles comprising a sequence of hinged, rigid units of various shapes and sizes, designed by evolution to click into unique, compact, space filling objects. If the bipartisan strategy is ever applied to solving these biologically important puzzles, projections that restore the shapes of the rigid units will play a major role.

Protein structures are defined by the positions of their atoms. In practice, hydrogen atoms are ignored because of their small size.[3] Groups of covalently bonded atoms form rigid structures, one of the largest being the 10-atom *side-chain* of the amino acid tryptophan, shown in the right panel of Figure 4.4. The six- and five-membered rings of tryptophan, and one additional atom, have collectively just three translational and three rotational degrees of freedom. These atoms participate in several competing constraints, often resulting in configurations like the one shown in the left panel, which we denote x. The projection to the "tryptophan constraint" finds the distance minimizing change to x such that the result, in the right panel, is some rigid-body motion applied to the true tryptophan structure.

Because the derivation of the projection is the same for all dimensions, we consider n "atoms" $x_i \in \mathbb{R}^m$, where $m = 3$ and $n = 10$ in the case of the tryptophan side chain. The true structure is defined by n constant vectors $Y_i \in \mathbb{R}^m$, corresponding to a particular orientation and centroid position of the rigid group of atoms. It will be mathematically convenient to put the

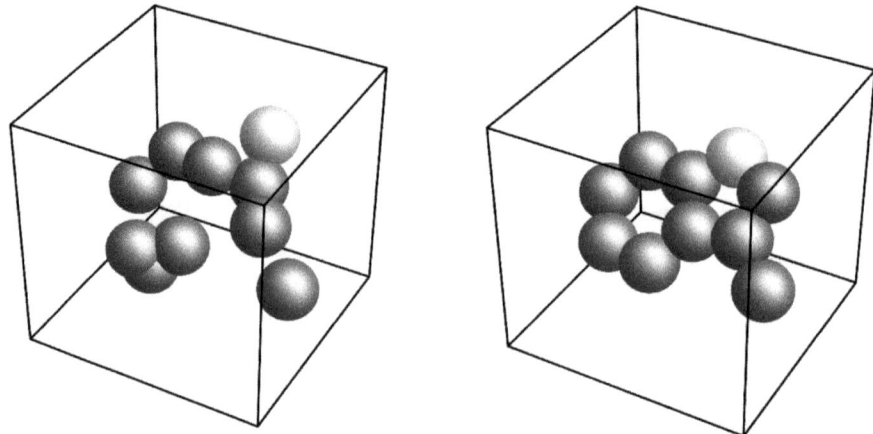

Figure 4.4 *Left*: The 10 (nonhydrogen) atoms of the tryptophan side chain (C_9H_6N) before projection to the rigid-body constraint. *Right*: A proper rotation and translation applied to the atoms of the true tryptophan structure that minimizes the total squared distance to their counterparts on the left.

centroid at the origin:

$$\sum_{i=1}^{n} Y_i = 0 \ .$$

Here is the constraint set defined by the rigid-body data of the atoms $Y \in \mathbb{R}^{n \times m}$:

$$A = \left\{ x \in \mathbb{R}^{n \times m} : \forall i \in \{1, \ldots, n\} : x_i = Y_i u + c, \ u \in \mathcal{O}^m, \ c \in \mathbb{R}^m \right\} . \quad (4.50)$$

The rigid-body relationship, between the true structure Y and a point x of the projection, is the application of the same rotation u and translation c to all n atoms of the true structure. Though it's not brought out in Figure 4.4, the matching of atoms x with their true counterparts Y is known.

When using the rigid-body constraints with proteins, we have to respect the fact that the true structures, when they have a handedness, always occur with one particular handedness (because of a handedness in the biochemical machinery that manufactures them). The allowed u are therefore restricted to proper rotations. Because the tryptophan side chain has no handedness, the handedness restriction can be ignored for this rigid body. The formula for the projection is transparent regarding this detail.

Theorem 4.16 (Rigid-body projection) *The projection of a configuration of atoms* $x \in \mathbb{R}^{n \times m}$ *to the handedness-preserving rigid-body constraint* (4.50), *defined by zeroed-centroid data* $Y \in \mathbb{R}^{n \times m}$, *is*

$$P_A(x) = Y\, P_{\mathcal{O}}\left(Y^{\mathsf{T}} x\right) + P_C(x)\,,$$

where $P_{\mathcal{O}}$ *is the projection to* $\mathcal{O}^m$ *(orientation-preserving rotations) and* P_C *is the concur projection. If A includes rotations that change handedness, $P_{\mathcal{O}}$ is replaced by P_O.*

Proof The projection is the configuration parameterized by $u \in \mathcal{O}^m$ and $c \in \mathbb{R}^m$ that minimizes the squared distance

$$\sum_{i=1}^{n}(x_i - Y_i\, u - c)(x_i - Y_i\, u - c)^{\mathsf{T}} = \sum_{i=1}^{n}\left(\|x_i - c\|^2 - 2x_i\, u^{\mathsf{T}} Y_i^{\mathsf{T}} + Y_i Y_i^{\mathsf{T}}\right),$$

where thanks to the convention $\sum_i Y_i = 0$, the c and u terms are decoupled. The minimizer for c is just the centroid

$$\bar{x} = \frac{1}{n}\sum_{i=1}^{n} x_i\,.$$

In the space of configurations, this would be expressed in terms of the concur projection:

$$(\bar{x}, \dots, \bar{x}) = P_C(x)\,.$$

The u minimizer is also not new:

$$
\begin{aligned}
u^* &= \underset{u \in \mathcal{O}^m}{\arg\min}\left(-2\sum_{i=1}^{n} x_i\, u^{\mathsf{T}} Y_i^{\mathsf{T}}\right) \\
&= \underset{u \in \mathcal{O}^m}{\arg\min}\left(-2\operatorname{Tr} u\left(x^{\mathsf{T}} Y\right)\right) \\
&= \underset{u \in \mathcal{O}^m}{\arg\min}\|u - Y^{\mathsf{T}} x\|^2 \\
&= P_{\mathcal{O}}\left(Y^{\mathsf{T}} x\right)\,.
\end{aligned}
$$

In the projection formula, this rotation is applied to Y and the entire configuration is translated by $P_C(x)$. The relaxation of the handedness restriction enters only through the projection that produced u^*. $\qquad\square$

4.5.2 Coplanarity

In Section 7.3, we will try out a divide-and-concur strategy for finding configurations of n equidistant lines in m dimensions. The idea is to represent each line with $n-1$ points, each point tasked with the distance constraint to one of the other lines, say in constraint A. Constraint B has to then impose colinearity of each of the n sets of $n-1$ points, and the projection P_B would do this in a distance-minimizing way.

More generally, it might be useful to know how to project n points in m dimensions so they lie in a hyperplane of dimension k, where $0 < k < m$. We make this restriction because $k = m$ is no constraint at all, while $k = 0$ is better known as the concur constraint, whose projection we already know. The general constraint is called k-coplanarity and the symbol for the constraint set is C_k.

We start with a simpler question: What happens to the centroid of the points in the projection?

Lemma 4.17 (k-coplanarity projection preserves the centroid) *If $x \in \mathbb{R}^{n \times m}$ is a configuration of n points in m dimensions, then the k-coplanarity projection $P_{C_k}(x)$ has the same centroid as x.*

Proof The n points can be orthogonally decomposed as

$$x_i = x_{i\parallel} + x_{i\perp} ,$$

where $\parallel$ denotes the k-dimensional subspace that P_{C_k} selects for the hyperplane that will contains all the points. If $x' = P_{C_k}(x)$ is the projected configuration, the k-coplanarity property implies

$$x'_{i\parallel} = x_{i\parallel} ,$$
$$x'_{i\perp} = c_\perp .$$

The $\parallel$ parts are unchanged because that is clearly distance minimizing, and the $\perp$ parts are the same (independent of i) because that is what it means to be coplanar. The distance minimizing $c_\perp$ is an instance of a concur projection, which we know is the mean of the inputs, $x_{i\perp}$. Summarizing, the projection has no effect on the $\parallel$ parts and changes the $\perp$ parts in a way that does not change their mean. The centroid is therefore preserved by the projection. □

Any subspace projector in $\mathbb{R}^m$, to a subspace of k dimensions, can be expressed as a real-symmetric matrix

$$u^\mathsf{T} d_k\, u ,$$

where $u \in O^m$ and d_k is any diagonal matrix with k elements equal to 1 and the rest 0. The rows of u that get multiplied by the 1's are a basis for the subspace. Because we will be using the spectral projection theorem, it is convenient to define the set of projectors to k-dimensional subspaces as symmetric matrices with a constraint on their spectrum,

$$S_k = \left\{ s \in \mathbb{S}^m : \operatorname{spec}(s) \in \Sigma_k \right\},$$

where

$$\Sigma_k = \left\{ \sigma \in \{0,1\}^m : \sum_{i=1}^m \sigma_i = k \right\}.$$

Recall that the projection of any $s \in \mathbb{S}^m$ to the set S_k starts with the computation of the eigen-decomposition of s. The resulting eigenvectors are unchanged, while the spectrum σ is projected to the set Σ_k. This means replacing the k largest elements of σ with 1 and setting the rest to 0.

We now have all the elements in place for the k-coplanarity projection:

Theorem 4.18 (k-coplanarity projection) *If $x \in \mathbb{R}^{n \times m}$ is a configuration of n points in m dimensions, then the projection to a configuration where all n points lie in a k-dimensional hyperplane has formula*

$$P_{C_k}(x) = \tilde{x}\, P_{S_k}\left(\tilde{x}^\mathsf{T}\tilde{x}\right) + P_C(x),$$

where

$$\tilde{x} = x - P_C(x).$$

Proof Since constraint C_k is invariant under translations,

$$P_{C_k}\left(x - P_C(x)\right) = P_{C_k}(x) - P_C(x).$$

In terms of centroid-centric variables

$$\tilde{x} = x - P_C(x),$$

the projection formula therefore takes the form

$$P_{C_k}(x) = P_{C_k}(\tilde{x}) + P_C(x).$$

By Lemma 4.17 the centroid of $P_{C_k}(\tilde{x})$ is zero, the same as the centroid of $\tilde{x}$. Using the notation used in the proof of the lemma, we therefore know that the projected points $\tilde{x}_i'$ satisfy

$$\forall i : \tilde{x}_i' = \tilde{x}_{i\parallel},$$

as any constant value for $\tilde{x}_{i\perp}$ other than zero would not have a configuration centroid of zero. Moreover, since P_{C_k} does not change the $\tilde{x}_{i\parallel}$, a distance-minimizing projector $s \in S_k$ is one that maximizes

$$\sum_{i=1}^n (\tilde{x}_i s)(\tilde{x}_i s)^\mathsf{T},$$

since that is the same as minimizing $\sum_{i=1}^{n} \|\tilde{x}_{i\perp}\|^2$. Since $P_{C_k}(\tilde{x}) = \tilde{x}_{\parallel}$ (the same projector s applied to all points of the configuration), we can confirm the formula for the projection:

$$P_{C_k}(\tilde{x}) = \tilde{x} \left(\underset{s \in S_k}{\arg\max} \sum_{i=1}^{n} (\tilde{x}_i s)(\tilde{x}_i s)^{\mathsf{T}} \right)$$

$$= \tilde{x} \left(\underset{s \in S_k}{\arg\max} \sum_{i=1}^{n} (\tilde{x}_i \, s \, \tilde{x}_i^{\mathsf{T}}) \right)$$

$$= \tilde{x} \left(\underset{s \in S_k}{\arg\max} \, \mathrm{Tr}\,(\tilde{x}^{\mathsf{T}} \tilde{x} \, s) \right)$$

$$= \tilde{x} \left(\underset{s \in S_k}{\arg\min} \, \|\tilde{x}^{\mathsf{T}} \tilde{x} - s\|^2 \right)$$

$$= \tilde{x} \, P_{S_k}(\tilde{x}^{\mathsf{T}} \tilde{x}) \, .$$

In this derivation, we used the identities $s^{\mathsf{T}} = s$, $s^2 = s$ and the fact that s has a constant trace. $\qquad\qquad\qquad\qquad\qquad\qquad\qquad\qquad\qquad\qquad\quad\square$

4.5.3 Covariance Spectrum Constraints

Any collection of n points in m dimensions counts as an "object" whose shape we may want to constrain. As data scientists know well, the most salient shape characteristics of such objects are their extent along principal axes – the magnitudes of their principal components. Because these are also the square roots of the eigenvalues of the covariance matrix associated with the n vectors from the object's center, we call this kind of constraint a constraint on the covariance spectrum. The special case of a spectrum constrained to be k-sparse corresponds to the k-coplanarity constraint. The covariance spectrum constraint differs from a spectrum constraint on symmetric matrices in that the projection does not act on symmetric matrices but sets of vectors that define, via the covariances, a symmetric matrix.

In Section 7.5 we use this constraint with the help of divide-and-concur, for $n = 3$ points in $m = 2$ dimensions, to design origami folds. The pattern of creases in origami can always be represented as a triangulation of the paper square. Avoiding skinny triangles, by imposing a lower bound on their covariance spectra, is a good design principle. When $n = m + 1$, the points define a simplex that possesses a handedness, or an orientation in the context of a topological cell complex. The triangles formed by the creases in the origami square also have this attribute, and the projection may have to flip their handedness to preserve the integrity of the triangulation.

Let $x \in \mathbb{R}^{n \times m}$ be the points of the configuration we wish to constrain. We may assume the n rows of x sum to zero, that is, correspond to positions relative to the centroid. That's because shape is independent of the centroid, so that the projection to the covariance spectrum constraint only acts on the factor of $x \in \mathbb{R}^{n \times m}$ with zero column sums.

The covariance matrix of the n points is the $m \times m$ symmetric, positive semi-definite matrix $x^\mathsf{T} x$. The principal axis u_i is the normalized eigenvector with eigenvalue (squared principal component) ρ_i :

$$u_i\, x^\mathsf{T} x = \rho_i\, u_i \,. \tag{4.51}$$

In this section we work out the projection to the set

$$A(\rho) = \left\{ x \in \mathbb{R}^{n \times m}, \sum_{i=1}^{n} x_i = 0 :\ \mathrm{spec}(x^\mathsf{T} x) = \rho \right\} \,.$$

We use the fiber approach introduced in Section 4.4.4, where $A(\rho)$ is viewed as the level set of the functions $\rho_1(x), \ldots, \rho_m(x)$.

The first step is to compute the gradients of the level-set functions. From (4.51) we obtain the relationships among the linear variations of u_i, x, and ρ_i :

$$du_i\, x^\mathsf{T} x + u_i\, dx^\mathsf{T} x + u_i\, x^\mathsf{T} dx = d\rho_i\, u_i + \rho_i\, du_i \,.$$

Multiplying on the right by u_i^T, using $du_i\, u_i^\mathsf{T} = 0$ (because $u_i\, u_i^\mathsf{T} = 1$), and also the transposed eigenvector relationship

$$x^\mathsf{T} x\, u_i^\mathsf{T} = \rho_i\, u_i^\mathsf{T} \,,$$

we obtain

$$
\begin{aligned}
d\rho_i &= u_i (dx^\mathsf{T} x + x^\mathsf{T} dx) u_i^\mathsf{T} \\
&= 2\,\mathrm{Tr}\left(dx^\mathsf{T}\, x\, u_i^\mathsf{T}\, u_i \right) \,.
\end{aligned}
$$

Since the $n \times m$ variations dx are independent, the gradient is the following $n \times m$ matrix:

$$\nabla \rho_i(x) = 2x\, u_i^\mathsf{T}\, u_i \,.$$

The span of the m gradients defines the fiber generated by x:

$$F(x) = \left\{ \sum_{i=1}^{m} \lambda_i\, x\, u_i^\mathsf{T}\, u_i :\ \lambda \in \mathbb{R}^m \right\} \,. \tag{4.52}$$

Since $\sum_{i=1}^{m} u_i^{\mathsf{T}} u_i$ is the identity matrix, we see that the point with fiber coordinates $\lambda_1 = \cdots = \lambda_m = 1$ is just the point x, so $x \in F(x)$.

The most general covariance spectrum constraint is the union

$$A = \bigcup_{\rho \in P} A(\rho) \, , \qquad (4.53)$$

where P is some distribution of the squared principal components. For example, if all we cared about was that our "object" x has a minimum size $\sqrt{\rho_0}$ along all m of its principal axes, we would use the distribution

$$P_0 = \{\rho \in \mathbb{R}^m : \rho_1 \geq \rho_0, \, \ldots, \rho_m \geq \rho_0\} \, . \qquad (4.54)$$

The general properties of fibers and projections to unions makes it easy to write down the projection to set A after having seen how this was done in the projection to singular-value distributions (Section 4.4.4).

Theorem 4.19 (Projection to covariance spectrum distribution) *The projection to set* (4.53), *for some distribution* P *of squared principal components, has formula*

$$P_A(x) = \underset{a \in \left\{ \sum_{i=1}^{m} \sqrt{\frac{\rho_i}{\rho_i(x)}} \, x \, u_i^{\mathsf{T}} u_i \, : \, \rho \in P \right\}}{\arg\min} \|a - x\| \, ,$$

where the $\rho_i(x)$ *are the eigenvalues of* $x^{\mathsf{T}} x$ *and* u_i *the corresponding normalized eigenvectors.*

Proof First consider the projection $a(\rho) = P_{A(\rho)}(x)$ for some element $\rho \in P$. Above we calculated the gradients of the functions $\rho_i(x)$ to establish formula (4.52) for the fiber $F(x)$. By the general relationship between fibers and projections (Section 4.4.4), $a(\rho) \in F(x)$ and we can write

$$a(\rho) = \sum_{i=1}^{m} \lambda_i \, x \, u_i^{\mathsf{T}} u_i$$

for some fiber coordinates $\lambda \in \mathbb{R}^m$. From this we find

$$a^{\mathsf{T}}(\rho) \, a(\rho) = \sum_{i=1}^{m} \lambda_i^2 \, \rho_i(x) \, u_i^{\mathsf{T}} u_i \, ,$$

or that the covariance matrix of the projection has the same eigenvectors as the covariance matrix of x, but the ith eigenvalue has been changed from $\rho_i(x)$ to

$$\lambda_i^2 \, \rho_i(x) \, .$$

Since the ith eigenvalue of the covariance matrix for $a(\rho)$ should be ρ_i, we must have

$$\lambda_i = \pm\sqrt{\frac{\rho_i}{\rho_i(x)}} \ . \tag{4.55}$$

To determine the distance-minimizing choice, we recall that x has all 1's for its fiber coordinates, with the result

$$\|a(\rho) - x\|^2 = \mathrm{Tr}\left((a(\rho) - x)^{\mathsf{T}}(a(\rho) - x)\right)$$
$$= \sum_{i=1}^{m}(\lambda_i - 1)^2\,\rho_i(x) \ . \tag{4.56}$$

Since the $\rho_i(x)$ are positive with probability 1, the distance is minimized with the positive root in (4.55) and

$$P_{A(\rho)}(x) = \sum_{i=1}^{m} \sqrt{\frac{\rho_i}{\rho_i(x)}}\, x\, u_i^{\mathsf{T}} u_i \ .$$

By the general formula for projecting to unions (Section 4.1.4),

$$P_A(x) = \underset{a \in \bigcup_{\rho \in P} P_{A(\rho)}(x)}{\arg\min}\ \|a - x\| \ ,$$

which is just a restatement of the theorem. $\square$

We now specialize Theorem 4.19 to the *positive-chirality simplex* constraint. This is where $n = m+1$ and the principal component magnitudes lie in the distribution P_0 (4.54) for some positive ρ_0 to keep the corresponding simplex full dimensional. The orientation (chirality) of a simplex x is defined by

$$\chi(x) = \mathrm{sgn}\left(\det(x_n - x_m\,,\ \dots\,,\ x_n - x_1)\right) \ .$$

For example, when the vertices x_1, x_2, x_3 of a triangle have a clockwise arrangement (by label) in the plane, $\chi(x) = +1$. In general, here is the constraint set,

$$A = \left\{ x \in \mathbb{R}^{(m+1)\times m},\ \sum_{i=1}^{m+1} x_i = 0 : \mathrm{spec}(x^{\mathsf{T}} x) \in P_0\ ,\ \chi(x) = 1 \right\} \ ,$$

and the corresponding projection:

Theorem 4.20 (Positive-chirality simplex projection) *Let $u_1, \rho_1(x),\ \dots$ be the eigenvectors/eigenvalues of $x^{\mathsf{T}} x$. Then*

$$P_A(x) = \sum_{i=1}^{m} \lambda_i\, x\, u_i^{\mathsf{T}} u_i \ , \tag{4.57}$$

where

$$
\lambda_i = \begin{cases} 1\,, & \rho_i(x) \geq \rho_0, \\ \sqrt{\dfrac{\rho_0}{\rho_i(x)}}\,, & \text{otherwise}, \end{cases} \tag{4.58}
$$

provided the resulting projection has positive chirality. If the chirality is negative, the coordinate λ_i associated with the smallest $\rho_i(x)$ is replaced by $-\lambda_i$.

Proof Since this is a special case of Theorem 4.19 we can adapt the proof for that theorem. Except for one detail, the minimizations are independent over the m terms in the sum because the eigenvalue distribution P_0 is a product over independent factors. This results in the minimizing fiber coordinates (4.58). However, if the resulting projection (4.57) has negative chirality, we need to reconsider taking only the positive roots in (4.55). The chirality is reversed whenever an odd number of roots are negative, since that corresponds to an odd number of reflections (in orthogonal planes defined by the eigenvectors).

Here is the squared projection distance (4.56) written explicitly in terms of the two roots,

$$
\begin{aligned}
d^2 &= \|a(\rho) - x\|^2 \\
&= \sum_{i=1}^{m} \left(\sqrt{\rho_i(x)} \mp \sqrt{\rho_i} \right)^2,
\end{aligned}
$$

where the negative sign is the distance minimizing choice without regard to chirality. The distance to a chirality-reversing projection is minimized when only a single term has the positive sign (a negative λ). For this term, the distance is minimized by choosing $\rho_i = \rho_0$. Depending on $\rho_i(x)$, there are two cases for the increase in distance:

$$
\Delta(d^2) = \begin{cases} \left(\sqrt{\rho_i(x)} + \sqrt{\rho_0} \right)^2, & \rho_i(x) > \rho_0, \\ 4\sqrt{\rho_i(x)\rho_0}\,, & \text{otherwise}. \end{cases}
$$

Because the first increase is greater than the second (arithmetic-geometric mean inequality), the second leads to a smaller d^2. If $\rho_i(x) < \rho_0$ for some i, the negative λ would be applied to one of them, and the least $\Delta(d^2)$ is attained with the i having the smallest $\rho_i(x)$. Otherwise, if $\rho_i(x) > \rho_0$ for all i, then $\Delta(d^2)$ is also minimized by selecting the i that minimizes $\rho_i(x)$. $\square$

Figure 4.5 shows the projection P_A applied to two triangles in the plane ($n = 3$, $m = 2$). Clockwise circulating arrows correspond to positive chirality. Because the thin triangle on the left has positive chirality, the projection only dilates the triangle, in this case along just one principal axis, to satisfy

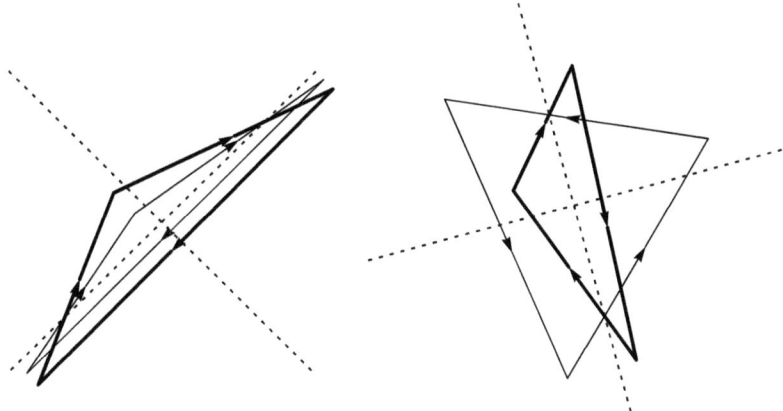

Figure 4.5 Orientation-preserving (left) and orientation-reversing (right) projections of triangles. The projected triangles, rendered with thicker lines, are also constrained to have both principal component magnitudes above a specified bound. Covariance spectrum projections, such as these, preserve the principal axes, shown as dashed lines.

the lower bound ρ_0 on the principal component magnitudes. Both principal component magnitudes of the triangle on the right exceed ρ_0, but the chirality is negative. In this case, the projection identifies the distance-minimizing axis for reflecting the triangle. The result is again a nondegenerate triangle, where the principal component magnitude along the reflection axis is the smallest possible, ρ_0.

4.5.4 Distance Constraints

The simplicity of the norm projection (Section 4.4.4) should not give the impression this constraint cannot play an important role in solving hard problems. The nonconvexity of the many norm constraints in phase retrieval is what makes that problem hard. In the protein folding problem, there are many lower bound constraints on the distances between pairs of atoms, all of which are nonconvex.

Because pairwise distance lower bounds between atoms in a protein or disks in a packing puzzle are so widely used, we feature that special case in this section. The configuration space is the one used by divide-and-concur, comprising two sphere centers in m dimensions. Colloquially, x_{ij} is the replica (copy) of sphere-center i that "watches out for" sphere j, while x_{ji} is the replica of sphere-center j that does the same for sphere i. Here is the constraint set, called *disjoint spheres*:

$$A_{ij} = \left\{ (x_{ij}, x_{ji}) \in \mathbb{R}^{2 \times m} : \|x_{ij} - x_{ji}\| \geq r_i + r_j \right\} .$$

The positive real numbers r_i and r_j, the radii of the spheres, are the constraint data. Here is the projection:

Proposition 4.13 (Disjoint sphere projection) *If* $(x_{ij}, x_{ji}) \in A_{ij}$ *(the inequality is satisfied) the two sphere centers are unchanged, otherwise*

$$P_{A_{ij}}(x_{ij}, x_{ji}) = \left(y + \left(\frac{r_i + r_j}{2\|z\|} \right) z \, , \, y - \left(\frac{r_i + r_j}{2\|z\|} \right) z \right) ,$$

where

$$y = \frac{1}{2}(x_{ij} + x_{ji}) ,$$

$$z = x_{ij} - x_{ji} .$$

Proof In terms of the mean and relative position variables, y and z, the projection distance takes the form

$$\|\delta x_{ij}\|^2 + \|\delta x_{ji}\|^2 = \|\delta y + \delta z/2\|^2 + \|\delta y - \delta z/2\|^2$$

$$= 2\|\delta y\|^2 + \frac{1}{2}\|\delta z\|^2 ,$$

where δ denotes the change in a variable. Only the relative position appears in the constraint:

$$A_{ij} = \left\{ (y, z) \in \mathbb{R}^{2 \times m} : \|z\| \geq r_i + r_j \right\} .$$

Since the y and z contribute independently to the distance, the projection only changes z, on which there is a simple norm constraint. Applying the same rescaling as in (4.39) to z (when the inequality is violated), and reexpressing the result in terms of the original variables, we obtain the projection formula above.   □

In geometrical language, the projection moves the two (non-disjoint) sphere replicas equally and oppositely, along the axis defined by their original positions, by the distance that makes them tangent.

4.6 Nearest and Optimizing Permutations

This section should have followed the section on distribution constraints, as the associated projections also do the seemingly impossible task of selecting one of $n!$ permutations with an amount of work that grows much more slowly. How this is possible for value-set distributions was easily understood because

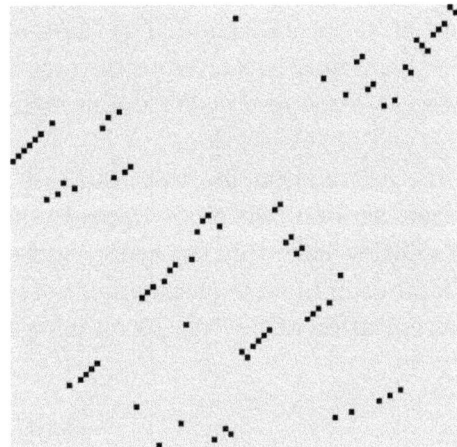

Figure 4.6 Jim Fienup (left) and the projection of his pixels (an 81×81 matrix) to the nearest permutation matrix (right). Reprinted with permission from © Jim Fienup.

the work of distance minimization turned out to be the same as the easy task of sorting a list of numbers. For the constraints in this section, the existence of an efficient projection is much more miraculous. Because the geometrical idea behind the projection involves a polytope in high dimensions, by ending the chapter with its visualization we also manage to close out the chapter in the same way it began!

In this section we study the following constraint set:

$$P_n = \left\{ x \in \mathbb{R}^{n \times n} : x_{ij} \in \{0, 1\}, \ \sum_{i=1}^n x_{ij} = \sum_{j=1}^n x_{ij} = 1 \right\}.$$

P_n is the set of all $n \times n$ permutation matrices. The projection $P_{P_n}(x)$ finds the distance-minimizing change to x that yields a single 1 in each row and column, zeroes elsewhere. An example[4] is shown in Figure 4.6. The small change,

$$G_n = \left\{ x \in \mathbb{R}^{n \times n} : x_{ij} \geq 0, \ \sum_{i=1}^n x_{ij} = \sum_{j=1}^n x_{ij} = 1 \right\},$$

defines the set of doubly stochastic matrices. This set is a convex polytope because it is the intersection of half-spaces and a hyperplane. We have given it the name *Garrett's polytope*, for the mathematician whose 1946 paper [12] introduced it to the world, and to avoid confusion with the mathematician father, George Birkhoff. The existence of an efficient algorithm for computing $P_{P_n}(x)$ is closely related to the fact that the *extreme points* of G_n, its vertices, are the same as P_n. We define extreme points and prove this statement at the end of this section.

As motivation for the study of constraint P_n, we present two applications: sudoku, where P_9 appears directly because the indicator variables in a solution have this structure, and edge-matching puzzles, where finding the distance-minimizing assignment of indicator variables, corresponding to some permutation of the tiles, is an instance of finding an optimizing permutation.

4.6.1 Sudoku

Already in Chapter 1 we pointed out that sudoku is a nice example of bipartisanship: two discrete sets A and B whose intersection (a unique point in a well-designed puzzle) is the solution. But how are these sets represented in a Euclidean space? Remarkably, the same representation is utilized in the real world by the software developer *Pseudo-Co*, in their very strict personal-space rules.

Pseudo-Co world headquarters is a nine-story all-glass building. Not only are the outside walls entirely glass, so too are all the internal walls and floors. The 81 cubicles on each of its nine floors are arranged into 3-by-3 offices, again demarcated only with glass.

The 81 employees at the headquarters are asked to respect the following cubicle occupation rules:

1. Looking directly up or down all nine floors, employees should not be able to see the shoes or head of another employee.
2. There should not be another employee in the same office.
3. Looking through the cubicle walls on the left and right it should not be possible to see another employee.
4. Looking through the cubicle walls to the front and back it should not be possible to see another employee.

A moment's reflection should convince you the intrepid *Pseudo-Co* employees invest (collectively) as much effort to get settled in their cubicles as it is to solve a sudoku puzzle! The employees on floor k are the k digits of a filled-in sudoku grid. The k's don't share grid space with other digits because of rule 1. By rule 2, there is a single k in each 3×3 block of the sudoku grid for all k (floors of the building). Rules 3 and 4 allow just a single k in each of the nine rows and columns of the grid, again for all k.

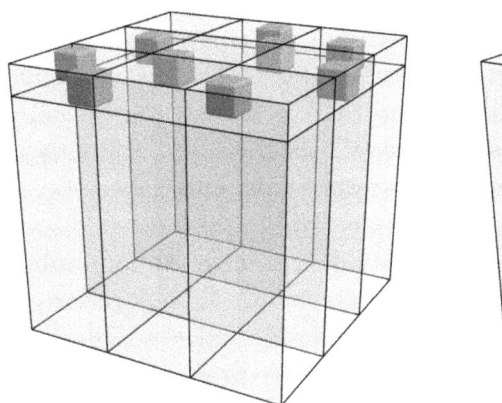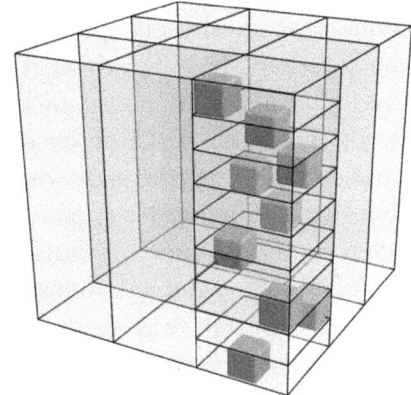

Figure 4.7 Allowed cubicle occupations on one floor of the *Pseudo-Co* building (left) and in one of the office stacks (right). The nine independent floor-wise constraints are set B, the nine independent office-wise constraints are set A.

Cubicle occupation can be represented by a $9 \times 9 \times 9$ tensor of indicator variables, and the personal-space (sudoku) rules are constraints on these indicators. In the bipartisan formulation of Chapter 1, the constraints are divided into sets A and B. Set A corresponds to cubicle occupation rules 1 and 2, set B takes care of rules 3 and 4.

Let's start with set B. This set is a product of nine sets,

$$B = B_1 \times \cdots \times B_9 \,,$$

corresponding to independent row-column constraints on each of the nine floors. The constraint for each k (floor) is that the corresponding 9×9 indicators are an element of P_9. An example of an allowed occupation of the top floor of the *Pseudo-Co* building is shown in the left panel of Figure 4.7.

Set A, or occupation rules 1 and 2, has the same product-of-permutations structure as set B:

$$A = A_1 \times \cdots \times A_9 \,.$$

Set A_1 constrains the occupations in office 1 (one of the 3×3 sudoku blocks) over all floors of the building (digits of the puzzle). The nine constraints in the product, imposed on the building's nine office stacks, are just as independent as were the floor-wise constraints B. Each factor has a 9×9 structure where a permutation pairs floors with office cubicles. An example of an allowed occupation of this kind is shown in the right panel of Figure 4.7.

To understand why occupations x that lie in the intersection $A \cap B$ are sudoku solutions, it's best to start with set A. If $x \in A$, then we know that each sudoku cell contains a single digit, and there is exactly one of each digit in all the 3×3 blocks. But constraint A allows rows and columns to be multiply occupied by the same digit. To prevent this, and have x be a true solution, we also require $x \in B$.

To find a random sudoku solution, one could start with random indicators $x \in \mathbb{R}^{9 \times 9 \times 9}$, say sampled uniformly between 0 and 1, and then use the Douglas–Rachford iteration rule, with projections P_A and P_B, to find a point in $A \cap B$. The projections act on the two kinds of product structure. For example,

$$P_A(x) = P_{P_9}(x_{A_1}) \times \cdots \times P_{P_9}(x_{A_9})$$

uses the floor-wise factors, while P_B acts on the office-stack factors. Both projections are built from the projection P_{P_9} to 9×9 permutation matrices.

Sudoku puzzles normally come with a set of *clues* $\mathcal{C}$ (also called *givens*) that force a unique solution. A clue $(i, j, k) \in \mathcal{C}$ indicates that the cubicle at row i, column j and floor k of the building must be occupied. The projections P_A and P_B can be modified to take clues into account, by fixing some elements of the distance-minimizing permutation. But there is a simpler way of imposing clues on a solution. Start by defining the puzzle's *clue point* $c \in \mathbb{R}^{9 \times 9 \times 9}$ as

$$c_{ijk} = \begin{cases} 1, & \text{if } (i, j, k) \in \mathcal{C} , \\ 0, & \text{otherwise.} \end{cases} \tag{4.59}$$

Then, to make sure that all the "cluebicles" are occupied, add a large multiple $\lambda > 0$ of the clue point to any x that you project. Assuming the clues $\mathcal{C}$ are consistent, the point $P_A(x + \lambda c)$ will then be distance minimizing to the floor-wise occupation constraints while respecting the occupations in $\mathcal{C}$. The same trick works for constraint B.

4.6.2 Edge-Matching Tiles

Edge-matching tile puzzles (Section 3.6) are a good example of how the permutation constraint can arise as a subproblem. Recall that we are given a set of n square tiles, $\mathcal{T} = \{T_1, \dots, T_n\}$, with instructions to place these at n positions forming a rectangular region. The challenge is to pair the n tiles with the n positions so the colored edges of adjacent tiles match, after also selecting one of four rotations for each tile.

The obvious bipartisan formulation is to have set A impose the constraint that all n tiles of $\mathcal{T}$ are used, that is, are matched 1-to-1 with the n positions. Constraint B is in charge of matching colors on adjacent edges. The projection to the tile-position matching will have to allow for tile rotations. This brings up the question of representing the placed, colored tiles with variables.

The obvious representation is to use variables x_{ie}, where $i \in \{1, \ldots, n\}$ labels the position of a tile's center and $e \in \{1, 2, 3, 4\}$ its four edges, say with a clockwise-increasing (mod 4) convention. Each x_{ie} should in fact be a k-component selector variable, since that is the correct way of representing one of k colors on the tile edges. The space of variables is therefore $x \in \mathbb{R}^{n \times 4 \times k}$.

It makes sense to represent the puzzle data in the same way as the variables. For example, the statement

$$\forall e : x_{ie} = T_{j\,e+r(j)} \tag{4.60}$$

is interpreted as " the four edge-colors of the tile placed at position i match those of tile j of the tile set after this tile is rotated by $r(j)$," where the edge-index addition is mod 4. The puzzle data is therefore a point $T \in \mathbb{R}^{n \times 4 \times k}$ with a particular combination of 0/1 coordinates.

If equation (4.60) holds for all tile positions i, and each i is matched with a unique tile j, then x satisfies the A constraint. Short of this, how do we modify (project) x in a distance-minimizing way to satisfy the A constraint?

The contribution to the squared distance, when position i is matched with tile j, has the form

$$(d^2)_{ij} = \min_r \left(\sum_e \|x_{ie} - T_{j\,e+r}\|^2 \right)$$

$$= u_i + v_j - 2w_{ij} ,$$

where

$$w_{ij} = \max_r \left(\sum_e (x_{ie} \cdot T_{j\,e+r}) \right)$$

and the dot product reminds us the variables and data are k-component selectors. We only care about the contribution that depends on both i and j because the distance-minimizing x ultimately hinges on the minimizing permutation $\pi \in P_n$ (an $n \times n$ permutation matrix) that matches positions

i with tiles j. The total squared distance

$$\text{Tr}\left(d^2\,\pi^\mathsf{T}\right) = \sum_{i=1}^{n} u_i + \sum_{j=1}^{n} v_j - 2\,\text{Tr}\left(w\,\pi^\mathsf{T}\right)$$

has two π-independent sums because every π has a single 1 in each row and column. The distance-minimizing permutation is therefore given by the simple expression

$$\pi^* = \arg\max_{\pi\in P_n} \text{Tr}\left(w\,\pi^\mathsf{T}\right) . \tag{4.61}$$

Interestingly, (4.61) is also a way to express the formula for projecting a matrix x to an element of P_n. Think of the rows of x as n selector variables. The squared distance of selecting column j in row i, relative to the all-zero setting, is $1 - 2x_{ij}$. The minimization of the total squared distance for all n selectors, subject to the constraint that the selectors select different columns, is exactly how the projection is defined,

$$P_{P_n}(x) = \arg\max_{\pi\in P_n} \text{Tr}\left(\tilde{x}\,\pi^\mathsf{T}\right) ,$$

where

$$\tilde{x}_{ij} = 2x_{ij} - 1 .$$

This formula is also nice because it is easily generalized for the metric

$$\|x' - x\|_g^2 = \sum_{i=1}^{n}\sum_{j=1}^{n} g_{ij}(x'_{ij} - x_{ij})^2 .$$

All that's required is a change in the definition of $\tilde{x}$:

$$\tilde{x}_{ij} = g_{ij}(2x_{ij} - 1) .$$

4.6.3 Optimized Matchings and Linear Programming

The optimization problem (4.61) is known as finding a maximum-weight bipartite matching [79]. The entries of the $n \times n$ matrix w can be interpreted as edge weights on a bipartite graph with n nodes on both sides. In a maximum-weight matching, n of the edges are selected that match nodes on the two sides and have the greatest combined weight.

The existence of an efficient algorithm for computing maximum-weight bipartite matchings is made plausible by the fact that the maximization over the $n!$ points of P_n can be replaced by the maximization over a convex set:

$$\arg\max_{\pi \in P_n} \mathrm{Tr}\left(w\,\pi^\mathsf{T}\right) = \arg\max_{x \in G_n} \mathrm{Tr}\left(w\,x^\mathsf{T}\right) \;.$$

The set G_n is not only convex, but a convex polytope (Garrett's). When the domain is a convex polytope, the maxima of a linear function, like $\mathrm{Tr}\left(w\,x^\mathsf{T}\right)$, can always be found at the polytope's vertices. Moreover, the maximum is unique with probability 1 when the coefficients w of the linear function are sampled from a full-dimensional continuous distribution. Since in our case w is derived from the iterates of an algorithm that has a continuously distributed initial point, it is safe to assume the maximum is unique. All that remains, then, for us to be able to optimize over G_n instead of P_n, is to check that the latter concides with the vertices of the former.

Before we prove the claim linking G_n and P_n, we should be cognizant of the advantages conferred by the convex domain. On a convex domain, it is possible to reach the maximum without ever having to go downhill. The greedy approach is not only safe, but is probably a good strategy!

Optimizing a linear function on a polytope, or a domain defined by linear equations and inequalities, is called linear programming (LP). Convexity suggests LP is a tractable problem and solutions – an optimizing polytope vertex – can be computed efficiently. This overlooks the possibility of a combinatorial explosion in the complexity of the polytope's boundary that even an uphill-guided algorithm may have to contend with. Fortunately, the LP algorithm known as the primal-dual method is not sidelined by this phenomenon in the case of Garret's polytope, where it is known as the *Hungarian algorithm* [66].

The vertices of a convex polytope, known as its extreme points in the LP context, are best characterized by a property they do *not* possess. A point x in a convex domain C is not extreme when there exists a line Λ through x such that an open neighborhood of x on Λ lies inside C. A generic linear function on C, when restricted to Λ, will then not have an extreme value at x. We now identify all the extreme points of G_n.

Theorem 4.21 (Extreme points of G_n) *The extreme points of G_n are the permutation matrices P_n.*

Proof The coordinates of G_n will be understood to be the matrix elements x_{ij}. All of these lie in the closed interval $[0, 1]$. Let $\mathcal{G}_n$ be the bipartite graph whose two sets of vertices correspond to the n rows and n columns of x. Place an edge between vertices i and j whenever x_{ij} is fractional (neither 0 or 1).

Because every row-sum and column-sum of x equals 1, no vertex of $\mathcal{G}_n$ can have only one incident edge (fractional element). In a bipartite $\mathcal{G}_n$ whose vertex degrees are all at least two, it is easy to find a cycle, assuming $\mathcal{G}_n$ has at least one edge. Just tracing random, previously untraversed edges from one side of $\mathcal{G}_n$ to the other always eventually produces a cycle. The cycle completes when a previously visited vertex is revisited, and this must happen because there are no dead ends when all the edge degrees are greater than one.

Let $\mathcal{C}$ be a cycle in $\mathcal{G}_n$, of even length because $\mathcal{G}_n$ is bipartite. Using $\mathcal{C}$ we construct a perturbation x' of x as follows. To the matrix elements corresponding to $\mathcal{C}$, alternate between adding and subtracting a small number ϵ, bounded in magnitude by the minimum (fractional) matrix element on $\mathcal{C}$. The new point, x', is clearly in G_n because $x' > 0$ and the alternations ensure that the row and column sums are unchanged. But ϵ can have either sign and parameterizes a line of points x' that passes though x (when $\epsilon = 0$). Since the only assumption we made was that $\mathcal{G}_n$ had at least one edge, the only candidate points where a line cannot be constructed are points with no fractional coordinates at all – the elements of P_n.   □

Garrett's polytope G_n, or the set of $n \times n$ doubly stochastic matrices, is a polytope in $(n-1)^2$ dimensions. To see this, observe that there are n linear equations constraining the rows, another n constraining the columns, and that there is one dependency due to the sum of the row-sums being equal to the sum of the column-sums. The hyperplane in which G_n resides therefore has $n^2 - (2n - 1)$ dimensions. What can we say about the $2n - 1$ coordinates orthogonal to this hyperplane? Start by asking, "What changes to a point $x \in \mathbb{R}^{n \times n}$ change *only* the orthogonal coordinates?" We can find the answer in Section 4.2.4, where we learned that the projection that adjusts the row and column sums of a rectangular table of numbers is built out of uniform shifts applied to all the elements of individual rows and columns. Allowing for the linear dependency, there are $2n - 1$ such shifts. And in being projections, the shifts do not change the $(n-1)^2$ coordinates in the orthogonal complement.

Given the central role of G_n for efficiently computing an important projection, it is normal to be curious about what this polytope looks like! G_2, in one dimension, is just an interval – the continuous interpolation between the identity permutation and the transposition – and too simple to satisfy that curiosity. On the other hand, G_3 is already in one dimension above what we can visualize. Can one of George Gamow's projections help?

Here is a symmetry-inspired map from $x \in \mathbb{R}^{3 \times 3}$ to $(z_1, z_2) \in \mathbb{C}^2$, a four-dimensional space:

$$z_1 = \sum_{i=1}^{3} \sum_{j=1}^{3} \omega^{i+j} x_{ij} \, ,$$

$$z_2 = \sum_{i=1}^{3} \sum_{j=1}^{3} \omega^{i-j} x_{ij} \, .$$

The coefficients are powers of $\omega = \exp(2\pi i/3)$. Notice that applying a uniform shift to any row or column of x has no effect on either z_1 or z_2 because $1+\omega+\omega^2 = 0$. From the remarks above, we therefore know that z_1 and z_2 are coordinates in G_3's hyperplane. We can also check that the corresponding 9-component vectors

$$e_{1ij} = \mathrm{Re}(\omega^{i+j})$$
$$e_{2ij} = \mathrm{Im}(\omega^{i+j})$$
$$e_{3ij} = \mathrm{Re}(\omega^{i-j})$$
$$e_{4ij} = \mathrm{Im}(\omega^{i-j})$$

form an orthogonal basis.

On mapping the vertices of G_3, the six permutation matrices, we find that the three transpositions map to $(3\,\omega^k, 0)$, and the three powers of the 3-cycle map to $(0, 3\,\omega^k)$, where $k \in \{1, 2, 3\}$. The two triples of permutations therefore form equilateral triangles in the two orthogonal complex planes. Pairs of permutations in the same complex plane are related by 3-cycles. This means that the edge joining such a pair comprises a one-parameter family of doubly stochastic matrices, as in the proof about the extreme points of G_n. The cycle $\mathcal{C}$ in the bipartite graph, that interpolates between the vertices, has length six.

Since G_3 is the convex hull of its vertices, the lines joining vertices in different complex planes are also edges of G_3. There are nine such edges and each of them fixes one of the nine matrix elements of x at 1. The endpoint permutations are related by a transposition, and the interpolating doubly stochastic matrices are 4-cycles in the bipartite graph.

Figure 4.8 is a rendering of G_3 based on our analysis. One of the complex planes was collapsed into a single axis by a Gamow projection. The positions of the three vertices on that axis move up and down when the collapsed complex plane is rotated.

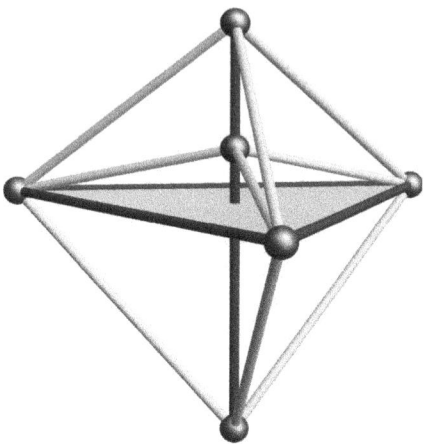

Figure 4.8 The four-dimensional polytope G_3 projected into three dimensions. Three vertices form an equilateral triangle in one two-dimensional plane, the other three a congruent triangle in an orthogonal two-dimensional plane. The second plane has been projected into a single axis. The six inter-triangle edges of G_3 (rendered dark) correspond to permutation pairs related by a 3-cycle. There are also nine edges between the two triangles, where the endpoint permutations are related by transpositions.

Exercises

4.1 A first step in projecting to a geometrical shape is transforming to coordinates best suited to that shape. What coordinates are best for the equilateral triangle,[5] the square?

4.2 Suppose you have n real-valued variables that need to satisfy the equality constraint:

$$x_1 = x_2 = \cdots = x_n \ .$$

Work out the projection to this constraint (concur) when using the 1-norm for the distance:

$$\|x' - x\|_1 = \sum_{i=1}^{n} |x'_i - x_i| \ .$$

As usual, you are not concerned about probability-0 inputs to the projection. However, you should be aware that your answer may depend on whether n is even or odd!

4.3 A standard procedure for "normalizing" a set of n points in m dimensions is to first apply a translation that puts their centroid at the origin, and then rescale their coordinates by a common factor to make the mean of their squared norms equal to 1. Write down the constraint set that corresponds to this procedure and show that the steps given above are a projection to this constraint.

4.4 Recall the two constraint sets of bit retrieval,

$$A = \left\{ x \in \mathbb{R}^m : x \in \{-1, 1\}^m \right\} ,$$
$$B = \left\{ x \in \mathbb{R}^m : |\hat{x}| = \sqrt{I} \right\} ,$$

where $\hat{x}$ is the Fourier transform of x and I is the intensity data. Both sets are nonconvex: A is discrete (an m-dimensional hypercube), and B is the product of m circles, one in each of the m complex dimensions.

The *convex hull* of a set X is the smallest convex set $\bar{X}$ that contains X. The convex hull is also called the *convex closure* because if x_1 and x_2 are any points of X, then $\bar{X}$ also includes the points

$$\alpha x_1 + (1 - \alpha) x_2$$

for any $\alpha \in [0, 1]$. A *convex relaxation* of a constraint problem is one where one or more of the nonconvex sets of the original problem are replaced by their convex hulls.

(a) Give explicit definitions of the convex hulls/closures $\bar{A}$ and $\bar{B}$.
(b) Consider the two convex relaxations $\bar{A} \cap B$ and $A \cap \bar{B}$ of bit retrieval. Use Parseval's theorem to show that these have the same set of solutions as the original problem, $A \cap B$. Why is $\bar{A} \cap \bar{B}$ not an acceptable relaxation?

4.5 The *Cut Above* pizzeria has given new meaning to the expression "serves k," where $k = 1, 2, 3, \ldots$ If you ordered a serves-four pizza, you would receive a pizza cut into six pieces having areas $\{1, 1, 2, 2, 3, 3\}$. The rationale for this way of cutting the pizza is explained by the 3×4 table (an order-2 tensor) below:

1	0	3	0
0	1	0	3
2	2	0	0

The nonzero entries are the areas of the six pieces. To "serve four," the pieces are combined by columns and everyone receives 3 units of pizza. But "serves four" really means "serves up to four." If we are serving just three, then the pieces are combined by rows for total area 4. When serving two we realize that two divides four and the column partition can be used (two columns per person). The serves-one case does not deserve comment.

(a) What order and size of tensor x would *Cut Above* use to design its serves-up-to-six pizza? Keep the order as small as possible. What are the sums-with-symmetry constraints on x?
(b) *Cut Above* takes pride in minimizing the number of pieces, and six pieces is the fewest possible for serves-up-to-four. What constraint would you use in a bipartisan formulation to limit the number of pieces at p?

4.6 By closely following the derivation of Section 4.4.3, derive the formula for projecting an arbitrary $m \times m$ complex matrix to the nearest unitary matrix. Why is there is no counterpart, for unitary matrices, of the rotation being proper or improper?

4.7 Your friend "Hamilton" has a thing about symmetric matrices. If you give him one, he will diagonalize it and admire its eigenvalues. For his birthday you would like to send him a matrix with three of his favorite numbers in its spectrum[6]:

$$0.406\ldots$$
$$1.618\ldots$$
$$3.141\ldots$$

To keep the greeting simple, the matrix elements should be single-digit integers $(0, \ldots, 9)$.

(a) Define sets A and B that will allow you to design such a matrix.
(b) Estimate the size of matrix needed to get the three eigenvalues above to three decimal digits.

4.8 Interpret $x \in \mathbb{R}^{n \times m}$ as a configuration of n vectors in $\mathbb{R}^m$. Suppose our information about these vectors is stronger than it was for the kissing spheres problem, in that their inner products are known, not just their bounds. The corresponding constraint set is

$$A = \left\{ x \in \mathbb{R}^{n \times m} : xx^{\mathsf{T}} = G \right\},$$

where the symmetric, positive semi-definite Gram matrix G is the data that defines the constraint. Derive a formula[7] for the projection $P_A(x)$. Check that your formula reduces to the projection P_O when $n = m$ and $G = I$.

4.9 In data science, one of the simplest schemes for interpreting a list of vectors v is to express them as combinations of a basis b with some coefficients c. Say there are n vectors of length m and the basis has size k. Then $v \in \mathbb{R}^{n \times m}$, $b \in \mathbb{R}^{k \times m}$, $c \in \mathbb{R}^{n \times k}$, and

$$v = c\,b. \tag{4.62}$$

The value of the interpretation is greatest when the size of the basis k is smallest. We have already learned how to project a matrix such as v to the nearest matrix having a given rank k using the SVD, and the result is exactly (4.62) once the singular values are absorbed into the coefficients.

For greater interpretability when the data vectors v are nonnegative, one would like the factors b and c to be likewise nonnegative. The basis vectors are then called *features*, the coefficients are *weights*, and instead of combinations one makes *mixtures*.

Projecting to the nearest *nonnegative matrix factorization* (NMF) of a given rank k is in general much harder than projecting without the nonnegative restriction. In fact, finding *any* NMF of a nonnegative v having a given rank k is by itself a hard problem [110]. In this exercise you will try your hand at formulating NMF along bipartisan lines.

Start by interpreting the matrix product as a sum of k rank-1 matrices:

$$v = \sum_{\ell=1}^{k} (c^{\mathsf{T}})_\ell \, b_\ell$$

$$= \sum_{\ell=1}^{k} x_\ell \, .$$

As usual, the unknowns in a bipartisan formulation are called x, this time with the structure

$$x = (x_1, \ldots, x_k) \, ,$$

where $x_\ell \in \mathbb{R}^{n \times m}$. Your job is now to define constraints A and B such that when x satisfies both constraints, the sum of its k components is a NMF of v. You will see that A and B work with different product-space factorizations of x, and both of the projections that get applied to the factors were covered in this chapter.[8]

4.10 A bipartisan formulation of the NMF problem that is very different from the one in the previous exercise (and also a nice application of divide-and-concur) is based on *bilinear constraints*. Focusing just on this constraint, and not how it is used for NMF, consider vectors $b \in \mathbb{R}^k$ and $c \in \mathbb{R}^k$. For any given $v \in \mathbb{R}$, what are the distance-minimizing vectors b' and c' such that $c' \cdot b' = v$? Since the weight vector c and feature vector b play very different roles, use the metric

$$\|b' - b\|^2 + g\|c' - c\|^2$$

when minimizing the distance.[9]

4.11 This exercise is another example of using the extreme parts of a set to completely characterize something, in this case the symmetries of NMF. Recall that the extreme points of doubly stochastic matrices turned out to be the permutation matrices.

Let $v = c\,b$ be a NMF with matrix dimensions as in Exercise 4.9.

The factorization $v = (c\,a)(a^{-1}b)$, for any invertible $a \in \mathbb{R}^{k \times k}$, is also nonnegative provided both $c\,a$ and $a^{-1}b$ are nonnegative. Expressed in terms of any of the rows of c or columns of b, written as vectors $y \in \mathbb{R}^k$, these conditions hold when the linear transformation a and its inverse map points in the nonnegative orthant, $y \geq 0$, to points in the nonnegative orthant: $y\,a \geq 0$, $y\,(a^{-1})^\mathsf{T} \geq 0$. The map a must therefore be a bijection of the nonnegative orthant, $\mathbb{R}^k_+$. The set of all linear transformations with this property form a group G under matrix multiplication. This is the symmetry group of NMF.

Now think about the elements of G geometrically. The set $\mathbb{R}^k_+$ has one extreme point, $(0,\ldots,0)$, and k extreme rays, the edges of $\mathbb{R}^k_+$ where just one coordinate is positive. Linear transformations map extreme sets to extreme sets (of the same type). Used this fact to completely characterize the elements of G.

Credit: Yoav Kallus

4.12 Suppose you have a sphere packing puzzle where the sphere radii may vary, as well as their centers. The nonintersecting constraint for a pair of spheres in m dimensions with centers x_1, x_2 and radii r_1, r_2 is

$$\|x_1 - x_2\|^2 \geq (r_1 + r_2)^2 \,.$$

Derive the projection to this constraint using the metric

$$\|x'_1 - x_1\|^2 + \|x'_2 - x_2\|^2 + g\left((r'_1 - r_1)^2 + (r'_2 - r_2)^2\right) \,.$$

The radii should be treated as any other type of variable in a bipartisan formulation: negative values are allowed (if the nonintersecting constraints are in set A, set B can be used to impose positive radii).

4.13 A line in any number of dimensions m is defined by a pair of points through which it passes. Suppose you have two lines, defined by point-pairs (y_1, z_1) and (y_2, z_2). Assuming an isotropic metric, derive the projection to pairs (y'_1, z'_1) and (y'_2, z'_2) such that the lines through them are separated by distance d. The distance between two lines (in any number of dimensions) is the minimum distance between a point on one line and a point on the other.

(a) First derive an explicit formula for the *proximal* pair of points on the two lines, that is, the pair that have minimum distance.

(b) Use the geometrical relationship between the proximal point-pair and the lines to help you limit the changes to the line-defining point-pairs that minimize the projection distance.

4.14 In equation (4.59) we defined the clue point c of a sudoku puzzle so that for large λ both $P_A(x+\lambda c)$ and $P_B(x+\lambda c)$ are consistent with the clues of the puzzle (in addition to satisfying their half of the sudoku rules). As you know, the clues very directly also eliminate a large number of co-occupations of the *Pseudo-Co* cubicles – not much logic required! Use this idea to define *anticlue points* to be used instead of clue points when performing projections. Not surprisingly, anticlue points are the superior way of imposing the clue constraints. New York Times "easy" puzzles are solved in only two Douglas–Rachford iterations when using anticlue points.

4.15 The extreme points of a polytope are also its vertices, and the convex hull of the vertices restores the polytope. The proof of Theorem 4.21, that the extreme points of Garrett's polytope G_n (doubly stochastic matrices) are the set P_n of permutation matrices, therefore also proves that the convex hull of P_n is G_n, or $\bar{P}_n = G_n$.

Consider the formulation of sudoku in Section 4.6.1, where both A and B are the product of nine P_9 factors.

(a) Do the convex relaxations, $\bar{A} \cap B$ or $A \cap \bar{B}$, also define sudoku solutions?

(b) Now consider the affine relaxations $\widetilde{A}$ and $\widetilde{B}$, where nonnegativity of the elements is also relaxed. Do $\widetilde{A} \cap B$ or $A \cap \widetilde{B}$ define sudoku solutions? What is the name of the projection to the affine relaxations?

4.16 Find explicit parameterizations of the 3×3 matrices corresponding to the points in the interiors of the two orthogonal triangles of G_3 shown in Figure 4.8. Consider using barycentric coordinates, which in this case are nonnegative triples of numbers that sum to 1. Of the doubly stochastic matrices in the two triangles, can you point to any that are definitely *not* the squared magnitudes of a 3×3 unitary matrix?

5

Reflect-Reflect-Relax

This is the chapter where the hard work of the previous chapters finally finds its reward. Now that you are able to express practically any problem in terms of sets A and B (Ch. 2), have some analytical tools for estimating the size of $A \cap B$ (Ch. 3), and have learned how to project to a bewildering variety of sets (Ch. 4), you are finally ready to confront the logic behind

$$x \mapsto x + P_B(R_A(x)) - P_A(x) . \tag{5.1}$$

At the risk of repetition, we remind you that the power behind the new method of search lies not so much in the formula above, but in the creative problem formulations the formula enables. You may even find the formula a bit of an anticlimax, once you understand what it is doing!

Before we get started, we need a better way to refer to formula (5.1). An obvious choice is to refer to searches based on iterating this particular combination of projections and reflections *the Douglas–Rachford algorithm.* Jim Douglas, Jr. and H. H. Rachford, Jr. were after all the first to write down the formula. In the original application (partial differential equations) the sets A and B were convex, and the rationale for the formula was its convergence properties. In many contemporary applications, the sets are also convex, and convergence is front and center.

But our primary interest is search problems where at least one of A or B is nonconvex. Typically, the iterations where the behavior is truly convergent represent just a blink of the eye in the solution process. Most of the time the iterates are doing the very opposite of convergence, by "moving on" when a solution is nowhere nearby. Though convergence takes place eventually, once a fixed point is within sight, we will see that it is really the process of selecting the next place to search that makes (5.1) special.

While *Douglas–Rachford, Sr.* would get across the point that the algorithm has come of age, we prefer a name that lays out in plain terms what each iteration is doing. First, let's introduce a very natural *time-step* parameter β:

$$x \mapsto \mathcal{R}(x) = x + \beta\Big(P_B(R_A(x)) - P_A(x)\Big) . \tag{5.2}$$

Starting with the introduction of the β parameter, we will use the symbol $\mathcal{R}$ for the iteration rule. Because the two projections (and corresponding reflections) act as maps with probability 1, we will also treat $\mathcal{R}$ as a map. The $\beta \to 0$ limit of $\mathcal{R}$ describes a flow, though it is not a gradient flow.

It's a simple exercise to rewrite $\mathcal{R}$ entirely in terms of the two reflections:

$$R_A(x) = 2P_A(x) - x ,$$
$$R_B(x) = 2P_B(x) - x .$$

The result is:

$$\mathcal{R}(x) = (1 - \beta/2)x + (\beta/2)R_B(R_A(x)) . \tag{5.3}$$

For $\beta = 1$, the formula tells us to reflect x, first in set A and then set B, and average the result with x. Jon Borwein called this case, which is also the Douglas–Rachford formula, *reflect-reflect-average*. Because the construction of the formula for general β is called a relaxation, we will use the name *reflect-reflect-relax* for formula (5.3) and also for (5.2), which is equivalent. Iterating the map $\mathcal{R}$ to solve problems – the subject of this chapter – is called the *RRR algorithm*.

5.1 Solutions and Fixed Points

The reader's first encounter with fixed point algorithms was probably *Newton's method*. To find a root of the function $f(x)$ you start with a guess x_0, approximate $f(x)$ by the line tangent to f at $f(x_0)$, solve the linear equation to find the next estimate x_1, and repeat. If the sequence $x_0, x_1, \ldots$ converges to a fixed point x^*, the same point is also a solution to the problem, that is, $f(x^*) = 0$. The tangent construction of the sequence corresponds to iterating

$$x \mapsto x - \frac{f(x)}{f'(x)} , \tag{5.4}$$

where f' is the derivative of f.

The reader's first encounter with chaos is almost surely also linked with Newton's method. Consider the following choice for f:

$$f(x) = \operatorname{sgn} g(x) \left| g(x) \right|^{\frac{1}{r-1}} , \tag{5.5}$$
$$g(x) = -r + (r - 1)/x .$$

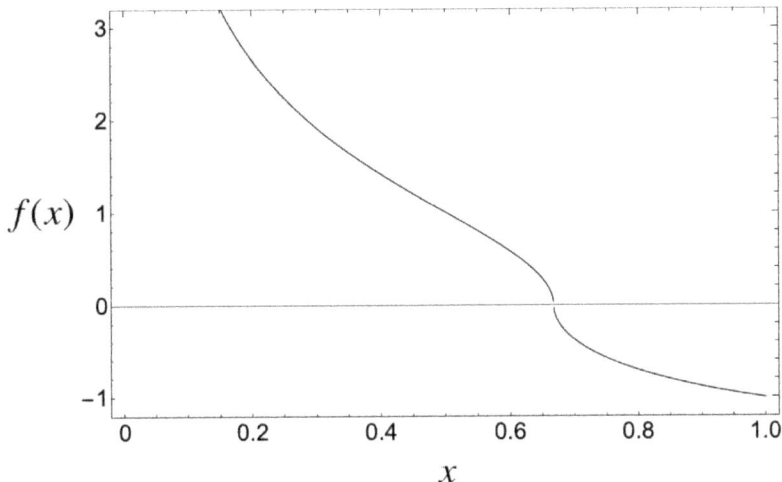

Figure 5.1 When Newton's root finding method is applied to the function
(5.5), shown here for $r = 3$, the iterates are given by the logistic map (5.6).
Unlike RRR, in this dynamical system chaos is the result of fixed points
becoming unstable.

There is a parameter r and Figure 5.1 shows the function for $r = 3$. From
the plot we see that the tangent estimates will oscillate above and below the
position of the root. For $r > 3$ the iterates not only oscillate but are *repelled*
from the root. Instead of converging on a fixed point, they settle into cycles
of lengths $2, 4, 8, \ldots$ until at $r = 3.56995 \cdots$ the dynamics is chaotic because
the cycle length has become infinite. If all this seems familiar, it is because
the application of (5.4) to (5.5) produces the *logistic map*, the simplest model
of chaos:

$$x \mapsto r\, x(1 - x) \,. \tag{5.6}$$

If RRR is a fixed point approach to solving problems that also exhibits
chaos, might it just be a multivariable generalization of Newton's method?
The point of this section is to convince you that the answer to this question
is a resounding "no."

Let's start with the relationship between solutions and fixed points. These
are the same in Newton's method. Now consider a fixed point x^* of $\mathcal{R}$. Using
the form (5.2), the fixed point property implies

$$P_B(R_A(x^*)) - P_A(x^*) = 0 \,,$$

from which we know that

$$P_B(R_A(x^*)) = P_A(x^*) = x_{\text{sol}} \tag{5.7}$$

is a solution because the projections are points in A and B and for them to be equal means x_{sol} is an element of $A \cap B$. But statement (5.7) is not the same as $x^* = x_{\text{sol}}$! Later, when we analyze the relationship between fixed points and solutions in detail, it will be very different from this.

Though the fixed points of $\mathcal{R}$ are not themselves solutions (elements of $A \cap B$), it's good to know – by (5.7) – that a solution is at hand whenever a fixed point is found. The alternating projections algorithm, which iterates

$$x \mapsto P_B(P_A(x)) \,, \tag{5.8}$$

does not have even this property. Consider the possibility of a *mutually proximal pair*. This is a pair $x_A \in A \setminus B$, $x_B \in B \setminus A$ with the properties

$$P_A(x_B) = x_A \,,$$
$$P_B(x_A) = x_B \,.$$

The point $x^* = x_B$ of such a pair would be a fixed point of (5.8) but not a solution. The alternating-projections map is very good at converging on proximal pairs, but that is all it will do reliably when either A or B is nonconvex!

The chaos mechanism in Newton's method is also very different from how this phenomenon arises in RRR. Chaos in Newton's method comes at the expense of fixed points. By tuning a parameter, fixed points that started out as stable become unstable. Chaos in the logistic map results when all the periodic orbits (fixed points of finitely many powers of the map) have become unstable. By contrast, and as we will see later in the chapter, RRR solutions are encoded by fixed points that are *always* stable – even when the dynamics leading to the discovery of the fixed point is chaotic.

RRR is chaotic in a strict sense only for infeasible instances, when $A \cap B$ is the empty set, and there are no fixed points. However, the question of feasibility is global in scope and sensitive to small changes in A and B (a single sudoku clue). If RRR is chaotic in an infeasible instance, a small change that makes it feasible will not change the character of the dynamics.

5.2 Generalizing the Metric

Often the variables most natural for a problem are hybrids of different variable types, instead of a single type (such as the pixel contrasts in an image). For example, suppose $x \in \mathbb{R}^m$ are the natural variables for constraint A_1, while variables $y \in \mathbb{R}^n$ are best suited for constraint A_2. The challenge is to satisfy both constraints when x and y are related in some way. If the relationship is linear, we have a bipartisan formulation with the following sets:

$$A = \left\{ (x, y) \in \mathbb{R}^{m+n} : x \in A_1, \, y \in A_2 \right\},$$
$$B = \left\{ (x, y) \in \mathbb{R}^{m+n} : xC = y \right\}.$$

The matrix $C \in \mathbb{R}^{m \times n}$ defines the linear relationship. Clearly there is some arbitrariness in the definition of C, including its overall scale. This arbitrariness should make us wary of using the standard Euclidean distance in $\mathbb{R}^{m+n}$ when computing projections. At a minimum we should address the scale arbitrariness by using the metric

$$\|x' - x\|^2 + g\|y' - y\|^2,$$

where g is a positive parameter. The value of g has an effect on the projection P_B of RRR and is therefore something that may improve the algorithm.

The most general form of metric we will encounter, for m real variables, is the diagonal generalization:

$$\|x' - x\|_g^2 = \sum_{i=1}^{m} g_i \left(x'_i - x_i \right)^2.$$

We've run across this already (Section 4.1.7), as the form of metric that applies to the projection of concur values to a side constraint, even when the isotropic metric is used for the original variables. Projecting with this metric to the two most common constraints can be found in Section 4.1.8. But in this chapter, about the analysis of RRR, we will avoid complications arising from the generalized metric with the help of a rescaling transformation.

Let r be the transformation

$$r(x_1, \ldots, x_m) = \left(\sqrt{g_1}\, x_1, \, \ldots, \, \sqrt{g_m}\, x_m \right)$$
$$= (\tilde{x}_1, \ldots, \tilde{x}_m).$$

Since

$$\|x' - x\|_g^2 = \|\tilde{x}' - \tilde{x}\|^2,$$

the isotropic metric applies to the transformed variables. Moreover, if A is some constraint for the original variables, then the transformed projection

$$r(P_A(x)) = r\left(\arg\min_{y \in A} \|y - x\|_g^2 \right)$$
$$= \arg\min_{\tilde{y} \in \tilde{A}} \|\tilde{y} - \tilde{x}\|^2$$
$$= P_{\tilde{A}}(\tilde{x})$$

is the same as the projection of the transformed variables to the transformed set $r(A) = \tilde{A}$ by the isotropic metric. This means that all the analysis in this chapter, for the isotropic metric, also applies to the diagonally generalized metric if we interpret the sets A and B as possibly being the transformed sets $\tilde{A}$ and $\tilde{B}$.

5.3 Generalizing the Rule

It is not nearly enough that $\mathcal{R}$ has the property that its fixed points translate to solutions. The alert reader will have noticed that the demonstration of this property in Section 5.1 made no reference at all to the reflection R_A! This operation could have been replaced by the identity operation without changing the conclusion. To bring out the significance of the reflection, we consider generalizations of $\mathcal{R}$, which we denote $\widetilde{\mathcal{R}}$, where the inclusion of the reflection is left open. To limit the scope of the generalizations, we start with a general property that is easy to take for granted.

5.3.1 Similarity Invariance

Two shapes are similar if one is obtained from the other by a combination of translation, rotation/reflection, and change of scale. Most generally, *similarity transformations* are invertible maps $s \colon \mathbb{R}^m \mapsto \mathbb{R}^m$,

$$s(x) = x\,u + t\,,$$

where $t \in \mathbb{R}^m$ is a translation and $u \in \mathbb{R}^{m \times m}$ is a scaled orthogonal matrix, or $u\,u^\intercal = \lambda I$ for some $\lambda \in \mathbb{R}_+$. Now consider iterates x_0, $x_1 = \widetilde{\mathcal{R}}(x_0)$, $x_2 = \widetilde{\mathcal{R}}^2(x_0)$, and so on for some iteration rule $\widetilde{\mathcal{R}}$. If the sequence $s(x_0)$, $s(x_1)$, $s(x_2)$, and so on is also generated by $\widetilde{\mathcal{R}}$, for initial point $s(x_0)$, we say that $\widetilde{\mathcal{R}}$ is *similarity invariant*. Since $\widetilde{\mathcal{R}}$ depends on sets A and B, the similarity of the iterate sequence should also involve replacing these with $s(A)$ and $s(B)$. In general, we say a map

$$x \mapsto f(x\,;\,A, B, \dots)$$

that depends parametrically on subsets $A, B, \dots$ of $\mathbb{R}^m$, is *similarity invariant* when

$$s\big(f(x\,;\,A, B, \dots)\big) = f\big(s(x)\,;\,s(A), s(B), \dots\big) \qquad (5.9)$$

for arbitrary similarity transformations s. Similarity invariance is a great convenience because it means any choice of origin, rotation, and scale of the constraint sets in the bipartisan formulation has no effect on the search for solutions.

Projections are for us the "atoms" from which similarity-invariant maps are built:

Proposition 5.1 (Similarity invariance of projections) *The projection P_A to the set A is similarity invariant.*

Proof Start by rewriting the formula for the projection:

$$P_{s(A)}(s(x)) = \arg\min_{y \in s(A)} \|s(x) - y\|^2$$
$$= s\left(\arg\min_{z \in A} \|s(x) - s(z)\|^2\right). \tag{5.10}$$

A defining property of similarity transformations is the preservation of inner products up to scale:

$$\|s(x) - s(z)\|^2 = \|(x\,u + t) - (z\,u + t)\|^2$$
$$= ((x - z)u)((x - z)u)^\mathsf{T}$$
$$= \lambda \|x - z\|^2.$$

Using this in (5.10),

$$P_{s(A)}(s(x)) = s\left(\arg\min_{z \in A} \|x - z\|^2\right)$$
$$= s(P_A(x)),$$

we confirm an instance of (5.9). □

We should not forget that the identity map (the projection to the whole space) is also similarity invariant!

There are two basic ways of building similarity invariant maps from simpler ones, and we will need both of them:

Proposition 5.2 (Similarity invariance of compositions) *The composition of two similarity invariant maps is similarity invariant.*

Proof Consider the map h defined by composing f with g:

$$h(x\,;\,A, \ldots, B, \ldots) = g\big(f(x\,;\,A, \ldots)\,;\,B, \ldots\big).$$

Using the action (5.9) of a similarity transformation s on the invariant maps

f and g,

$$s\Big(h(x\,;\,A,\ldots,B,\ldots)\Big) = g\Big(s(f(x\,;\,A,\ldots))\,;\,s(B),\ldots\Big)$$

$$= g\Big(\big(f(s(x)\,;\,s(A),\ldots)\big)\,;\,s(B),\ldots\Big)$$

$$= h\big(s(x)\,;\,s(A),\ldots,s(B),\ldots\big)\,,$$

we confirm that h is also similarity invariant. □

Proposition 5.3 (Similarity-invariant combinations) *If the maps*

$$f_1(x\,;\,A_1,\ldots)\,,\ \ldots\,,\ f_n(x\,;\,A_n,\ldots)$$

are similarity invariant, then the linear combination with real coefficients

$$\sum_{i=1}^{n} \alpha_i\, f_i$$

is similarity invariant if and only if the coefficients are "barycentric coordinates" :

$$\sum_{i=1}^{n} \alpha_i = 1\,.$$

Proof   Let $s(x) = x\,u + t$ be an arbitrary similarity transformation, then

$$s\left(\sum_{i=1}^{n} \alpha_i\, f_i(x\,;\,A_i,\ldots)\right) = \left(\sum_{i=1}^{n} \alpha_i f_i\right) u + t$$

$$= \sum_{i=1}^{n} \alpha_i\,(f_i\,u + t) + \left(1 - \sum_{i=1}^{n} \alpha_i\right) t$$

$$= \sum_{i=1}^{n} \alpha_i\, s\big(f_i(x\,;\,A_i,\ldots)\big) + \left(1 - \sum_{i=1}^{n} \alpha_i\right) t$$

$$= \sum_{i=1}^{n} \alpha_i\, f_i\big(s(x)\,;\,s(A_i),\ldots\big) + \left(1 - \sum_{i=1}^{n} \alpha_i\right) t\,.$$

The left side and the first term on the right is an instance of (5.9). Consequently, the linear combination is similarity invariant if and only if the second term on the right is zero for arbitrary t, or that the sum of the coefficients equals 1. □

5.3.2 General Criteria

We are now ready to state five criteria that the generalized map $\widetilde{\mathcal{R}}$ should adhere to:

Constraints implemented as projections

$\widetilde{\mathcal{R}}$ may only make reference to the sets A and B through the projections P_A and P_B.

Similarity invariance

$\widetilde{\mathcal{R}}$ should be similarity invariant.

Linkage of feasible and fixed points

Guided by what we learned in Section 5.1, fixed points will be linked to the feasible points $A \cap B$ by giving $\widetilde{\mathcal{R}}$ the structure

$$\widetilde{\mathcal{R}}(x) = x + \beta\Big(P_B\big(g(x)\big) - P_A\big(f(x)\big)\Big), \qquad (5.11)$$

where $f(x)$ and $g(x)$ are similarity invariant maps that may depend on A and B through their projections. By propositions 5.1, 5.2, and 5.3, this $\widetilde{\mathcal{R}}$ is similarity invariant. The real parameter β is the "time-step."

Set-wise attraction

It is not enough that feasible and fixed points are linked: the fixed points must also be attractive. But as we have seen, fixed points can implicate solutions without being identified with them. We therefore allow for the possibility that there is a continuous set of fixed points X^* associated with each feasible point. Fixed-point attraction is replaced by a slightly weaker property. Let x' be any perturbation of $x^* \in X^*$, then the iterates $(\widetilde{\mathcal{R}})^k(x')$, $k = 1, 2, \ldots$, converge to another point in X^*, and this new fixed point is near x^*.

Finite step size

Convergence to fixed points should hold for finite β. That is, "time-step" should not suggest we are trying to approximate continuous-time dynamics.

5.3.3 Two-Parameter Family

Similarity invariance and feasible/fixed-point linkage still admit too many possibilities for $\widetilde{\mathcal{R}}$! To narrow the field of candidates in a way that admits the Douglas–Rachford iteration rule, we consider only those $\widetilde{\mathcal{R}}$ where at most one of the maps f and g in (5.11) is not the identity. Since the two cases are related by a swap of A and B and changing the sign of β, without loss of generality, we set $f(x) = x$. For g we take the simplest similarity-invariant map built from just two "atoms" and one real parameter:

$$g(x) = (1 - \lambda)P_A(x) + \lambda\,x\,.$$

We could have used P_B instead, but then $P_B(g(x))$ collapses to $P_B(x)$ for the case $\lambda = 0$. By using P_A we have three interesting cases:

$$g(x) = \begin{cases} R_A(x)\,, & \lambda = -1\,, \\ P_A(x)\,, & \lambda = 0\,, \\ x\,, & \lambda = 1\,. \end{cases}$$

The RRR rule corresponds to $\lambda = -1$. We will refer to the $\lambda = 1$ case (no reflection) as the *difference-of-projections* rule.

Three of our five general criteria are satisfied by the generalization

$$\widetilde{\mathcal{R}}(x\,;\beta,\lambda) = x + \beta\Big(P_B\big((1 - \lambda)P_A(x) + \lambda\,x\big) - P_A(x)\Big)\,.$$

To address the final two criteria, we next analyze the local behavior of $\widetilde{\mathcal{R}}$. An interesting thing that will come out of this analysis is a picture of what RRR is doing when it is *not* converging on a fixed point set.

5.4 Local Behavior

Our local analysis of $\widetilde{\mathcal{R}}$ assumes that A and B are both close to a point x, whose neighborhood defines what we mean by local. Moreover, A and B are both assumed to be so close to x that they can be approximated by affine sets in that neighborhood. In a regular formulation this property holds whenever x is near a solution. For simplicity, in this section we use the same symbols, A and B, for these affine approximations.

5.4.1 Affine Geometry

The affine sets A and B have associated linear subspaces comprising the set of all differences:

$$\bar{A} = A - A \,,$$
$$\bar{B} = B - B \,.$$

Using X to denote the whole space, these subspaces define an orthogonal decomposition

$$X = W \times Y \times Z$$

as follows. First, we define the orthogonal complement of the first factor as the span of the linear subspaces:

$$W_\perp = Y \times Z = \bar{A} + \bar{B} \,. \tag{5.12}$$

Over most of the course of a search, the point x will not be able to reach a solution via steps in either $\bar{A}$ or $\bar{B}$ (for the current affine approximations). The space W will therefore emerge as the space in which the search does its most productive "walking." Second, the space Z is defined to be the intersection of the linear subspaces:

$$Z = \bar{A} \cap \bar{B} \,.$$

Z corresponds to the local space of symmetries: The joint constraining power of A and B is unchanged with any step of x in Z. The remaining space Y, orthogonal to both the walking and symmetry spaces, is called the *transverse space*. Contractive behavior in the transverse space is the key to understanding RRR.

The parts of $\bar{A}$ and $\bar{B}$ in the transverse space have special names:

$$C = \bar{A} \cap Y \,,$$
$$D = \bar{B} \cap Y \,. \tag{5.13}$$

Note that

$$C + D = Y \tag{5.14}$$

since

$$(\bar{A} \cap Y) + (\bar{B} \cap Y) = (\bar{A} + \bar{B}) \cap Y = Y \,.$$

Two key properties of these subspaces of the transverse space are summarized here:

Lemma 5.1 (Transverse-space subspace intersections) *The subspaces $C \subset Y$ and $D \subset Y$, and their orthogonal complements in Y, have trivial intersections:*

$$C \cap D = \{0\} \, , \tag{5.15}$$

$$C_{\perp} \cap D_{\perp} = \{0\} \, . \tag{5.16}$$

Proof From the definitions (5.13), any element of $C \cap D$ is an element of $\bar{A} \cap \bar{B} = Z$ on the one hand, but also an element of Y. Statement (5.15) follows from $Z \cap Y = \{0\}$ since these subspaces are orthogonal. Statement (5.16) follows from the general rule for the orthogonal complement of an intersection and (5.14):

$$\left((C_{\perp} \cap D_{\perp})_{\perp} \right)_{\perp} = (C + D)_{\perp} = Y_{\perp} = \{0\} \, .$$

$\square$

Using the above definitions, the original affine sets have the following structure:

$$A = w_A + (C + y_A) + Z \, ,$$
$$B = w_B + (D + y_B) + Z \, ,$$

where $w_A, w_B \in W$ and $y_A, y_B \in Y$. Since C and D span Y, we can write

$$y_A = c_A + d_A \, ,$$
$$y_B = c_B + d_B \, ,$$

for some $c_A, c_B \in C$ and $d_A, d_B \in D$. But since

$$C + c_A + d_A = C + c_B + d_A \, ,$$
$$D + c_B + d_B = D + c_B + d_A \, ,$$

we can write the affine sets more simply as

$$A = w_A + (C + y_0) + Z \, ,$$
$$B = w_B + (D + y_0) + Z \, , \tag{5.17}$$

where y_0 is a general element of Y. This representation of the local affine geometry of A and B (minus the symmetry space Z) is rendered in Figure 5.2.

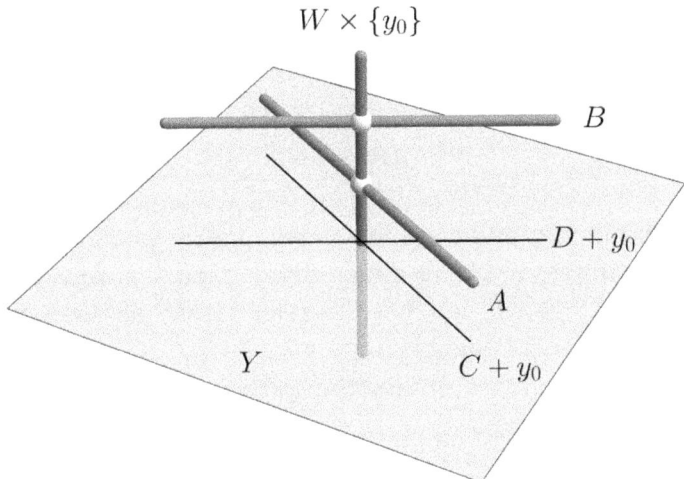

Figure 5.2 Cartoon of the sets A and B, approximated as affine sets. The transverse space Y is shown as two-dimensional and contains shadows of A and B as the translated subspaces $C + y_0$ and $D + y_0$. When (locally) there is no solution, RRR "walks" in the W space, shown here as one-dimensional. The two white balls are the proximal points on A and B. In typical RRR applications, A, B, and W have many hundreds of dimensions.

5.4.2 Projection Formulas

To keep track of the three components of our orthogonal decomposition, points $x \in X$ will be written as

$$x = (w \, , \, y \, , \, z) \, .$$

By the representation (5.17) of A and B, we know that $P_A(x)$ simply replaces w with w_A while for $P_B(x)$ the replacement is w_B. The action of these projections on z is even simpler. Neither changes z because the entire factor space Z is a factor of both A and B.

The only interesting projections are in Y, the transverse space. For these we can make use of the translational (similarity) invariance of projections:

$$P_{C+y_0}(y) = P_C(y - y_0) + y_0 \, .$$

There is a similar formula when projecting to $D + y_0$. Since the difference $y - y_0$ will appear repeatedly, we define

$$\tilde{y} = y - y_0 \, .$$

The projections P_C and P_D to the subspaces C and D can be written more explicitly in terms of *projection matrices* p and q:

$$P_C(y - y_0) = \tilde{y}\,p\,,$$
$$P_D(y - y_0) = \tilde{y}\,q\,.$$

If Y is m-dimensional, p and q are $m \times m$ symmetric matrices with "projector spectra." For example, if C is k-dimensional, then k eigenvalues of p are 1 and the rest are 0. We will not have to refer to any of these dimensions in our analysis of the local behavior.

With these definitions we obtain the following explicit formulas for the two constraint projections:

$$P_A(w\,,\,\tilde{y} + y_0\,,\,z) = (w_A\,,\,\tilde{y}\,p + y_0\,,\,z)\,,$$
$$P_B(w\,,\,\tilde{y} + y_0\,,\,z) = (w_B\,,\,\tilde{y}\,q + y_0\,,\,z)\,.$$

We get an interpretation of y_0 when we look into the possibility of a proximal pair. Applying P_B to the formula for P_A,

$$P_B\Big(P_A(w\,,\,\tilde{y} + y_0\,,\,z)\Big) = (w_B\,,\,\tilde{y}\,p\,q + y_0\,,\,z)\,,$$

we see that we get back to the original point – one half of the proximal pair – provided $w = w_B$ and

$$\tilde{y} = \tilde{y}\,p\,q\,.$$

But $\|\tilde{y}\| = \|\tilde{y}\,p\,q\|$ can only be true if $\tilde{y} = 0$ or $C \subset D$ (for $\tilde{y}\,p \in C$ to not decrease its length when projected into D). By (5.15) the second case implies $C = \{0\}$, so in either case we must have $\tilde{y} = 0$. A proximal pair must therefore have the form

$$x_A = (w_A\,,\,y_0\,,\,z)\,,$$
$$x_B = (w_B\,,\,y_0\,,\,z)\,.$$

This is a true proximal pair only when the points are distinct, or $w_A \neq w_B$. The distance between the proximal pair, $\|x_A - x_B\| = \|w_A - w_B\|$, is called the *gap*. Figure 5.2 also shows the proximal pair. When the points coincide they belong to $A \cap B$, a solution.

5.4.3 Iterating the Generalized Rule

We are finally ready to analyze the local behavior of $\widetilde{\mathcal{R}}$. Applying P_B to the generalized reflection instead of P_A, we obtain

$$
\begin{aligned}
P_B\big(g(x)\big) &= P_B\Big((1-\lambda)P_A(x) + \lambda\,x\Big) \\
&= P_B\Big((1-\lambda)(w_A\,,\ \tilde{y}\,p + y_0\,,\ z) + \lambda\,(w\,,\ \tilde{y} + y_0\,,\ z)\Big) \\
&= \Big(w_B\,,\ (1-\lambda)\tilde{y}\,p\,q + \lambda\,\tilde{y}\,q + y_0\,,\ z\Big)\,.
\end{aligned}
$$

Subtracting $P_A(x)$ from this expression, multiplying the result by β, and finally adding x, we arrive at

$$
\widetilde{\mathcal{R}}(w\,,\ \tilde{y} + y_0\,,\ z) = \Big(w + \beta(w_B - w_A)\,,\ \tilde{y} + \beta\,\tilde{y}\,r + y_0\,,\ z\Big)\,,
$$

where

$$
r = (1-\lambda)p\,q + \lambda\,q - p\,. \tag{5.18}
$$

Here is the same result expressed as changes in the three components:

$$
\Delta w = \beta(w_B - w_A)\,, \tag{5.19a}
$$
$$
\Delta\tilde{y} = \beta\,\tilde{y}\,r\,, \tag{5.19b}
$$
$$
\Delta z = 0\,. \tag{5.19c}
$$

The symmetry component z does not change, as it should. Steps in the W space are uniform and proportional to the gap and β. If $\widetilde{\mathcal{R}}$ stumbles into a region where the affine approximation has zero gap, it is also near a fixed point set. That's because all three of the changes (5.19) are zero when $\tilde{y} = 0$. The local, affine approximation of the fixed point set is therefore

$$
X^* = W \times \{y_0\} \times Z\,.
$$

Even without symmetry, fixed point sets can be huge! Consider our favorite formulation of sudoku, where both A and B are discrete. In the $n \times n$ puzzle, $\dim X = n^3$, but that means $\dim W = n^3$ since $\dim A = \dim B = 0$. This is an extreme case, where the fixed point sets form a partition of the whole space. It is also not representative with regard to the question of fixed-point-set *attraction*. That's because $Y = \{0\}$, and the behavior $\tilde{y} \to 0$ never comes up. The question of attraction is taken up in Section 5.4.4, where we assume $\dim Y > 0$. This is where we will first come to appreciate the role of the reflection in RRR, since that detail only appears in the λ-dependence of the matrix r (5.18) behind the $\tilde{y}$ behavior. Why $\lambda = -1$ (RRR) is the right choice even when $\dim Y = 0$ is addressed in Section 5.5.6.

5.4.4 Fixed-Point-Set Attraction

To study whether there is contracting behavior in Y, to the point y_0, we define a quadratic form that gives us the change in the norm,

$$Q(\tilde{y}) = \|\tilde{y} + \Delta\tilde{y}\|^2 - \|\tilde{y}\|^2$$
$$= Q_1(\tilde{y}) + Q_2(\tilde{y}) \,,$$

where

$$Q_1(\tilde{y}) = 2\beta\, \tilde{y}\, r\, \tilde{y}^\mathsf{T} \,,$$
$$Q_2(\tilde{y}) = \beta^2\, \tilde{y}\, r\, r^\mathsf{T}\, \tilde{y}^\mathsf{T} \,.$$

Are there parameters β and λ for which the behavior is always *contractive*, that is, $Q(\tilde{y}) < 0$ for any (nonzero) $\tilde{y}$? Parameters where $Q(\tilde{y}) > 0$, for even one $\tilde{y}$, should be avoided because then there is expansion in Y away from the fixed point set. We use the term *expansive* when the behavior is not contractive.

Using only the fact that p and q are symmetric and idempotent, the matrix appearing in Q_2 takes the following form:

$$r\,r^\mathsf{T} = (\lambda^2 - 1)p\,q\,p + \lambda^2(q - q\,p - p\,q) + p \,. \tag{5.20}$$

Even without knowing this structure, because $Q_2 = \beta^2\|\tilde{y}\,r\|^2$, this part of Q is always expansive. Interestingly, (5.20) reduces to something simple for two cases, $\lambda = \pm 1$:

$$r\,r^\mathsf{T}\Big|_{\lambda=\pm 1} = (p - q)^2 \,. \tag{5.21}$$

These special cases differ in the matrix that appears in Q_1:

$$r = \begin{cases} q - p\,, & \lambda = 1\,, \\ 2p\,q - p - q\,, & \lambda = -1\,. \end{cases}$$

The next theorem tells us which combination of parameters to avoid.

Theorem 5.2 (Transversely expansive cases of $\widetilde{\mathcal{R}}$) *If the transverse subspaces C and D have positive dimension, then the map $\widetilde{\mathcal{R}}$ is expansive in the transverse space when $|\beta - 1| > 1$ or, if $|\beta - 1| < 1$, when $\lambda > 0$ or $\lambda < -2/\beta$.*

Proof Since $\dim(C) > 0$, there exists a nonzero $\tilde{y} \in C$. By (5.15) this implies $\tilde{y} \notin D$, or that $\tilde{y} \in D_\perp$. For this nonzero $\tilde{y} \in C \cap D_\perp$, the formulas (5.18) and (5.20), together with $\tilde{y}p = \tilde{y}$, $\tilde{y}q = 0$, imply

$$Q(\tilde{y}) = (-2\beta + \beta^2)\|\tilde{y}\|^2$$
$$= (|\beta - 1|^2 - 1)\|\tilde{y}\|^2 \,. \tag{5.22}$$

Similarly, since $\dim(D) > 0$ there exists a nonzero $\tilde{y} \in C_\perp \cap D$ for which $\tilde{y}p = 0$, $\tilde{y}q = \tilde{y}$ and

$$
\begin{aligned}
Q(\tilde{y}) &= \left(2\beta\lambda + (\beta\lambda)^2\right) \|\tilde{y}\|^2 \\
&= \left(|\beta\lambda + 1|^2 - 1\right) \|\tilde{y}\|^2 .
\end{aligned} \tag{5.23}
$$

From (5.22) we see that $\widetilde{\mathcal{R}}$ is transversely expansive when $|\beta - 1| > 1$. On the other hand, if $|\beta - 1| < 1$, we can turn to (5.23) and the transverse expansion for $|\beta\lambda + 1| > 1$. Since $|\beta - 1| < 1$ implies $\beta > 0$, we get expansion for $\lambda > 0$ or $\lambda < -2/\beta$, in that case. $\qquad\square$

Theorem 5.2 rules out many candidate algorithms, including difference-of-projections ($\lambda = 1$). Since we are limited to $|\beta - 1| < 1$, and $\lambda = -1$ is not ruled out, let's focus on the case of RRR. Interestingly, when $\lambda = -1$, the two parts of Q are magically related and give a rule that has fixed-point-set attraction.

Theorem 5.3 (RRR is transversely contractive) *The RRR rule, or $\widetilde{\mathcal{R}}$ with $\lambda = -1$, is contractive in the transverse space provided $|\beta - 1| < 1$.*

Proof By the identity

$$
\tilde{y}\,q\,p\,\tilde{y}^\mathsf{T} = \tilde{y}\,p\,q\,\tilde{y}^\mathsf{T} ,
$$

the order of the projection matrices in r may be reversed when they appear in the norm:

$$
\begin{aligned}
Q_1(\tilde{y}) &= 2\beta\,\tilde{y}\,(2\,p\,q - p - q)\,\tilde{y}^\mathsf{T} \\
&= 2\beta\,\tilde{y}\,(p\,q + q\,p - p - q)\,\tilde{y}^\mathsf{T} \\
&= -2\beta\,\tilde{y}\,(p - q)^2\,\tilde{y}^\mathsf{T} .
\end{aligned} \tag{5.24}
$$

Using (5.21) in Q_2 we see that the two parts of Q have the same form and can be combined:

$$
\begin{aligned}
Q(\tilde{y}) &= (-2\beta + \beta^2)\,\tilde{y}\,(p - q)^2\,\tilde{y}^\mathsf{T} \\
&= -\left(1 - (1 - \beta)^2\right) \|\tilde{y}\,(p - q)\|^2 .
\end{aligned} \tag{5.25}
$$

This shows $Q(\tilde{y}) \leq 0$ for $|\beta - 1| < 1$. To eliminate the possibility of $Q(\tilde{y}) = 0$ for a nonzero $\tilde{y}$, we note that the expression above would then imply

$$
\tilde{y}\,p = \tilde{y}\,q \in C \cap D = \{0\}
$$

after using (5.15). But

$$
\begin{aligned}
\tilde{y}\,p &= \{0\} , \\
\tilde{y}\,q &= \{0\}
\end{aligned}
$$

imply

$$\tilde{y} \in C_\perp \cap D_\perp$$

and (5.16) contradicts the claim that $\tilde{y}$ could be nonzero. □

We get transverse contraction, and fixed-point-set attraction, only for $\beta > 0$. To get attraction for the other sign of β, the maps f (identity) and g (reflection) in (5.11) must switch places. The positive-β form, where g is the reflection, will be our convention in this book. Notice that switching A and B (while keeping the reflection in the first term) produces another fixed-point-set attractive map and another way to find elements of $A \cap B$. In Section 5.5 we argue that swapping A and B in the RRR rule is a "quasi-time-reversal" transformation.

From (5.25) we see that transverse contraction is maximized for $\beta = 1$. This choice of β is a good default value for this parameter, and is clearly the best possible when the affine sets are not approximations, but the true (convex) constraints. In this book our main interest is in hard problems, where at least one of A or B is highly nonconvex. For these applications we will see that small β is typically superior to $\beta = 1$.

We conclude this section with the actual goal of the search, a nonempty intersection $A \cap B$ of the affine approximation. When this happens $w_A = w_B = w_{\text{sol}}$ and

$$A \cap B = \{w_{\text{sol}}\} \times \{y_0\} \times Z .$$

By contrast, the fixed point set is the superset

$$X^* = W \times \{y_0\} \times Z .$$

Both of these are, of course, only the local approximations (by affine sets). The presence of the space Z is explained by this being the space of continuous symmetries. It is tempting to speculate that RRR works as well as it does because the set of fixed points is a larger target than the set of solutions, especially when W has many dimensions. In Section 5.5 we make the opposite argument, that searches are more efficient when it is Y that has many dimensions!

5.5 Global Behavior

We cannot claim to understand how RRR manages to solve difficult problems until we at least have a picture of what it is doing globally. For this part of the analysis, there are no rigorous results, and indeed there probably should *not* be any if we wish to use the method on NP-complete problems! On the other hand, the local analysis has revealed what RRR is doing as it is taking the "next step" in the search, and that should allow you to form an informed opinion on whether RRR is at least a good search heuristic.

5.5.1 Transverse Contraction and Time-Reversal

The affine approximations to A and B shown in Figure 5.2 evolve over the course of the search. For small β, when the steps are small, the affine approximation of a smooth set evolves smoothly except at special times when a different part of the set takes over as the best approximation. When the set is discrete, or the union of many affine sets (such as the sparsity constraint), the evolution is discontinuous. In either case, when β is small, RRR makes several iterations before the affine approximations change significantly. Over these periods of continuous evolution, RRR is doing two simple things simultaneously: taking steps in the W space and contracting in the Y space (equations 5.19).

If we think of the iteration counter as "time," then only one part of the local behavior is time-reversal symmetric. Reversing time corresponds to changing the sign of β. Instead of taking w-steps from constraint A to constraint B (equation 5.19a), the reversed-time dynamics moves in the opposite direction. But with that sign-reversal of β, contraction in Y becomes expansion. To recover fixed-point-set attraction, one also has to move the reflection from the argument of P_B to the argument of P_A. Here is a summary of these observations.

Quasi-time-reversal

> Interchanging sets A and B in the RRR rule is almost like a reversal of time. In the W space, each step is β times the step between the proximal pair on A and B. As written, RRR steps from A to B, so interchanging A and B reverses the direction of the step in W. However, instead of contraction in Y being replaced by expansion in Y (as in a true time reversal), the dynamics remains contractive in the transverse space.

An analogy may be useful when thinking about the two parts of the local behavior. Firefighters perform a *primary search* for occupants of a burning building when they first arrive on the scene. The smoke may be so thick that even with lights, the firefighters are searching in the dark. In order for the search to be both efficient and exhaustive in these conditions, pairs of firefighters do the following. Crawling on the ground while staying in physical contact with one pair of hands, one sweeps the ground to the right, the other to the left, both with outstretched arms (augmented by the handle of a fire axe) to maximize their chances of locating a nonresponsive occupant as they comb the building.

The Y space corresponds to the space explored by the firefighter-pair's arms. When the set of feasible Y candidates has contracted to a single point (y_0), but without that being a fixed point because $w_A \neq w_B$, the search continues with a step in the W space. It's a good strategy for firefighters because they can search a two-dimensional space one-dimensionally. If the analogy holds, that might be the logic behind RRR. Naively, when Y has many dimensions, a large part of the whole space is being searched very efficiently. Mostly what RRR is doing, when it is not converging on a fixed point set, is stepping through the smaller W space. Having solved the problem to the greatest extent possible in the Y space – for the current approximation of the constraints – the algorithm moves in the only remaining way that can make a difference: the W space (Z is a symmetry).

The firefighter analogy breaks down when it comes to moving into the next room, or when exploring a large room. Firefighters then use walls and right-hand-turn rules in order to cover the whole building efficiently. RRR has no such guides or protocols and instead relies on chaos. Each discontinuous change in the affine approximation is like the firefighters moving between rooms in a random way. While that is a poor strategy for firefighters, random exploration is not that different from a systematic exhaustive search when space has many dimensions. The real problem faced by this picture of the global behavior is that RRR appears to be doing something much more clever than picking random rooms!

In some applications where RRR works very well, the high-dimensional-transverse-space argument falls flat even more spectacularly. A good example is our favorite bipartisan formulation of sudoku, where both A and B are discrete so that $\dim(Y) = \dim(Z) = 0$ and the orthogonal decomposition is trivial: $X = W$. Search efficiency cannot be explained by a large transverse space when this space has zero dimensions! Also, when both A and B are discrete, one should worry about RRR becoming indistinguishable from the difference-of-projections algorithm. To see this, note that $P_B(R_A(x)) = P_B(x)$ when x

is close to both some point $a \in A$ and some point $b \in B$. But difference-of-projections cannot solve even the easiest sudoku puzzles!

Both of the above concerns (random room switching, no transverse space) are the motivation behind most of what remains in this chapter. We first gain some insights from the $\beta \to 0$ limit of RRR, the *flow limit*. From the apparent random evolution of the affine approximations (as currently defined), we will be able to define a generalization of affine approximation that has more global scope. An interesting outcome of this generalization is that we recover contracting transverse spaces even when originally there were none!

5.5.2 The Flow Limit

The $\beta \to 0$ limit of RRR is the system of ordinary differential equations

$$\frac{dx}{dt} = \lim_{\beta \to 0} \frac{\Delta x}{\beta} = v(x) \,,$$

where Δx is the change in x between iterations and

$$v(x) = P_B(R_A(x)) - P_A(x)$$

is the *RRR velocity* at x. Iterates x_0 and $x_i = \mathcal{R}^i(x_0)$ are separated by i iterations and

$$t = i\beta$$

units of *RRR time*. In the flow limit, x is a continuous function of the time, and we write $x_0 = x(0)$, $x_i = x(t)$. If RRR finds a fixed point in i iterations, we say $t^* = i\beta$ is the *solution time*.

At a regular point x, the projection $P_A(x) = a$ is unique as is the projection $P_B(2a - x) = b$. Not only is the velocity well defined, $v(x) = b - a$, so is its gradient. That's because at regular points the constraint sets can be approximated as affine with projections

$$P_A(y) = a + (y - a)p_A \,,$$
$$P_B(y) = b + (y - b)p_B \,.$$

Here p_A and p_B are projection matrices for the corresponding subspaces and the argument of P_A should be close to a, while the argument of P_B, $2a - x$, should be close to b. The velocity therefore has the approximation

$$v(x) = b + \Big(2\big(a + (x - a)p_A\big) - x - b\Big)p_B - \big(a + (x - a)p_A\big)$$

with gradient

$$\nabla v(x) = 2p_A\, p_B - p_B - p_A\ .$$

Since $p_A = p_B = 0$ when A and B are discrete, the velocity is piecewise constant in that case.

Since the velocity is well defined at regular points, by Proposition 4.1 it may only have discontinuities on sets of positive codimension. For example, if $r = 2a - x^*$ is the reflection in A of an irregular point x^*, where $P_B(r) = \{b_1, b_2\}$, then $v(x)$ switches between $b_1 - a$ and $b_2 - a$ as x passes through x^*. The existence of nonunique projections, in nonconvex problems, is reframed in the flow limit as a velocity function with discontinuities.

The magnitude of $v(x)$, being the distance between points $P_B(R_A(x)) \in B$ and $P_A(x) \in A$ on the two constraint sets for the current iteration, is what we have called the gap. The gap and velocity both vanish at the solution time. If we had the power to visualize $v(x)$ on a global scale for our problem, we would know how any initial point evolves and whether it will arrive at a fixed point. There wouldn't be much point in running RRR, because solutions can be found much more directly by looking for x^* where $v(x^*) = 0$ (or even more directly by visualizing $A \cap B$). But in the high-dimensional worlds where RRR is used, and lacking stupendous powers of visualization, evolving a single point $x(t)$ in time is our only recourse for exploring $v(x)$.

5.5.3 Flows in Two Dimensions

While it is hopeless to visualize $v(x)$ for the problems RRR is used for, some general characteristics can be gleaned from examples in lower dimensions. All the examples in this section are in two dimensions, and streamlines tangent to $v(x)$ are used to convey at least the direction of the velocity. With the convention that A is rendered black and set B is rendered white, the flow is always in the direction black to white.

The flow that is most often identified with RRR – usually rendered as discrete iterates – is the spiraling pattern produced by a pair of intersecting lines, shown in Figure 5.3. The two lines could be the sets A and B in a two-dimensional X, or the subspaces C and D in a transverse Y. Spiraling is a trademark of RRR (and Douglas–Rachford) and can be traced to the fact that the matrix r in the transverse evolution (5.19b) is nonsymmetric for the case $\lambda = -1$, giving it complex eigenvalues. Because $v(x)$ only depends on the geometry of the sets, the rate of contraction can only depend on the angle of intersection of the lines in this two-dimensional example. It should not come as a surprise that contraction is instantaneous for perpendicular

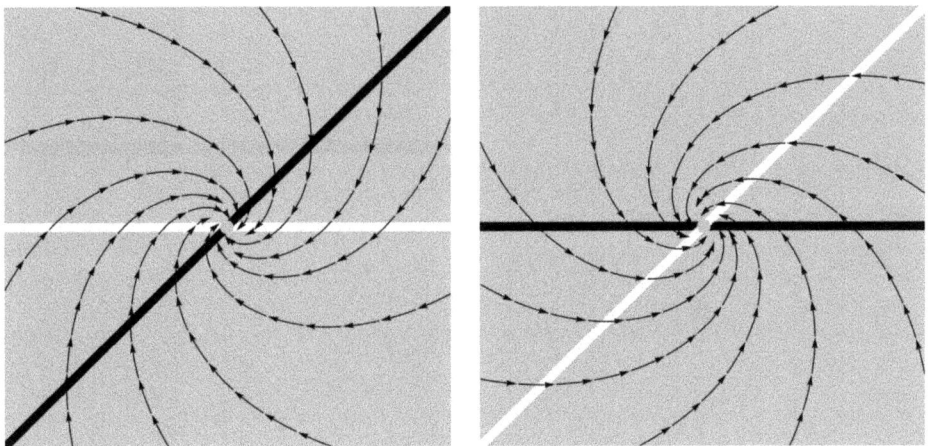

Figure 5.3 Some streamlines in the RRR flow for the case of two intersecting lines in a two-dimensional X, or in a two-dimensional transverse Y, with an orthogonal W (not shown). The flow-spiral is clockwise when B (white) is rotated clockwise by an acute angle relative to A (black), and counterclockwise in the other case.

lines, since for that case a single DR iteration takes x to the intersection. The derivation of explicit formulas for the spiral orbits and the rate of contraction is left as an exercise.

If Figure 5.3 is showing only what is happening in the transverse space Y, and the sets A and B do not in fact intersect because they have coordinates $w_A \neq w_B$ in an orthogonal W space, by equation (5.19a) we know the W-coordinate evolves as

$$
\begin{aligned}
w(t) &= w_i \\
&= w_0 + i\beta(w_B - w_A) \\
&= w(0) + t(w_B - w_A) \,.
\end{aligned}
$$

We see that the velocity in W is constant and equal in magnitude to the distance $\|w_B - w_A\|$ between the proximal points.

The second flow every student of RRR should know is where A is a point and B a line, or vice versa. Both are rendered in Figure 5.4. Because both sets are convex (as they were for the two lines), $v(x)$ has no discontinuities and the flow is smooth. These flows are even simpler than the flow for two lines. If B is the line, then D is one-dimensional, C is zero-dimensional, and $Y = C + D$ is also just one-dimensional. The subspace orthogonal to Y, W, is one-dimensional and the flow contracts to the affine set $W \times \{y_0\}$.

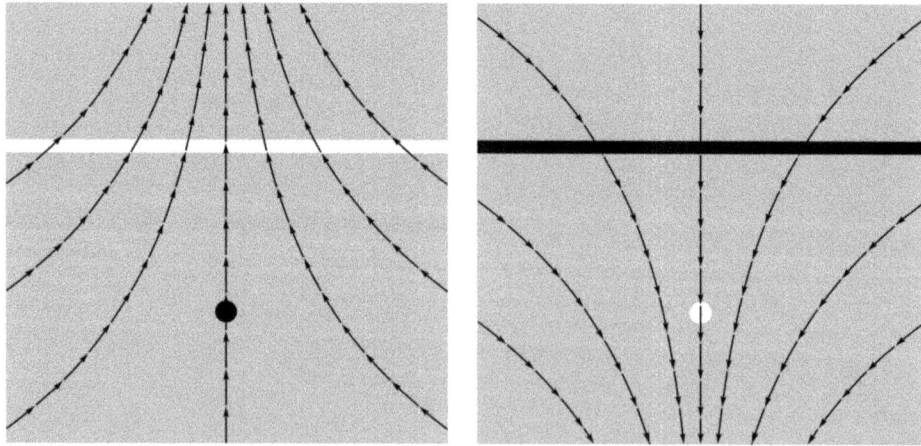

Figure 5.4 Flow streamlines when one set is a line and the other is a point.
By our convention, the direction of the flow is always from A (black) to B
(white).

For the one-dimensional Y that applies to both of the flows in Figure 5.4,
the matrix r in the local analysis is just a 1×1 matrix with element -1.
The flow in Y, with (one-dimensional) coordinate $\tilde{y}$ relative to the proximal
position y_0, is the $\beta \to 0$ limit of (5.19b) with $r = -1$,

$$\frac{d\tilde{y}}{dt} = -\tilde{y}$$

with solution

$$\tilde{y}(t) = \tilde{y}(0)e^{-t} .$$

The uniform velocity of the flow in $W \times \{y_0\}$ is equal to the distance between
the line and the point, $\|w_B - w_A\|$. When the point and line intersect, all of
$W \times \{y_0\}$ becomes a fixed point set. The contracting flow in Y is independent
of this detail.

In our final example, shown in Figure 5.5, one of the sets is a line and the
other is nonconvex because it comprises three points. The case where A is
the set of points, shown on the left, is easier to understand. That's because
the three points divide the plane into three regions, for which each is the
nearest point (two half-planes and an infinite strip). The flow in the two
regions where the point does not intersect the line is just the point-line flow
of Figure 5.4. In the strip region, where the point intersects the line, the
flow converges on a set of fixed points (there is no stepping in the W space).
It's easy to see that no matter where one starts in the plane, the flow-limit

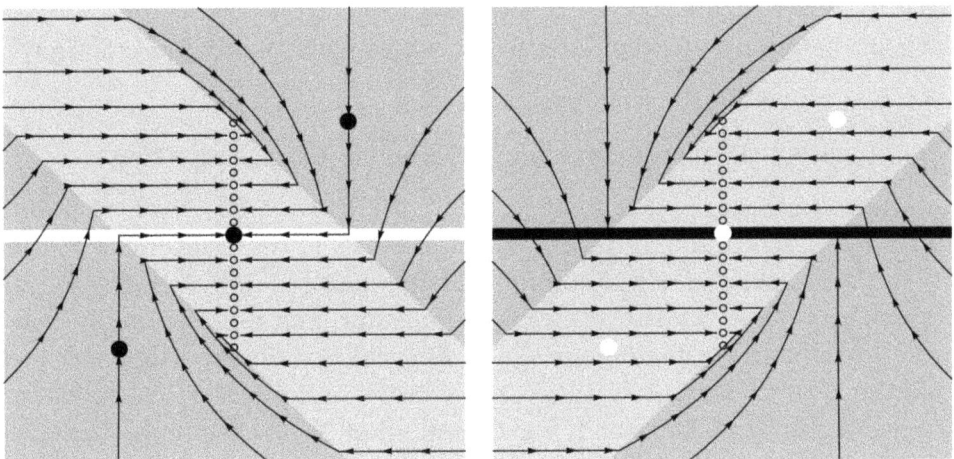

Figure 5.5 Flows where either A (left panel) or B (right panel) is nonconvex – a set of three points. Both flows have a spiraling character, not unlike
the two-line flows of Figure 5.3.

iterates eventually arrive at the one-dimensional fixed point set associated
with the solution point $A \cap B$.

To understand the other case, where B is the set of three points, we start
by constructing the regions of the plane where each of the three points is the
value of P_B. In these regions the flow will again be a point-line flow. Call the
three points b_1, b_2, and b_3, and let Y_i be the region defined by $P_B(Y_i) = b_i$.
The regions Y_1, Y_2, and Y_3 are the same regions (half-planes and strip) shown
in the left panel of Figure 5.5. The corresponding regions X_i, that determine
$v(x)$ for $x \in X_i$, are defined by $R_A(X_i) = Y_i$. But R_A, a reflection in a line,
is an involution so $X_i = R_A(Y_i)$. The three point-line flow regions in the
right panel are therefore just the reflections in the line of the regions in the
left panel. When the three flow-regions are stitched together, we again get a
flow that arrives at the fixed point set for any starting point in the plane.

In one of the exercises, you are asked to analyze the flow when the middle
point (in either case) is removed, so that $A \cap B$ is empty. You will find that
the orbit gets ever closer to the position of the missing point, trapping the
RRR iterates (with finite β) there. This is bad news if you had set your
hopes on RRR always finding an intersection if one exists, since A and B
might intersect outside the region shown!

Squinting at Figure 5.5 may bring to mind the spiraling two-line flows
of Figure 5.3. It appears that "the other line" can be approximated in some

sense by as few as three points! Before we develop this notion we first consider flows in the most nonconvex case of all, where both A and B are discrete.

5.5.4 Discrete Sets and Piecewise Constant Flow

Let's now confront the flow most fraught with discontinuities, where both A and B are discrete sets of points. For almost all x, the projections P_A and P_B in the formula for $v(x)$ produce a unique pair (a, b), $a \in A$, $b \in B$, and $v(x) = b - a$. The velocity is piecewise constant, and representing it comes down to working out the domains of x over which it takes particular values.

In the local analysis of Section 5.4, the nearest (approximating) affine sets are just a pair of points $a \in A$, $b \in B$, and all of X is taken up by the space W (since $Y = Z = \{0\}$). Extending this very trivial local analysis to a global one amounts to partitioning the global space X into W-domains over which $v(x)$ is constant. When there's a solution, so $A \cap B$ contains a point $a = b$, there will be a fixed point set W-domain wherein $v(x) = 0$. RRR is a finite algorithm in this setting: it makes a finite number of steps and stops when x lands in a domain where $v(x) = 0$. When RRR is applied to the formulation of sudoku where both A and B are discrete (permutations), the puzzle is solved in a finite number of iterations.

There are some familiar constructions that will help us construct the less familiar W-domains. First, recall the *Voronoi domain* associated with a point $a \in A$:

$$V_A(a) = \{x \in X : P_A(x) = a\} \ .$$

The Voronoi domains $V_A(a)$, where a ranges over A, form a familiar partition of X. There is an analogous partition of X by the Voronoi domains $V_B(b)$ of the points $b \in B$. We need not concern ourselves with the boundaries of these domains other than to note that it is only there that $v(x)$ changes (discontinuously). Another familiar construction is the reflection

$$R_a(x) = 2a - x \ ,$$

now with respect to a particular point a. If $Y \subset X$ is a general subset of X, we use the notation $R_a(Y)$ for the set formed by applying the reflection to all points in Y.

The partition of X into constant velocity domains has the form of a disjoint union,

$$X = \bigsqcup_{a \in A, \ b \in B} W(a, b) \ ,$$

and

$$v(x) = b - a, \qquad \forall \, x \in W(a,b) \, .$$

Not all differences $b - a$ need arise in the velocity field because the domains $W(a,b)$ may be empty. It's possible to write a general formula for W-domains:

Proposition 5.4 (W-domain formula) *Given discrete sets A and B, the W-domain associated with $a \in A$, $b \in B$ is*

$$W(a,b) = R_a \Big(V_B(b) \cap R_a\big(V_A(a)\big) \Big) \, .$$

Proof Consider an element $x \in W(a,b)$, that is, a point where $v(x) = b-a$. We know that $x \in V_A(a)$ if the second term of the velocity field, $P_A(x)$, equals a. That the first term equals b constrains the possible inputs to P_B, that is, the set of reflections $R = R_a\big(V_A(a)\big)$. This constraint is simply that if $r \in R$, then it should also be true that $r \in V_B(b)$. The set of reflections giving b for the first velocity term is therefore

$$V_B(b) \cap R_a\big(V_A(a)\big) \, .$$

To recover the x that corresponds to any point of this set, we apply the inverse of the reflection R_a, which is the same as applying R_a.   □

Figure 5.6 shows the W-domains when the two sets combined comprise $1 + 5$ points. Because the singleton set has just one Voronoi domain, and it is all of X, the W-domains are determined by the Voronoi domains of the set with five points. When A is the set with five points (right panel), the W-domains are the same as the Voronoi domains of those points. When B is the set with five points (left panel), the W-domains are the reflections of these Voronoi domains in the single point of the other set. The corresponding velocity $b - a$, uniform in each domain, is rendered as gray arrows in Figure 5.6. Because $A \cap B$ is empty, $v(x) \neq 0$ in all the domains and there is no fixed point set.

RRR trajectories in piecewise-constant flows exhibit an interesting phenomenon, examples of which are highlighted in Figure 5.6 (black arrows). When the RRR update steps across a domain boundary, and the flow normal to the boundary on the other side has the opposite direction, the trajectory takes a crisscrossing path that follows the boundary. *Domain-boundary-attraction* happens in higher dimensions as well, whenever the velocity normal to the boundary is convergent. Codimension-1 discontinuities arise even if only one of the constraint sets is discrete, such as in the bit retrieval

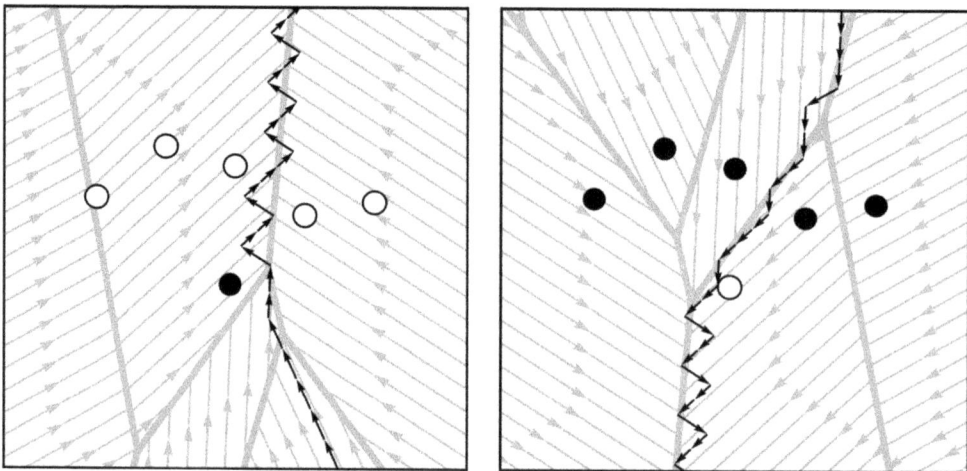

Figure 5.6 Flow (gray arrows) is uniform inside W-domains when both A and B are sets of points. The RRR trajectory (black arrows) for small β crisscrosses domain boundaries when the flow on opposite sides is convergent. The resulting attraction to a domain boundary, in both flows, is qualitatively similar to the transverse contraction in Figure 5.4, where the five points are replaced by a line.

problem. Moreover, in higher dimensions the attraction can be to higher-codimension sets, provided the flow in the adjacent domains converges on the set. In bit retrieval, with ± 1 bits, there is attraction to sets where as many as 15% of the variables are near zero [37] – apparently keeping their options open!

The flows in Figure 5.6 are our inspiration for what RRR is doing in nonconvex problems when there are many dimensions. By squinting and comparing with Figure 5.4, you get a hint of where we are going. Domain-boundary attraction may not be all that different from the transverse-space contraction we saw earlier, in the locally affine model.

5.5.5 Flows in Several Dimensions

The RRR flows in two dimensions, when both A and B are discrete, have a stochastic character but also hint at something regular. When the flow is convergent at a domain boundary, the RRR velocity orthogonal to the boundary switches sign from one iteration to the next. But there is also a constant velocity along the boundary, from the average of the velocities on the two sides. When there are more dimensions, it is reasonable to expect rapid fluctuations in some but not all of the velocity components.

The stochastic and regular parts of the velocity are not independent. In higher dimensions, a fluctuating velocity indicates the projections are sampling a diverse set of points on the constraint sets. A regular contribution to the velocity arises when these fluctuations have a nonzero average. The regular part of the velocity evolves in a regular way when the sampling of the constraints evolves slowly, and that is achieved when the time step β applied to the velocity is small.

Because flows in several dimensions are beyond our power to visualize, we need another way to observe them. The simplest thing we can do is to track a few of the coordinates of the iterate x. "Few" may be as small as 10, and instead of embedding the 10 variables in a 10-dimensional space, we plot them on a common axis. As we will see, this *one-dimensional sparse sampling* of the RRR dynamics reveals a regular flow and also does a good job in conveying the complexity and other characteristics of the search.

In this section we apply one-dimensional sparse sampling to sudoku and 16×16 Hadamard matrices. For sudoku we use the formulation where A and B are permutation-matrix constraints on two factorizations of the $9 \times 9 \times 9$ cube of 729 variables. We use a New York Times hard-rated puzzle, where 228 of the cube's variables are not directly implicated by the puzzle's 23 givens. In the Hadamard-matrix search, set A is the ± 1 constraint on the 256 elements, while a continuous B imposes orthogonality on the matrix. The two flows are respectively in 228 and 256 dimensions, and the A constraint is discrete in both.

The left panel of Figure 5.7 shows the $\beta = 0.02$ evolution of 10 sudoku variables in a search with solution time $t^* = 3.2$. Recall that when both A and B are discrete and space is partitioned into $W(a, b)$-domains, RRR is a finite algorithm because it encounters a fixed point x^* as soon as the iterate x steps into a domain where $a = b \in A \cap B$. The fixed point structure of the flow is emphasized by subtracting x^* from $x(t)$ in the plot. The right panel shows the evolution of the same variables, and the same $x(0)$, but with $\beta = 0.002$.

The evolution with the larger time step on the left helps us understand the more regular-appearing trajectories on the right. All the velocities are 3-valued because the components of $b - a$ are elements of $\{-1, 0, 1\}$ when A and B have only 0 and 1 elements. The appearance of a continuum of slopes comes from the three discrete velocities arising with different frequencies over the course of the puzzle's solution. The evolution in the right panel

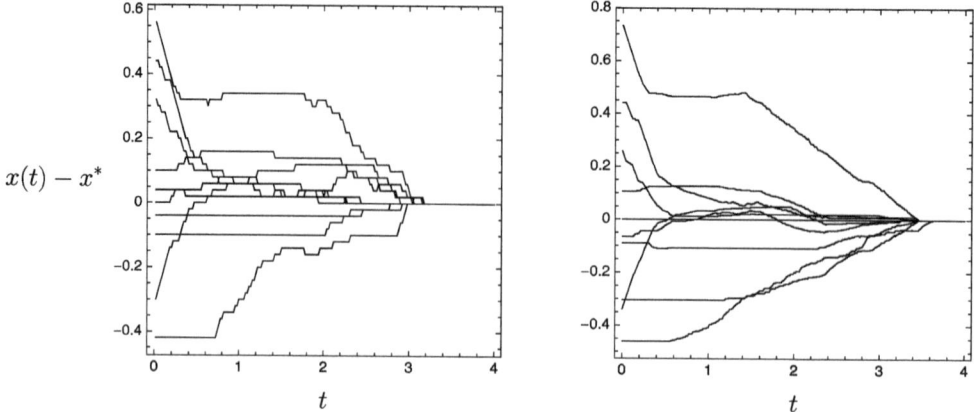

Figure 5.7 One-dimensional sparse sampling of the flow $x(t)$ in the solution of a NYT-hard sudoku puzzle with $\beta = 0.02$ on the left, $\beta = 0.002$ on the right. The convergent nature of the flow is brought out by subtracting the fixed point, x^*.

looks more regular because reducing β by a factor of 10 improves the quality of the running average of the discrete velocity.

From the mostly monotonic changes in the variables shown, it looks as if the algorithm did not have to "try out" too many options when solving this puzzle. When sampling different $x(0)$, the variables rise and fall in value just a few times before converging to the fixed point. The solution time t^* for this puzzle is always below 15.

The sudoku flows can be compared with Hadamard flows in Figure 5.8, now with $\beta = 0.1$ and $\beta = 0.01$. In Hadamard flow, only the A contribution to the velocity is discrete. Unlike sudoku, these flows have much greater sensitivity to the initial point because there are nearly 10^{32} (symmetry-related) solutions [83], each with its own domain for $x(0)$.

Though the trajectories on the right in Figures 5.7 and 5.8 look regular, we know this is just because β is too small to resolve the individual RRR iterates and the fluctuating velocity. But when solving a sudoku puzzle or constructing a Hadamard matrix, the net change in x, from $x(0)$ to x^*, is not small at all. And because the evolution of $x(t)$ on large scales looks regular, the evolution of the regular part may be enough to understand why RRR works on even highly nonconvex problems.

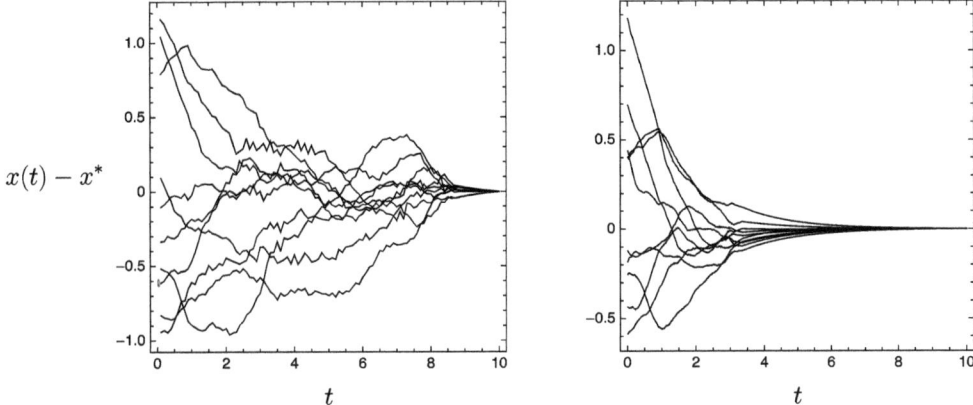

Figure 5.8 Same as Figure 5.7, but for 16×16 Hadamard matrix searches. The time steps β are five times as large as in the corresponding sudoku flows.

5.5.6 The Continuum Model

The existence of a regular part of the RRR dynamics in highly nonconvex problems can be interpreted with the help of a continuum model. Though models are no substitute for rigorous mathematical constructions, they are useful because they help us formulate hypotheses.

Just as in statistical mechanics, where it is hopeless and unproductive to study the evolution of microstates, in the case of (highly nonconvex) RRR we will settle for understanding the evolution of averages. In the flow limit, the RRR iterate $x(t)$ has small amplitude fluctuations, but its average $\langle x(t) \rangle$ appears to have a regular evolution. Instead of defining "average" by sampling perturbations of the initial point, we use this term in the sense of a "running average." The goal of the model is to explain the regular evolution of $\langle x(t) \rangle$. Because the modeling involves continuous sets, even when A and B are discrete, it is a *continuum model*.

Let $x_i = x(t + i\beta)$, $i = 0, 1, 2, \ldots$, be the stream of iterates generated by RRR and $a_i = P_A(x_i)$ the corresponding stream of projections to the A constraint. Section 5.5.7 shows how to construct a best, empirically defined, *affine* set $\tilde{A}$ with the property

$$\begin{aligned} \langle P_A(x) \rangle &= \langle P_{\tilde{A}}(x) \rangle \\ &= P_{\tilde{A}}(\langle x \rangle) \ . \end{aligned}$$

In words, the average of the projections to the true constraint can be interpreted as the average of projections to the affine set $\tilde{A}$. The second equality follows because the projection to an affine set is a linear function, and commutes with averaging. Projection fidelity, or the least-squares minimization of the residuals between true and approximated projections, is the principle that determines the best $\tilde{A}$. It is interesting that this principle also determines the dimension of $\tilde{A}$.

The same idea applied to the stream of reflections $r_i = 2a_i - x_i$ and their projections $b_i = P_B(r_i)$ determines the best empirically defined affine approximation $\tilde{B}$ of set B. With both definitions in place,

$$\langle v(x) \rangle = \tilde{v}(\langle x \rangle) \,,$$

where

$$\tilde{v}(x) = P_{\tilde{B}}(R_{\tilde{A}}(x)) - P_{\tilde{A}}(x)$$

is the RRR flow associated with the affine approximations $\tilde{A}$ and $\tilde{B}$. Because these sets are defined empirically, the model does not offer a way to evolve the average $\langle x \rangle$ by applying RRR to some simpler affine sets. After all, the affine approximations evolve as well. The most the model can offer is some insights why RRR works well even when the original sets are very far from affine.

Recall that RRR is indeed very efficient when A and B are affine. In the firefighter analogy (Section 5.5.1), the contraction in the transverse Y space means that many variables (dimensions) can be disposed of quickly, shifting the focus of the search to the orthogonal space, W. Both Y and W evolve when these are defined empirically, by the stream of samples that determine $\tilde{A}$ and $\tilde{B}$. Though the initial affine approximation may be infeasible, the gap $\|w_{\tilde{A}} - w_{\tilde{B}}\|$ evolves with time and becomes zero when eventually the affine approximations define a feasible problem.

5.5.7 Empirically Defined Affine Sets

Consider a stream of inputs x and projection outputs $a = P_A(x)$ generated by RRR:

$$(x_1, a_1) \,, \ \ldots \,, \ (x_n, a_n) \,.$$

These correspond to times $t + \beta, \ldots, t + n\beta$, that is, samples near a time t in the flow limit. If $\Delta t = n\beta$ is held fixed, then the number of samples $\Delta t/\beta$ grows and their distribution approaches a stationary distribution as $\beta \to 0$. Stationarity follows from the fact that the iterate x spans a fixed range of size

$$\|\Delta x\| = \Delta t \, \|\langle v(x) \rangle\| \,,$$

where $\langle v(x) \rangle$ is the average of the bounded RRR velocity in this limit. In this section we ask how the same distribution could have been generated by the projection to an affine constraint $\tilde{A}$, and what determines the parameters of $\tilde{A}$. Exactly the same analysis would be applied to constraint B, except that the inputs to P_B are the reflections $r = 2a - x$, which are stochastic as well.

The continuum model interprets the flow $v(x)$ as the result of projections to continuously evolving affine sets. This imposes the following constraint on the definition of $\tilde{A}$:

$$\frac{1}{n}\sum_{i=1}^{n} P_{\tilde{A}}(x_i) = \frac{1}{n}\sum_{i=1}^{n} P_A(x_i) \, . \tag{5.26}$$

Because there are infinitely many $\tilde{A}$ that have this property, we want to identify the $\tilde{A}$ that best approximates the projected points in a least-squares sense. Defining the current "center" of $\tilde{A}$ by

$$a_0 = \frac{1}{n}\sum_{i=1}^{n} a_i \, ,$$

we are interested in optimizing the affine projection

$$P_{\tilde{A}}(x) = (x - a_0)p + a_0$$

with respect to the projection matrix p. Our parameterization of $P_{\tilde{A}}$ is consistent with constraint 5.26 provided

$$(x_0 - a_0)p = 0 \, , \tag{5.27}$$

where

$$x_0 = \frac{1}{n}\sum_{i=1}^{n} x_i$$

is the current center of the x distribution. To satisfy constraint (5.27) we express p as the product

$$p = p_0 \, p_{\|} \, ,$$

where

$$p_0 = I - \frac{(x_0 - a_0)^{\mathsf{T}}(x_0 - a_0)}{\|x_0 - a_0\|^2}$$

projects into the corresponding codimension-1 subspace and $p_{\|}$ projects into a subspace of that subspace.

Optimizing the projection fidelity now means minimizing the function

$$f(p_{\|}) = \sum_{i=1}^{n} \|P_A(x_i) - P_{\tilde{A}}(x_i)\|^2 \tag{5.28}$$

whose argument includes the rank k of the projection matrix $p_{\|}$ as a parameter. When f is minimized by a projection of rank k, the best affine approximation is k-dimensional. As it is this characteristic that motivated the continuum model, our main result is the following formula for k:

Theorem 5.4 (Dimension of empirically defined affine sets) *The rank k of the projection $p_{\|}$ that minimizes (5.28) is equal to the number of positive eigenvalues of the $m \times m$ matrix*

$$w = u^{\mathsf{T}} u - v^{\mathsf{T}} v \,,$$

where the rows of u and v comprise the following n samples in $\mathbb{R}^m$:

$$u_i = (a_i - a_0)p_0 \,,$$
$$v_i = (a_i - x_i)p_0 \,.$$

Proof   We use the notation

$$y = y_{\|} + y_{\perp}$$

for the orthogonal decomposition of vectors $y \in \mathbb{R}^m$, where

$$y_{\|} = y \, p_0 \, p_{\|} \,.$$

We start by writing f only in terms of the $\|$ component:

$$f(p_{\|}) = \sum_{i=1}^{n} \left\|a_i - \left((x_i - a_0)p_0 p_{\|} + a_0\right)\right\|^2$$
$$= \sum_{i=1}^{n} \left(\left\|(a_i - a_0)_{\|} - (x_i - a_0)_{\|}\right\|^2 + \left\|(a_i - a_0)_{\perp}\right\|^2\right)$$
$$= f_0 - \sum_{i=1}^{n} \left(\left\|(a_i - a_0)_{\|}\right\|^2 - \left\|(a_i - x_i)_{\|}\right\|^2\right) , \tag{5.29}$$

where

$$f_0 = \sum_{i=1}^{n} \|a_i - a_0\|^2$$

is a constant independent of $p_{\|}$. Expressing (5.29) in terms of the matrices

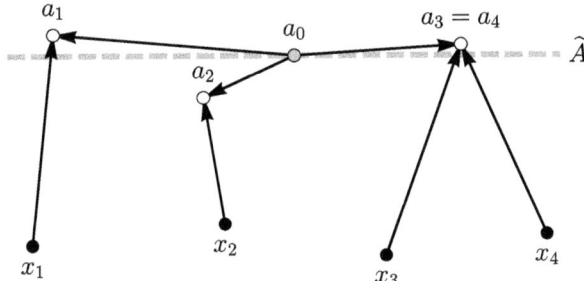

Figure 5.9 Cartoon showing four samples (x_i, a_i) used in the empirical definition of the affine approximation $\tilde{A}$.

defined above, we obtain

$$f(p_{\parallel}) = f_0 - \mathrm{Tr}\left((u\,p_{\parallel})^{\mathsf{T}}\,(u\,p_{\parallel}) - (v\,p_{\parallel})^{\mathsf{T}}\,(v\,p_{\parallel})\right)$$
$$= f_0 - \mathrm{Tr}\left(w\,p_{\parallel}\right).$$

Recalling the k-coplanarity projection (Section 4.5.2), we recognize the f-minimizing $p_{\parallel}$ as the projection of the symmetric matrix w to have a spectrum of k 1's and 0 for the other eigenvalues. Writing the eigen-decomposition of w as

$$w = s^{\mathsf{T}}\,\sigma\,s$$

with eigenvalues ordered as

$$\sigma_1 > \cdots > \sigma_m \,,$$

the minimizing projection is

$$p_{\parallel}^* = s^{\mathsf{T}}\,d_k\,s \,,$$

where d_k is diagonal with 1 in the first k elements, 0 elsewhere. Evaluating the least-squares function, we obtain

$$f(p_{\parallel}^*) = f_0 - \sum_{j=1}^{k}\sigma_j \,.$$

To minimize this expression with respect to k, the sum should contain all the positive eigenvalues and no more. $\qquad\square$

Figure 5.9 tries to convey the competition at work when defining the affine approximation $\tilde{A}$. While the vectors $a_i - a_0$ define what the subspace centered on a_0 should try to include, the vectors $a_i - x_i$ define what its orthogonal complement should be. The most interesting outcome of the competition is

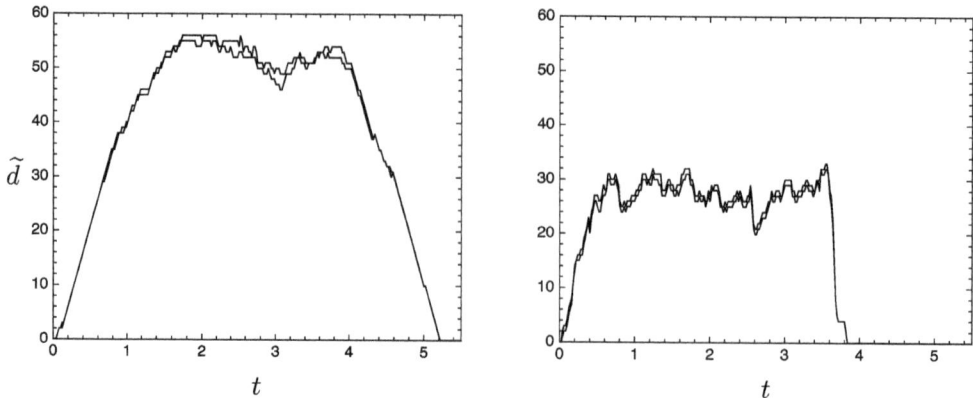

Figure 5.10 Evolution of the dimensions $d_{\tilde{A}}$ and $d_{\tilde{B}}$ of the empirically defined affine sets in the sudoku solutions shown in Figure 5.7. The left and right panels correspond to the same β values in that figure.

the dimension of $\tilde{A}$ where both of these objectives are achieved to the greatest extent possible. Cartoon 5.9 is misleading in one respect: In practice, the number of dimensions m of the ambient space is typically as large as the number of samples n.

The evolution from a stream of samples, of the dimensions $d_{\tilde{A}}$ and $d_{\tilde{B}}$ of the empirically defined affine sets for sudoku, are plotted in Figure 5.10. These are for the same puzzle and initial point whose flows are rendered in Figure 5.7. As in that figure, the two panels show the effect of decreasing β by a factor of 10. With the smaller time step, on the right, $d_{\tilde{A}}$ and $d_{\tilde{B}}$ are quite a bit smaller than the number of sudoku variables ($m = 228$). That $d_{\tilde{A}} \approx d_{\tilde{B}}$ is not surprising because the original A and B are similarly structured discrete sets in this formulation of sudoku.

The rise of the dimensions from zero, and their return to zero, are artifacts of the running-average process. To represent an affine set of $\tilde{d}$ dimensions, a minimum of $\tilde{d} + 1$ samples are needed. The evolution shown in Figure 5.10 used $n = 200$ samples, a number comfortably greater than what the dimensions turned out to be. However, it is only after the 199th RRR iteration that 200 samples are available. This explains the rise from zero. Similarly, the fall back to zero is not complete until all 200 samples are the singleton fixed point.

The dimension evolutions corresponding to the Hadamard flows of Figure 5.8 are plotted in Figure 5.11, with the same two β's and sample number that were used for sudoku. With the smaller β, the dimensions are about one-tenth the number of Hadamard variables (256).

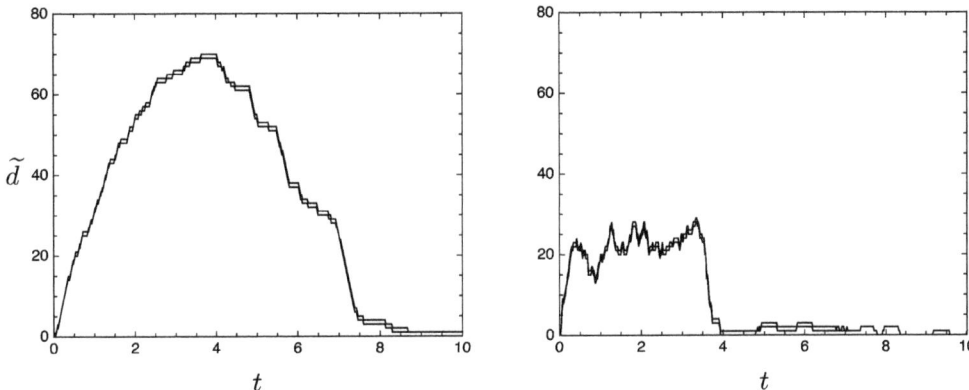

Figure 5.11 Same as Figure 5.10, but for the Hadamard matrix searches whose flows are rendered in Figure 5.8.

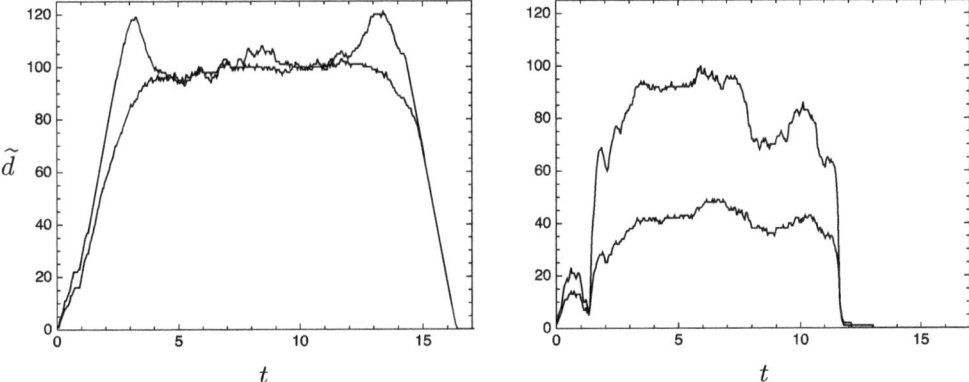

Figure 5.12 Same as Figure 5.11, but where the discrete set of the Hadamard matrix problem is now the B constraint of RRR. This makes the two empirical dimensions different, with the smaller one being the dimension of $\tilde{B}$.

Figure 5.11 might lead one to conjecture that the two empirically defined dimensions are always nearly the same, even when the original sets A and B are quite different. In our Hadamard matrix formulation, B is the set of 16×16 orthogonal matrices with 120 dimensions and very unlike the discrete A. The qualitative difference in the sets makes itself known when they are swapped in the definition of the RRR rule. Figure 5.12 shows the evolution of the empirically defined dimensions when B is the discrete set. The two dimensions are now quite different. The larger dimension corresponds to the approximation $\tilde{A}$ of the continuous set, and is not all that much smaller than the dimension of A (120), even with the smaller β (right panel).

5.5.8 The Flow Limit Conjecture

Let's review what the renderings of flow and the evolution of empirical dimensions, all in the limit of small β, have taught us. The least contentious lesson is that the flow, even for highly nonconvex problems, is regular in this limit. The continuum model provides a nice interpretation, in terms of empirically defined affine approximations $\tilde{A}$ and $\tilde{B}$ of the original sets. These have positive dimensions even when A and B are discrete. A thought experiment, that maps out the evolving $\tilde{A}$ and $\tilde{B}$ for all possible initial points, could define a much more regular pair of RRR constraints that explains the regular part of the flow. Since we have a good understanding of how RRR works with nice constraints – a combination of transverse contraction and steps in the orthogonal space – it is good to know the same intuition can be applied to highly nonconvex problems in the flow limit.

When learning *the three Rs* of RRR, the last one – relax – is the most challenging. Probably the best argument for $\beta < 1$ in nonconvex problems (deviating from Douglas–Rachford), is that it is only with this relaxation that the flow exhibits regularity in the most extreme case of discrete constraints. But when selecting the best β, or deciding whether A or B should be the more nonconvex of the two constraints, the answers are not that clear.

Before we considered the flow limit, the firefighter analogy suggested a large transverse space was good. Fewer dimensions in the orthogonal W space meant less searching. But our experience in rescuing Hadamard matrices from burning buildings tells a different story. Comparing Figures 5.11 and 5.12, we see that the search with the *smaller*-dimensioned affine approximations was faster ($t^* = 4$ vs. $t^* = 12$). Perhaps our naive conclusion at the end of Section 5.4.4 was not that far off the mark: fewer Y dimensions and more W dimensions translates to a bigger and easier target for the search.

Reducing the number of dimensions of the empirical approximations, of the sets that explain the regular flow, also appears to be a good thing when we compare the left and right panels of the same two Figures. Searches with small β generally have a smaller solution time t^*. Because this is a useful thing to know (if true) and easy to test, here is a formal statement:

Flow limit conjecture

> The RRR solution time t^* has a decreasing limiting behavior in the $\beta \to 0$ limit.

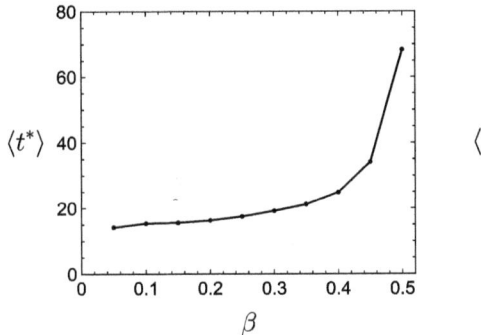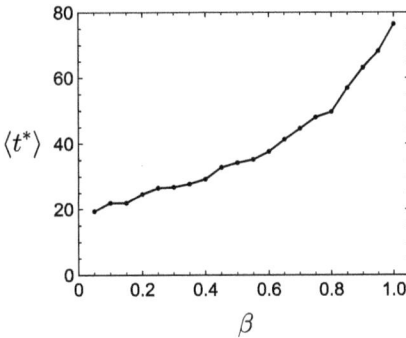

Figure 5.13 β-dependence of solution times averaged over the initial point for 16×16 Hadamard matrices (left) and a sudoku puzzle (right).

Figure 5.13 confirms the conjecture for 16×16 Hadamard matrices and the sudoku puzzle used previously. The solution times have been averaged over $1\,000$ random initial points.

Because there is a limiting solution time, the number of iterations t^*/β grows inversely with β for small β. Since one is always interested in minimizing the number of iterations, an intermediate β accomplishes this. These turn out to be $\beta = 0.35$ for 16×16 Hadamard matrices and $\beta = 0.7$ for the sudoku puzzle.

5.6 RRR and ADMM

In the alternate universe, where the author is invited to a conference on "Distributed Computing with Nonconvex Constraints" instead of a phase retrieval workshop, this chapter might have a very different acronym.

The *alternating direction method of multipliers* or ADMM algorithm, like RRR, has strong bipartisan origins and its three-term update rule looks remarkably like RRR when applied to the problem of finding a point in $A \cap B$. There is even a time-step parameter α analogous to the β of RRR. In this section we briefly provide the background for ADMM most relevant for its relationship to RRR and also show that ADMM with $\alpha = 1$ is equivalent to RRR with $\beta = 1$ (Douglas–Rachford).

The bipartisan character of ADMM is immediate from an application called the *consensus problem*:

$$
\begin{aligned}
\text{minimize:} \quad & f(x) + g(z) \\
\text{subject to:} \quad & x = z \, .
\end{aligned}
\tag{5.30}
$$

For the most part, our notation matches the notation in the ADMM review by Boyd et al. [13]. Like Douglas–Rachford, ADMM is mostly applied to convex problems and the functions f and g are convex. But also like RRR, the ADMM "rule" makes no reference to this fact.

The odd way of splitting the objective function into two parts signals the intention of splitting the optimization into independent subproblems. This is a much-valued feature in distributed computing, but also makes ADMM a nice framework for some standard applications. For example, if g is the indicator function for a set A,

$$g(z) = \left\{ \begin{array}{ll} 0, & z \in A \\ \infty, & \text{otherwise} , \end{array} \right.$$

then (5.30) is the constrained optimization problem

$$\text{minimize:} \quad f(x)$$
$$\text{subject to:} \quad x \in A .$$

The many mysterious terms in the acronym hint that the setup for ADMM is elaborate. A key ingredient is the *augmented Lagrangian*:

$$L_\rho(x, z; y) = f(x) + g(z) + (x - z)y^\mathsf{T} + (\rho/2)\|x - z\|^2 .$$

Here y is the dual or Lagrange multiplier variable that imposes equality of x and z. The quadratic penalty function with parameter $\rho > 0$ is introduced to improve convergence. This has no effect on the optimum, since its value is zero when the dual variable imposes $x = z$.

A second key ingredient is the update rule. Instead of jointly minimizing L_ρ with respect to x and z, the minimizations are performed sequentially, and followed by an update of the dual variable:

$$x' = \arg \min_x L_\rho(x, z; y) ,$$
$$z' = \arg \min_z L_\rho(x', z; y) , \qquad (5.31)$$
$$y' = y + \tilde{\alpha}(x' - z') .$$

The sign of $\tilde{\alpha} > 0$ follows from the fact that ADMM is also solving the dual problem, which is a maximization with respect to y of the dual objective, whose gradient is $x - z$. When performing one ADMM update, the three variables must be updated in the order given above.

The meaning of "alternating direction" is best understood by specializing to the case where both functions are indicator functions: f for set B and g for set A. After combining the dual-variable term with the penalty function and defining the rescaled variable $u = y/\rho$,

$$(x - z)y^\mathsf{T} + (\rho/2)\|x - z\|^2 = (\rho/2)\|x - z + (1/\rho)y\|^2 - (1/2\rho)\|y\|^2$$
$$= (\rho/2)\|x - z + u\|^2 - (\rho/2)\|u\|^2 ,$$

the x update takes on a simple form:

$$x' = \arg\min_x \left(f(x) + (\rho/2)\|x - z + u\|^2 \right)$$
$$= P_B(z - u) .$$

This is just the projection to the set B from the point $z-u$ and is independent of ρ. The z update is similar

$$z' = P_A(x + u) ,$$

but differs in the sign applied to u. Summarizing, $x' \in B$, $z' \in A$, and the two projections take each variable to the other but only after first applying shifts u with "alternating directions."

To facilitate the comparison with RRR, we write the ADMM update rule in terms of variables $x_B = x$, $x_A = z$, and with a shift in the definition of the start of the update cycle:

$$
\begin{aligned}
x_A &= P_A(x_B + u) , \\
u' &= u + \alpha(x_B - x_A) , \\
x_B' &= P_B(x_A - u') .
\end{aligned}
\tag{5.32}
$$

We've also defined a rescaled dual-variable update rate, $\alpha = \tilde\alpha/\rho$. The updates in (5.32) have the same cyclic order as the updates in (5.31). This way of writing the rule also shows there are really just two independent ADMM variables, x_B and u, because these explicitly define x_A. At a fixed point, $u' = u$ and $x_B - x_A = 0$ means we have found a point in $A \cap B$.

To see that ADMM with $\alpha = 1$ is equivalent to Douglas–Rachford, start by defining

$$\tilde{x} = x_B + u .$$

Now using the first two updates of (5.32) and $\alpha = 1$, the dual variable update takes the following form:

$$
\begin{aligned}
u' &= (u + x_B) - x_A \\
&= \tilde{x} - P_A(\tilde{x}) .
\end{aligned}
$$

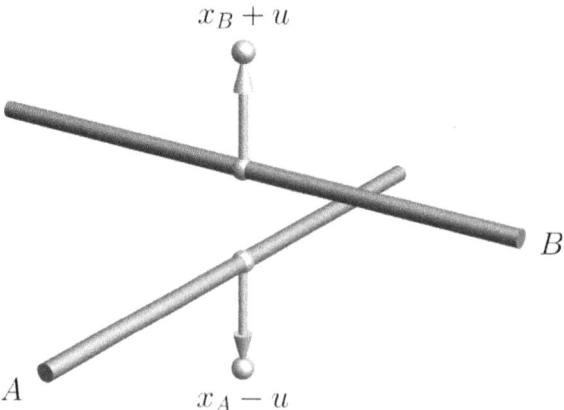

$x_B + u$

B

A

$x_A - u$

Figure 5.14 The ADMM dual vector u grows at near solutions, pushing the projection inputs away from the near solution in opposite directions.

Using this and the update of x_B, we can obtain the update of $\tilde{x}$:

$$\begin{aligned}
\tilde{x}' &= x'_B + u' \\
&= P_B(x_A - u') + \tilde{x} - P_A(\tilde{x}) \\
&= \tilde{x} + P_B\big(P_A(\tilde{x}) - \tilde{x} + P_A(\tilde{x})\big) - P_A(\tilde{x}) \\
&= \tilde{x} + P_B(R_A(\tilde{x})) - P_A(\tilde{x}) \ .
\end{aligned}$$

This is the Douglas–Rachford update of the variable $\tilde{x}$. ADMM still differs from DR in having two initial vectors, $x_B(0)$ and $u(0)$. It is only for $\alpha = 1$ that the evolution of the sum, $x_B + u$, depends only on $x_B(0) + u(0)$.

For nonconvex problems, ADMM with $\alpha < 1$ is interesting for the same reasons RRR is interesting for $\beta < 1$. Figure 5.14 shows what ADMM is doing at a near solution, when the two affine approximations of A and B (rendered as rods) do not intersect. The projections x_A and x_B have converged on a pair of proximal points while the inputs to the two projections, $x_B + u$ and $x_A - u$, move ever further from these points in opposite directions. The scaled dual vector u can be interpreted as the "accumulated discrepancy," which grows with rate α in the direction of the constraint discrepancy $x_B - x_A$. The algorithm extricates itself from the proximal-point trap when the discrepancy u has accumulated to a degree such that x_A is no longer the point on A nearest to $x_B + u$ or similarly for x_B.

Of the two variables whose initial values can be specified, x_B and u, the scenario just described suggests that x_B has the more direct interpretation as a search variable. The role of the dual variable u is to track the relative position of the proximal point on A. When sampling initial points for ADMM, it therefore suffices to sample $x_B(0)$ and set $u(0) = 0$.

In highly nonconvex problems, such as when A or B are discrete, the cartoon sets shown in Figure 5.14 represent the best affine approximations of distributions that get sampled stochastically by the algorithm's chaotic dynamics. In the $\beta \to 0$ limit of RRR, the distributions being sampled approach stationary distributions and the quality of the affine approximations is improved. We expect the same phenomenon for ADMM when the growth of u is slowed with a small α.

Exercises

5.1 Use an iterative fixed point scheme to fill in the blanks with numbers:

In this sentence,

the digit 0 appears ____ times;
the digit 1 appears ____ times;
the digit 2 appears ____ times;
the digit 3 appears ____ times;
the digit 4 appears ____ times;
the digit 5 appears ____ times;
the digit 6 appears ____ times;
the digit 7 appears ____ times;
the digit 8 appears ____ times;
the digit 9 appears ____ times.

For some initializations you may find that your iteration scheme ends up trapped in a cycle!

Credit: N.D. Mermin

5.2 If A is an affine set, then the reflection R_A is an involution. Are there nonaffine sets whose reflections are also involutions?

 (a) Show that R_A is *not* an involution both when A is a (nonconvex) sphere and a (convex) ball.
 (b) Show that reflection in the nonconvex set

$$A_k = \left\{ x \in \mathbb{R}^{m \times m} : \operatorname{rank}(x) = k \right\}, \ k > 0$$

is an involution with probability 1 (ignoring elements in the set of measure zero whose reflections are not unique).

5.3 The fact that iterates under the Douglas–Rachford algorithm move away from a proximal pair $a \in A$, $b \in B$ (along the axis joining them), no matter how small their separation $\|a - b\|$, did not sit well with applied mathematicians concerned about noise. To mitigate the problem, Russell Luke proposed the *relaxed averaged alternating reflections* (RAAR) iteration rule [71]:

$$x \mapsto \operatorname{RAAR}(x) = (\beta/2)\big(x + R_B(R_A(x))\big) + (1 - \beta)P_A(x) \,.$$

The parameter β is unrelated to the β of RRR and has the restriction $0 < \beta < 1$.

 (a) Show that RAAR is similarity invariant.
 (b) For the simple case $A = \{a\}$, $B = \{b\}$ (a pair of points in m dimensions), obtain an explicit formula[1] for $x_n = \operatorname{RAAR}^n(x_0)$.

5.4   Consider the iteration rule[2]

$$x \mapsto x + \beta\left(P_B\big((1-\lambda)P_A(x) + \lambda x\big) - P_A\big((1+\lambda)P_B(x) - \lambda x\big)\right),$$

which reduces to RRR for $\lambda = -1$.

(a) Is this generalization similarity invariant?

(b) Repeat the local analysis of Section 5.4 and show that the only thing that changes is the matrix

$$r = pq - qp + \lambda(p-q)^2$$

that acts on the vector $\tilde{y}$ in the transverse space.

(c) Next show that the first part of the quadratic form for the change in the norm is unchanged, with λ replacing the -1 in (5.24):

$$Q_1(\tilde{y}) = 2\beta\,\tilde{y}\,r\,\tilde{y}^\mathsf{T}$$
$$= 2\beta\lambda\,\tilde{y}(p-q)^2\tilde{y}^\mathsf{T}\ .$$

(d) The second part of the change in the transverse norm is itself a norm:

$$Q_2(\tilde{y}) = \beta^2\|\tilde{y}\,r\|^2\ .$$

Using the identity

$$pq - qp = (p-q)(p+q-I)\,,$$

and the triangle inequality, show that

$$\|\tilde{y}\,r\| \le \|v(p+q-I)\| + |\lambda|\,\|v(p-q)\|\,, \qquad (5.33)$$

where $v = \tilde{y}(p-q)$.

(e) The first term in (5.33) has the bound

$$\|v(p+q-I)\| \le \sigma\|v\|\,,$$

where

$$\sigma = \max\left(|\operatorname{spec}(p+q-I)|\right)\,.$$

Show that $\sigma \le 1$.

(f) Use the triangle inequality on the second term of (5.33) to derive the bound

$$\|v(p-q)\| \le 2\|v\|\,.$$

(g) Combine the previous parts to show

$$Q(\tilde{y}) = Q_1(\tilde{y}) + Q_2(\tilde{y}) \le \left(2\beta\lambda + \beta^2(1+2|\lambda|)^2\right)\|v\|^2\ .$$

(h) As in the proof of Theorem 5.3, there are no nonzero $\tilde{y}$ for which $v = 0$. Transverse contraction, or $Q(\tilde{y}) < 0$, is therefore ensured by the inequality

$$2\beta\lambda + \beta^2(1 + 2|\lambda|)^2 < 0 .$$

Show that there are two branches, one of which is

$$\lambda < 0, \quad \beta \in \left(0, \frac{2|\lambda|}{(1 + 2|\lambda|)^2}\right) .$$

The other branch, obtained by reversing the signs of β and λ, is equivalent to swapping A and B in the definition of the rule.

5.5 Construct the RRR velocity field $v(x, y)$ for the case of two intersecting lines in two dimensions. Let A be the line $(\sin\alpha)x - (\cos\alpha)y = 0$, and B the line $y = 0$.

(a) Start by working out explicit forms for the two projections,

$$P_A(x, y) = \cdots ,$$
$$P_B(x, y) = \cdots ,$$

and from these obtain $v(x, y)$.

(b) Integrate v to obtain explicit orbits $(x(t), y(t))$ for arbitrary $(x(0), y(0))$. Compare your results with Figure 5.3.

(c) Work out the behavior of the norm,

$$x^2(t) + y^2(t) .$$

Interpret your result in terms of the matrix r (5.18) in the local analysis for these sets. For which intersection angle α is transverse contraction greatest?

5.6 Construct the RRR flow in two dimensions (x, y) where A is the pair of points $(0, \pm 1)$ and B is the line $y = 0$. Your streamlines will be a hybrid of those in Figure 5.4.

(a) By using the analytical forms of the orbits in the two regions of point-line flow, show that all orbits converge to the point $(0, 0)$.

(b) Show that $\|v(x)\|$ (the gap) converges to a positive number.

(c) Repeat for the case where the point-pair, A, is $\pm(1, 1)$. This is the flow in the left panel of Figure 5.5 when the middle point is removed.

5.7 Construct the simplest example of flow for discrete constraints in two dimensions where there is attraction not just to one-dimensional domain boundaries (edges), but also the zero-dimensional boundary (point) where the one-dimensional boundaries meet. In any number of dimensions, if the iterates are ever attracted to a zero-dimensional domain boundary, the search is effectively over.

5.8 Suppose A and B are discrete sets in the space X, but that for x in some domain $X^* \subset X$ only the two-element subsets $A = \{a_1, a_2\}$, $B = \{b_1, b_2\}$ are relevant for the RRR dynamics because all other elements of A and B have a much greater distance to x. Moreover, there is a local symmetry in that the point-pairs have a common centroid:

$$\frac{1}{2}(a_1 + a_2) = c = \frac{1}{2}(b_1 + b_2).$$

Assume that the vectors $u = a_1 - c$ and $v = b_1 - c$ are not co-linear and span a two-dimensional subspace W of X.

(a) If we write $X = W \times Z$, where Z is the orthogonal complement of W in X, show that Z is no different from the Z we defined for the locally affine case, that is, $\mathcal{R}(w, z) = (w', z)$ for all $z \in Z$.

(b) Construct the piecewise-constant flow in the W-plane. Without loss of generality, assume $u \cdot v > 0$. Show that for all initial points, the flow converges on a line segment centered on c.

(c) Show that for finite β the RRR iterates eventually end up in a 2-cycle in which the velocity v is reversed in every iteration (and the gap is constant). Finally, show that this holds even for the special case $u \cdot v = 0$.

6

The User's Guide

Let's say the easy part is behind you. You've found a novel bipartisan formulation with sets A and B and can show that an element of $A \cap B$ solves your problem. You've also worked out the projections to these sets and feel confident that your projection code does what it's supposed to do. The last step, of coding an RRR update with your P_A and P_B, is the same as in your previous, presumably successful, application of the algorithm. What can go wrong?

Users of RRR learn to live with two forms of anxiety. The first stems from novel sets and their projections. It's quite possible no one has ever needed to project to covariance-distribution-constraints-on-chiral-simplices. One error in your projection code will derail the whole project. On a higher level, there's the nagging fact that there are no convergence guarantees when constraints are nonconvex. When the gap is not decreasing with every iteration, as you've come to expect in hard problems, there's always the possibility of trapping.

This chapter tries to help you manage both of these anxieties. Observation plays a major part. Rendering x and its projections is an art, but an indispensable tool for detecting bugs in code. Observing how x changes from iteration to iteration will also let you assess, in qualitative terms, whether the search is going well. And by observing changes in the evolution induced by modified constraints and parameters, you will be in a better position to improve the algorithm.

6.1 Getting Started

The piece of code you should always write first is the function that generates the initial point. That will force you to declare the data structure that gets iterated, the object denoted by x throughout this book. As the applications in Chapter 7 will make clear, x can be much more elaborate than the rectangular arrays (vectors, matrices) that suffice for most "textbook" problems. As you establish the shape of x in the initial-point code, you will be looking ahead to the two projections that act on x. Some shapes make more sense than others. Also, the shape that works best for P_A may complicate P_B, and vice versa. In any case, the exercise of generating the initial point will prepare you for what lies ahead.

Adding the option for the initial-point code to use an x from input provides a convenient way to test your projections. If the x you input is the solution x_{sol} to a designed problem, and the initial point code creates $x(0)$ by applying a perturbation to x_{sol}, you can let RRR compute some iterations and look for convergence. From the local analysis in Section 5.4.4 we know that if x_{sol} is a regular point and the perturbation is sufficiently small (so the locally affine approximations of A and B are good), the gap will shrink exponentially in the number of iterations. This critically tests both projections, not only that their outputs are on the true sets, but that these are distance minimizing to the inputs.

The not completely random starting point of this guide is the application that launched this book (Section 2.1).

6.1.1 Bug Retrieval

You've been given a 128×128 array I of diffraction intensities obtained by shining a laser on "a bug." By making a rough contour plot of its Fourier transform $\hat{I}$, shown in Figure 6.1, you have an idea what size of bug you are dealing with. This is the bug's autocorrelation, and the radius of the autocorrelation support is an estimate of the bug's diameter. Precise numbers are not required here. You mostly want to confirm that the bug will fit inside a modest support region without wrap-around (on the 128×128 grid). It looks like a 60×60 support $\mathcal{S}$ will be generous.

Though the x for this problem is a real, 128×128 array of contrast values, you might as well define your initial-point function so that it puts positive contrast only in the support region $\mathcal{S}$. Since the contrast grid is periodic, it does not matter where $\mathcal{S}$ is placed. The upper left-hand corner is a convenient choice. The left panel of Figure 6.2 shows a random initial point, $x(0)$. In the large, uniformly gray region, $x(0)$ is zero. Constraint B will continue to impose this property, as well as nonnegativity of the contrast in $\mathcal{S}$.

The initialization of the contrast in the support region seems outrageous in its simplicity: independent, uniform samples in each pixel. Surely the contrast should drop off near the edges of $\mathcal{S}$, and adjacent pixels ought to be strongly correlated. While these are things we expect in the true contrast, we are going

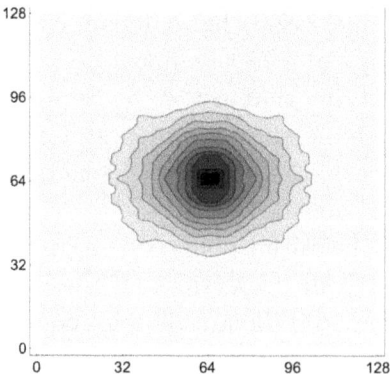

Figure 6.1 A bug autocorrelation $\hat{I}$, obtained by Fourier transforming a bug diffraction intensity I. Since the support of $\hat{I}$ does not span the 128×128 grid, the bug's contrast will fit inside a support region $\mathcal{S}$ of size smaller than 64×64.

to rely entirely on constraints A and B to make that happen. Indeed, "rough" initial points will give us greater confidence in reconstructions, assuming they are reproduced in different runs of the algorithm.

The middle panel of Figure 6.2 shows $P_A(x(0))$, the result of projecting $x(0)$ to be consistent with the diffraction intensities. We can think of this as a forward-backward virtual experiment. In the forward experiment, we Fourier transform $x(0) \to \hat{x}(0)$ to see what the diffraction amplitude would be if $x(0)$ was indeed the true bug. Of course we only know the magnitudes $\sqrt{I}$ of the true amplitudes, so the most we can do is fix that part of $\hat{x}(0)$ by rescaling the magnitudes to agree with $\sqrt{I}$. The rescaling is followed

 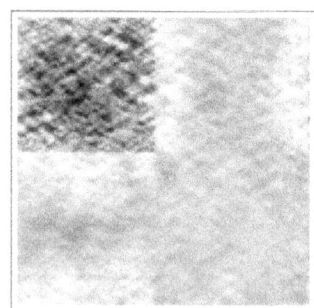 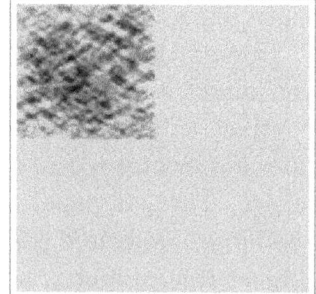

Figure 6.2 Left to right: $x(0)$, $P_A(x(0))$, $P_B(P_A(x(0)))$. The gray contrast outside the support in the left and right panels is zero; negative contrast in the middle panel is rendered white.

by the backward experiment, an inverse Fourier transform, that gives us a new bug contrast which we identify with $P_A(x(0))$. It's a projection because magnitude rescaling is the distance-minimizing way to impose the intensity constraints.

Notice that $P_A(x(0))$ has strong violations of the B constraint, and therefore is not our bug. Not only is the contrast nonzero outside the support, there are negative pixels inside S (white regions). Turning to a good thing about $P_A(x(0))$, notice that the contrast is smoother than $x(0)$. That comes from the fact that the imposed Fourier magnitudes are very small for the high spatial frequencies.

We can test the other projection by applying P_B to $P_A(x(0))$. This is shown in the right panel. The support constraint is restored, and the negative contrast inside S is set to zero (white → gray). RRR does *not* do alternating projections, or $P_B(P_A(x(0)))$, but generating the three images in Figure 6.2 is a good way to test the projection code.

Phase retrieval with known support is not considered to be all that hard, so we will try RRR with $\beta = 1$, or Douglas–Rachford. One question remains: How should we monitor the progress of the algorithm? The time series of the gap will let us know when things are working, but is not very helpful when diagnosing problems. In phase retrieval, we are lucky in that x and its associated projections $x_A = P_A(x)$ and $x_B = P_B(R_A(x))$ are directly meaningful. After all, we are doing microscopy and either x_A or x_B, when they converge, are the contrast of our bug. We will use x_B because then we only have to render the contrast in the support.

Figure 6.3 shows x_B at 50-iteration intervals. After 500 iterations (first 10 frames), the contrast shows no sign of settling down, at least on what appear to be the "legs." The "body" itself is relatively constant but not well centered in the support. We also know, from general principles, that reconstructions are translationally nonunique. Could it be that the legs on the left side need more room to properly reconstruct, and that the body is centering itself far too slowly to allow that to happen?

Acting on the poor-centering hypothesis, we make the following intervention. First, we enlarge the support from 60×60 to 70×70 (keeping it in the upper-left corner). Second, we translate x so the region defined by the original S is centered in the new S. The last five frames in Figure 6.3 show how the reconstruction develops with the enlarged support. After two frames (100 iterations) we see the legs too have settled down, confirming our hypothesis. Our bug is a spider that appears to be defying the laws of gravity. We expect anti-gravity spiders in half of our reconstructions because of inversion nonuniqueness.

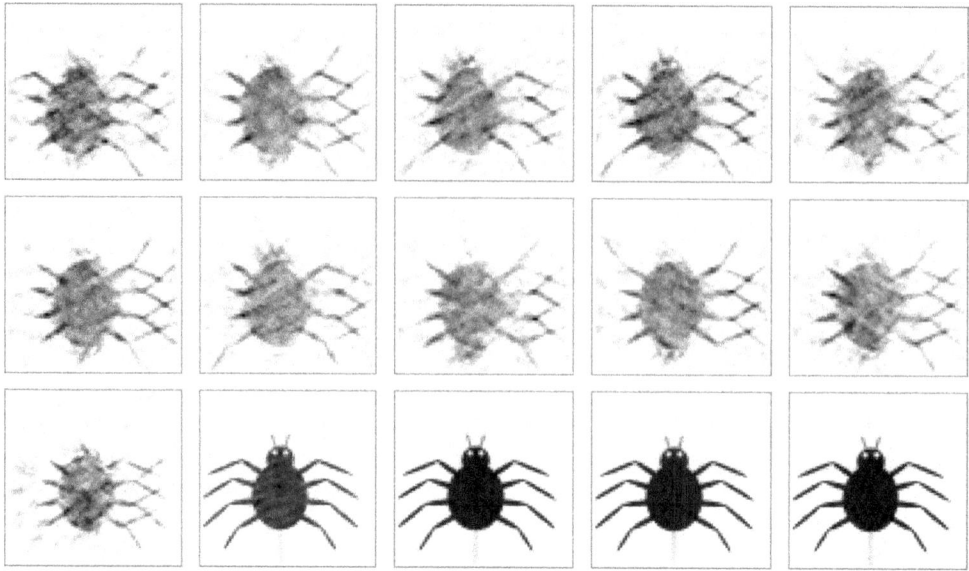

Figure 6.3 Reconstructing a spider from its diffraction intensities with RRR. All frames show the contrast within the support, whose size in pixels was slightly expanded in the last five frames.

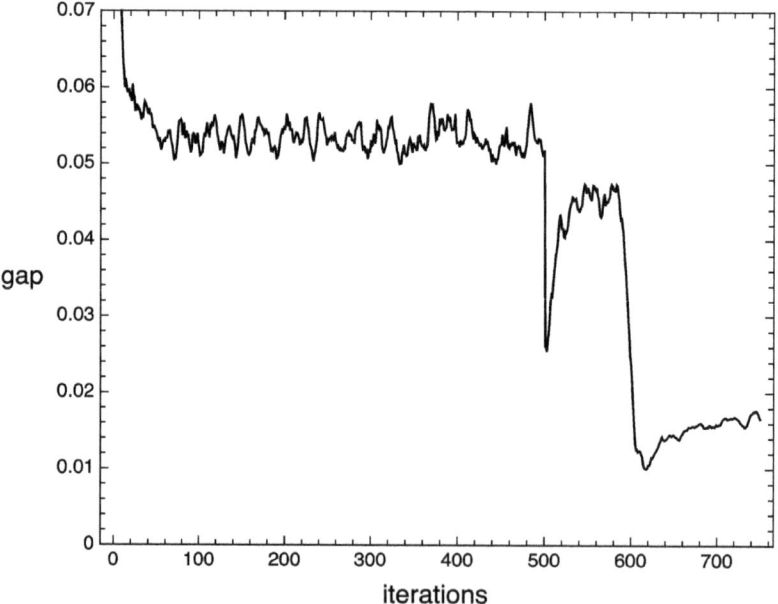

Figure 6.4 Evolution of the gap that goes with the spider reconstruction above, where frames are separated by 50 iterations. The support was expanded at iteration 500.

The evolution of the reconstruction is reflected in the plot of the gap in Figure 6.4. The sharp drop in the gap right after the intervention comes from the weakening of the B constraint. That it then rebounds is a nice display of bipartisanship: finding the true solution remains a negotiation between competing interests. Noise in the data – or a subversion of the support constraint by the spider silk! – might explain why the gap settles down to a small positive value.

The refinement of the support region can be automated. A popular method, called *shrink-wrap* [72], does this by updating $\mathcal{S}$ using the low-pass filtered contrast. To preserve the fixed-point convergence of RRR, the $\mathcal{S}$-updates must be adiabatic, or small over one RRR iteration.

We were lucky that in bug retrieval we had a convenient way to diagnose the origin of trapping (a spider literally trapped against the edge of the support). The evolution of the gap was no help in that regard. But as we will see in Section 6.2, some forms of trapping are obvious from the behavior of the gap.

6.2 Tuning the Constraints

Recall the bipartisan formulation of the set-cover problem (Section 2.4). The variables live on the edges of a bipartite graph and the A constraints are imposed independently on edges incident to each of the antenna nodes $a \in \mathcal{A}$. In addition to taking values 0 or 1, at least one incident edge must have value 1 in order for antenna a to be "covered." On the side of the broadcast nodes, $\mathcal{B}$, there are two constraints that comprise B. First, all edges incident to each $b \in \mathcal{B}$ must have the same value, $\bar{x}(b)$. The second constraint is a side constraint on these concur values that bounds the size c of the cover.

Let $m = |\mathcal{B}|$ be the number of broadcast nodes, and

$$\bar{x} = \big(\bar{x}(b_1), \ldots, \bar{x}(b_m)\big)$$

the vector of concur values. In the formulation of Section 2.4, the side constraint was the inequality

$$B_{\text{sum}} = \big\{\bar{x} \in \mathbb{R}^m : \textstyle\sum_{i=1}^{m} \bar{x}(b_i) \leq c\big\}.$$

Here is a stronger form of this constraint:

$$B_{\text{binary}} = \big\{\bar{x} \in \mathbb{R}^m : \bar{x} \in \{0,1\}^m, \ \textstyle\sum_{i=1}^{m} \bar{x}(b_i) \leq c\big\}.$$

And here is yet another proposal for the side constraint:

$$B_{\text{sparse}} = \big\{\bar{x} \in \mathbb{R}^m : \bar{x} \in \mathbb{R}_c^m\big\}.$$

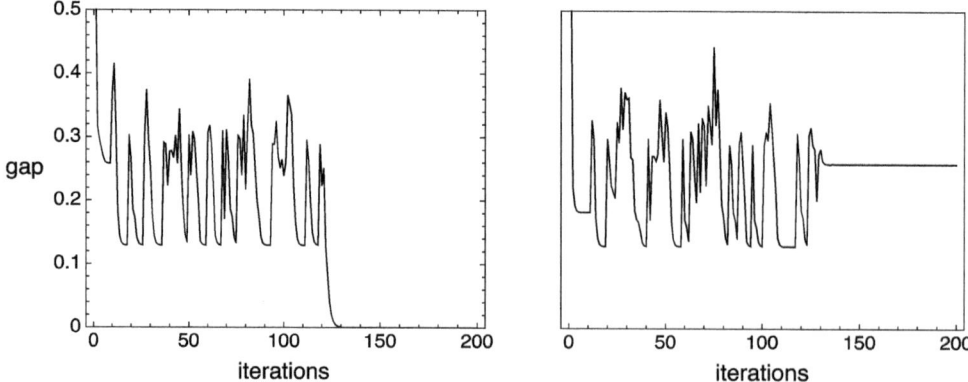

Figure 6.5 Two runs with constraint B_{sparse} on the $n = 6$ Turán instance and $\beta = 0.5$. The "flatlined" (positive) gap in the run on the right is a sign of symmetric trapping.

Recall that $\mathbb{R}^m_c$ is our notation for c-sparse vectors in $\mathbb{R}^m$, the union of m-choose-c Cartesian subspaces. Because the A constraint imposes indicator values on all the variables, in a solution the concur vector $\bar{x}$ is binary and all three options, B_{sum}, B_{binary}, and B_{sparse}, are the same constraint on that binary vector. However, these three variants of constraint B are quite different geometrically, and for that reason may have a profound effect on the *search* for solutions.

When comparing constraints, it's good to experiment with a family of instances that span a range of difficulty. For comparing the set-cover B constraints, we will use the graphs $\mathcal{G}(n, 4, 3)$ for the Turán numbers, with conjectured minimum cover sizes (2.14) when n is a multiple of three. Already for the easiest instance in this family, $n = 6$, constraint B_{sparse} has a trapping problem. Figure 6.5 shows two runs with $\beta = 0.5$. In one run the solution is found in 130 iterations, while the gap has "flatlined" at a positive value in the other.

A repeating positive gap can arise when there is a local symmetry in the constraints. In the simplest example, $\tilde{A} = \{a_1, a_2\}$ comprises two points and $\tilde{B} = \{(a_1 + a_2)/2\}$ is the midpoint. Think of these as subsets of A and B that apply in a small domain, far from all other points of A and B. The piecewise constant flow is orthogonal and convergent to the hyperplane H that separates the Voronoi domains of the two points of $\tilde{A}$. After arriving at H, x hops from one side of H to the other while reversing the sign of the velocity, $v = \pm(a_1 - a_2)/2$. Since $\|v\|$ is the gap, the explanation of the flatlined gap is trapping on a symmetric cycle of period two.

Something similar happens when, again locally, the point $(a_1 + a_2)/2 = b$

| | $\beta = 0.2$ | | $\beta = 0.5$ | |
	iterations	success (%)	iterations	success (%)
B_{sum}	145	96	56	98
B_{binary}	400	99	175	99
B_{sparse}	81	4	144	96

Table 6.1 Comparison of the three options for the B constraint in the $\mathcal{G}(6,4,3)$ set-cover problem. B_{sparse} is plagued with trapping for small β.

of $\tilde{B}$ is replaced by a hyperplane of positive dimension that passes through b. Since this $\tilde{B}$ is continuous, the trapping is on a *limit cycle* of period two. The limiting flatness of the gap is again a manifestation of the symmetry of the local constraints.

A different kind of local constraint symmetry is responsible for the positive flatlined gap in Figure 6.5. The trapping dynamics plays out in just the $m = 20$ concur values, and in fact just a $k = 7$ dimensional subspace (the other 13 components of $\bar{x}$ are static). The relevant part of the B constraint is therefore $\tilde{B} = \mathbb{R}^k_{k-1}$, since $c = 6 = k - 1$ in this set-cover instance. The relevant part of A is just a single point, which happens to satisfy the concur constraint with value 1 on all the variables singled out by the k-dimensional subspace. Consequently, $\tilde{A}$ is just one k-tuple of 1's.

The RRR dynamics for the $\tilde{A}$ and $\tilde{B}$ just defined is analyzed in Exercise 6.3. The iterates fall on a limit cycle of period 7. Over the course of one cycle, each of the seven variables in turn is set to zero in the 6-sparsity constraint. But since all are nonzero (with value 1) in $\tilde{A}$, the local constraints offer only a near solution. Though the velocity v executes a 7-cycle, the gap $\|v\|$ is constant because of symmetry and hides the true period of the cycle.

The extent of the trapping problem with constraint B_{sparse} is exposed at small β. Table 6.1 gives the solution success rates and mean iteration counts (of the successful runs) in 1 000 trials for the three side constraints. Less than 100% success implies trapping because the iteration limit was set at 10^5. Whereas the trapping rate is just a few percent for all three with $\beta = 0.5$, trapping is a near certainty in the case of B_{sparse} with $\beta = 0.2$. By contrast, constraints B_{sum} and B_{binary} have negligible trapping at this β and exhibit the β-behavior we like to see: mean iterations scaling inversely with β (a β-independent solution time). Since $t_{\text{sum}} = 29$ is smaller than $t_{\text{binary}} = 80$, constraint B_{sum} is slightly favored.

We could declare constraint B_{sparse} acceptable, provided β is not too small. But that goes against the logic of the flow limit conjecture (Section 5.5.8). In

	$\beta = 0.2$		$\beta = 0.5$	
	iterations	success (%)	iterations	success (%)
B_{sum}	5 900	100	1 400	100
B_{binary}	3 500 000	87	3 200 000	93
B_{sparse}	–	0	4 100 000	59

Table 6.2 Comparison of the three options for the B constraint on the next set-cover instance, $\mathcal{G}(9, 4, 3)$. Constraint B_{sum} is favored by a large margin.

case you need reminding, that's the conjecture that the most direct route to the solution is achieved when the empirically defined affine approximations of A and B (responsible for the evolution of $\langle x \rangle$) are derived from near-stationary distributions. The latter are realized in the $\beta \to 0$ limit of the actual algorithm with sets A and B. However, because of trapping all flow limit arguments are void for B_{sparse}. In fact, problematic constraints like B_{sparse} are made worse in the flow limit because the local subsets $\tilde{A}$ and $\tilde{B}$, that account for the trapping, are valid local models precisely when the iterate makes small steps.

As stated earlier, we should withhold judgment on constraints until a range of problem instances have been tested. Table 6.2 gives results for the next Turán number instance, with $n = 9$ and cover size $c = 30$. The iteration limit was increased to 10^7, though we see this was hardly necessary for constraint B_{sum}. Whereas B_{binary} continues to have no trapping problem, B_{sum} is clearly the better constraint!

If we restrict B_{sum} and B_{binary} to the equality case on the cover size – surely the most active part of these constraints – then the former is the *affine relaxation* of the latter. This and the clear superiority of B_{sum} in finding covers suggests the reasonable general principle that affine (or convex) relaxations should be used whenever this does not change the feasible set $A \cap B$. Our next example shows that this principle is not at all reliable!

Figure 6.6 shows two sudoku puzzles that are difficult for RRR for very different reasons. The puzzle on the left, nicknamed `escargot` by its creator Arto Inkala, was claimed to be the hardest puzzle ever designed when first published [57]. Not everyone agrees, mostly because there is no universally accepted yardstick for hardness. Still, no one disputes that `escargot` is much more difficult than the hardest newspaper puzzles. It is also very challenging for RRR. In the formulation where both A and B are discrete, there is no trapping and the solution-time minimum, realized at small β, is 140. The mean solution time for NYT "hard" puzzles is only about 24.

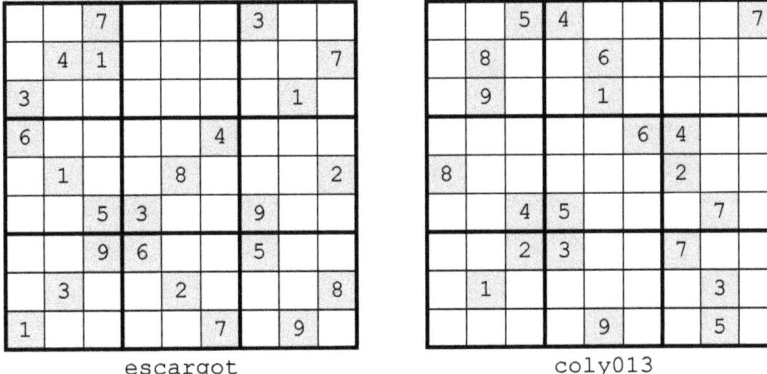

escargot   coly013

Figure 6.6 Two difficult sudoku puzzles.

In the fully discrete sudoku formulation, each factor of A and B has the following form:

$$P_9 = \left\{ x \in \mathbb{R}^{9\times9} : \ \forall(i,j) : \ x_{ij} \in \{0,1\}, \ \textstyle\sum_{k=1}^{9} x_{ik} = 1, \ \sum_{k=1}^{9} x_{kj} = 1 \right\} .$$

These are the 9×9 permutation matrices. The affine relaxation[1] is the set of matrices with all row and column sums equal to 1:

$$\bar{P}_9 = \left\{ x \in \mathbb{R}^{9\times9} : \ \forall(i,j) : \ \textstyle\sum_{k=1}^{9} x_{ik} = 1, \ \sum_{k=1}^{9} x_{kj} = 1 \right\} .$$

Suppose we use the relaxation $\bar{P}_9$ in all the factors of B, and call the resulting constraint set $\bar{B}$. It's easy to see that $A \cap B = A \cap \bar{B}$, because A imposes $0/1$ values on the $9 \times 9 \times 9$ cube of variables. The general principle, that affinely relaxed constraints (for the same feasible set) are superior to their discrete counterparts, would then predict a reduction in solution time when using $\bar{B}$ instead of B. However, escargot convincingly demonstrates that such a principle does not extend to sudoku. The solution time when using $\bar{B}$ is 34 000 – more than a 200-fold increase over the time when both A and B are discrete!

The hard set-cover and hard sudoku examples show that the only reliable guide for choosing constraints is experiment. For set-cover, an argument for a continuous B is that the concur vector $\bar{x}$ holds valuable information that is lost when projected to a binary vector. But when applied to sudoku, the same heuristic recommends continuous arrays (with row/column sums equal to 1) over binary arrays – which turned out to be a poor choice. The affine relaxation is probably too generous in this case, because it allows, and the projection exploits, negative array elements. The weaker relaxation, which replaces $x_{ij} \in \{0, 1\}$ with $x_{ij} \geq 0$, defines a convex set and would be a better continuous alternative.[2]

Trapping in the fully discrete formulation of sudoku occurs in some, but not all, of the very hard puzzles. The second puzzle[3] in Figure 6.6, coly013, is an extreme example. Trapping, with a flatlined gap, occurs in about 94% of solution attempts. The mean solution time in the lucky successful attempts is 45 (far below escargot). It seems a shame to modify our discrete A and B, which work so well in most puzzles, just to avoid trapping in a handful of cases. Fortunately, a very different kind of life-saving intervention is introduced in Section 6.3.

6.3 Tuning the Metric

The metric can have a huge effect on the behavior of RRR. A good metric can mean the difference between never finding solutions and finding them consistently. In this section we look at some examples of the easiest kind of metric intervention, the *diagonal generalization*.

Recognizing when a diagonally generalized metric makes sense is the first step. In some problems, it's not even possible to apply this idea. The best example is phase retrieval with a known support. If we used nonuniform metric coefficients on the contrast pixels, we could still project to the support constraint (with or without nonnegativity). In fact, that projection is completely unaffected by the modified metric. On the other hand, it would be very difficult to project to the diffraction intensity constraint. That's because this involves a Fourier transform and its inverse, both of which fundamentally preserve the isotropic metric (Parseval's theorem). The rescaling trick (Section 5.2) to restore the isotropic metric is also no help, because the Fourier magnitude constraint is not simply rescaled by this action.

A simple criterion for considering metric modifications is the feasibility of computing the two projections. If these are only minimally changed by the diagonal generalization, then that's usually a good reason to explore that option. Because our favorite formulations of set-cover and sudoku easily satisfy this criterion, we demonstrate diagonal metric tuning for these applications of RRR.

6.3.1 Asymmetry

The symmetry of the variables in a problem should always be the first consideration when deciding on a metric. In set-cover this translates to the

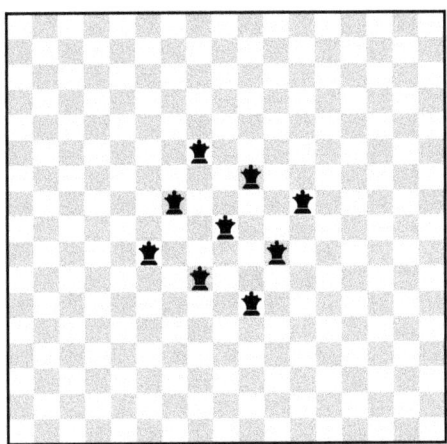 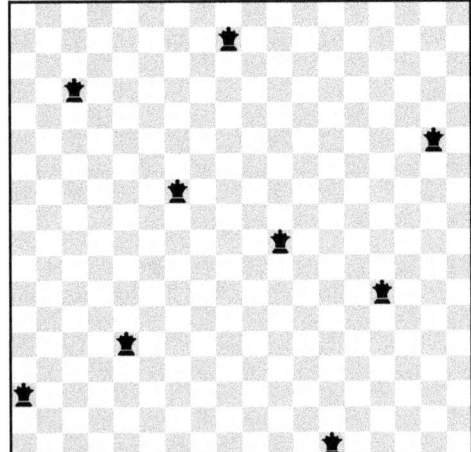

Figure 6.7 Nine queens dominating a 17×17 chessboard on the left, and a 18×18 board on the right. The queens on the right also dominate a 17×17 board (remove the unoccupied squares in the top and right rows).

symmetry of the problem instance, that is, the symmetry of the bipartite graph.

The graphs $\mathcal{G}(n, k, r)$ associated with the Turán numbers exemplify one extreme, where the graph has so much symmetry that an isotropic metric seems like the only reasonable option. Any permutation of the n symbols from which the nodes are built induces a symmetry transformation of the graph. As you can work out in one of the exercises, this group action is transitive on the edges of the graph (and the associated variables). All the variables in a Turán set-cover instance are, in a technical sense, the same kind of variable.

Bipartite graphs with much lower symmetry arise in the problem of *dominating sets*. This problem is defined for general graphs and makes no direct reference to a bipartite structure. But before we develop the relationship with bipartite graphs, let's look at a dominating set problem from Queensland.

Figure 6.7 shows two configurations of nine queens, one where the queens "dominate" a 17×17 chessboard, and another where the size of the board is 18×18. Every square of the boards is dominated, either by the attack or the occupation of a queen. These board sizes cannot be dominated by fewer than nine queens.

The *queen's graph* Q_n has n^2 nodes, corresponding to the squares of the $n \times n$ chessboard, and has edges between all nodes related by a queen-move. A dominating set, more generally for a graph $\mathcal{G}$ with nodes $\mathcal{N}$ and edges $\mathcal{E}$,

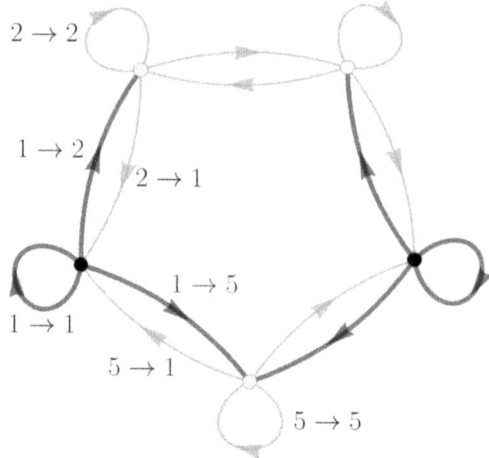

Figure 6.8 The doubled pentagon graph, with a loop added at each node, can be interpreted as a bipartite graph of five antenna and five broadcast nodes, and with three out-edges at each broadcast node. In this set-cover solution, for a cover of size two, all the out-edges at two nodes are black. Edge variables are 0 on the gray edges, 1 on the black edges.

is a subset $\mathcal{D} \subset \mathcal{N}$ such that every element of $\mathcal{N}$ is either an element of $\mathcal{D}$ or adjacent, by one of the edges in $\mathcal{E}$, to an element of $\mathcal{D}$. The minimum size of $\mathcal{D}$ is the *domination number* of $\mathcal{G}$.

The construction of the bipartite graph $\mathcal{G}_2$ associated with a general graph $\mathcal{G}$ is demonstrated for the case of the pentagon graph in Figure 6.8. There are two copies of each node $i \in \mathcal{N}$ in the bipartite graph, with symbol $\to i$ on the antenna side, and $i \to$ on the broadcast side. The edges $(i, j) \in \mathcal{E}$ are also doubled, and given opposite orientations, $i \to j$ and $j \to i$, in the edges $\mathcal{E}_2$ of $\mathcal{G}_2$. Starting at any broadcast node and following an out-arrow, one arrives at an antenna node. In addition to the doubled edges from $\mathcal{E}$, $\mathcal{E}_2$ also has edges $i \to i$ from the broadcast copy to the antenna copy of each node $i \in \mathcal{N}$.

The set-cover interpretation of the bipartite graph $\mathcal{G}_2$ is equivalent to the dominating-set interpretation of $\mathcal{G}$. A node i of $\mathcal{G}$ can be dominated by itself, $i \in \mathcal{D}$, when the edge $i \to i$ of $\mathcal{G}_2$ is in the set cover, or $i \to i \in \mathcal{E}_2^*$. Node i can also be dominated by an adjacent node $j \in \mathcal{D}$, which corresponds to $j \to i \in \mathcal{E}_2^*$ being in the set cover.

When $\mathcal{G}$ is the pentagon graph, the bipartite graph $\mathcal{G}_2$ has 15 directed edges. In our formulation each of these is given a variable. The cover constraint at antenna node $\to 1$, using the node labels in Figure 6.8, is

$$x_{1 \to 1} + x_{2 \to 1} + x_{5 \to 1} \geq 1 \, .$$

	c	$\mathcal{N}$	$\mathcal{E}$	$g = 0.1$	0.2	0.5	1.0	2.0	5.0	10.
Q_{17}	9	289	7616	95	93	95	95	97	94	91
H_9	62	512	2304	0	0	72	98	82	16	4

Table 6.3 Dependence on the metric coefficient g of the number of success-ful searches for optimal dominating sets in 100 trials.

This is in constraint A, which also imposes indicator values on these variables. The concur constraint at the corresponding broadcast node $1\rightarrow$ is

$$x_{1\rightarrow 1} = x_{1\rightarrow 2} = x_{1\rightarrow 5} \; .$$

In constraint B the sum of the concurring values at the five nodes is addition-ally constrained to be at least the cover size c. The figure shows a solution with domination number 2, or a set-cover of size $c = 2$. Each node in the cover contributes three edges to $\mathcal{E}_2^*$.

Even when the graph $\mathcal{G}$ has symmetry (like the pentagon graph), the edges of the associated bipartite graph $\mathcal{G}_2$ are always inequivalent, symmetry-wise. In particular, edges of type $i \rightarrow i$ are never in a symmetry orbit with edges of type $i \rightarrow j$, $i \neq j$. From a symmetry perspective, there is no reason to give the associated variables the same metric coefficients. This absence of symmetry compels us to use a metric with at least one free parameter:

$$\|x' - x\|_g^2 = \sum_{i \in \mathcal{N}} (x'_{i\rightarrow i} - x_{i\rightarrow i})^2 + g \sum_{(i,j) \in \mathcal{E}} \left((x'_{i\rightarrow j} - x_{i\rightarrow j})^2 + (x'_{j\rightarrow i} - x_{j\rightarrow i})^2 \right) .$$

$$(6.1)$$

Table 6.3 compares RRR dominating-set searches for a range of the metric parameter g on two instances. In addition to the order-17 queen's graph problem (Q_{17}) of Figure 6.7, results are also given for the 9-dimensional hypercube graph (H_9). Always with $\beta = 0.5$, 100 trials of 10^7 iterations were performed for each g. The two cover sizes (domination numbers) are known to be optimal, either by clever analysis [112] or lengthy exhaustive search [86].

Only the experiments with H_9 have a strong metric-parameter preference, and it is the isotropic value, $g = 1$. The g-dependence for Q_{17} is very weak, and reflects the fact that very little domination is of the self-domination variety. The dominating set in Q_{17} comprises only 3% of all nodes, compared to 12% for H_9. For Q_{17} it would have made more sense to base the metric on the variety of edges (long vs. short, diagonal vs. nondiagonal). Of course, for anything that elaborate, the tuning of the metric has to be automated. As all the edges in the hypercube are related by symmetry, the one parameter

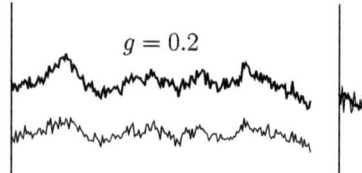 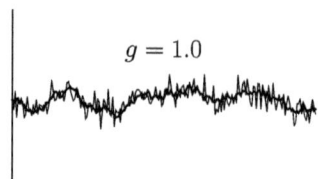 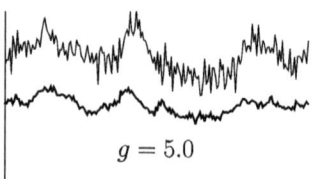

Figure 6.9 Detail (200 iterations) of the gaps $\Delta_{\rightarrow}$ (thick) and $\Delta_{\circlearrowleft}$ (thin) for the H_9 instance of the dominating set problem, for three settings of the metric parameter g.

metric (6.1) for the edges $\mathcal{E}_2$ in the bipartite graph for H_9 was the only metric generalization consistent with the symmetry of the problem.

When the variables in a bipartisan formulation have low symmetry, and the projections are easily modified to accommodate a metric consistent with that low symmetry, the task of tuning the many metric parameters quickly becomes unmanageable. But Figure 6.9 suggests a simple basis for automating the tuning process. Plotted are short time series of the gap in dominating set searches, where the gap has been decomposed into two contributions:

$$\Delta_{\circlearrowleft}^2 = \frac{1}{|\mathcal{N}|} \sum_{i \in \mathcal{N}} (x_{i \rightarrow i}^B - x_{i \rightarrow i}^A)^2 \,,$$

$$\Delta_{\rightarrow}^2 = \frac{1}{2|\mathcal{E}|} \sum_{(i,j) \in \mathcal{E}} (x_{i \rightarrow j}^B - x_{i \rightarrow j}^A)^2 + (x_{j \rightarrow i}^B - x_{j \rightarrow i}^A)^2 \,.$$

The thicker plot shows $\Delta_{\rightarrow}$, which is larger than $\Delta_{\circlearrowleft}$ when the corresponding variables are given the smaller metric coefficient ($g = 0.2$). Reversing the relative metric compliance of the variable types ($g = 5.0$) reverses the relative sizes of the two gaps. These plots are for the H_9 hypercube-graph instance, where the solution rates are poor (0% and 16%) at these settings of g. Since $g = 1$ is empirically the best setting, for which $\Delta_{\rightarrow} \approx \Delta_{\circlearrowleft}$, a feedback scheme based on gap equality might be a good way to automate the tuning of the metric.

6.3.2 Automated Tuning

Often the metric is adjusted with the restriction that it remains constant on groups of variables. Suppose we are packing molecules or some other rigid objects that we represent by sets of points (the constituent atoms). The projection that imposes the geometry of each molecule (Section 4.5.1) would be greatly complicated if each atom (and each Cartesian coordinate) was given its own metric coefficient. To avoid this complication, the metric is kept constant on each molecule. The packing of the molecules is imposed with distance constraints between replicated atom variables. Unlike the rigid-body constraint, projecting an atom pair to a minimum distance is easily generalized when the variables are given different metric coefficients. When atom replicas are forced to be equal, in the other constraint, the "concur" projection simply uses these coefficients in a weighted average.

In some problems every variable can be given its own metric coefficient without complicating the projections. Ironically, this turns out to be a smart thing to do even when the variables are related by a symmetry! To define things for the most general case, let $\mathcal{P}$ be a partition of the variables such that all variables $i \in \mathcal{I}$ in part $\mathcal{I} \in \mathcal{P}$ have the same metric coefficient:

$$\|x' - x\|_g^2 = \sum_{\mathcal{I} \in \mathcal{P}} g_{\mathcal{I}} \left(\sum_{i \in \mathcal{I}} (x_i' - x_i)^2 \right) .$$

For each group of variables $\mathcal{I}$, we can define an average gap:

$$\Delta_{\mathcal{I}}^2 = \frac{1}{|\mathcal{I}|} \sum_{i \in \mathcal{I}} (x_i^B - x_i^A)^2 .$$

If our groups of variables have sensible normalizations, such as 0 and 1 for indicators, we might be concerned if two groups, such as variables of type $\circlearrowleft$ and $\rightarrow$ in the dominating set problem, have very different gaps. We also know that decreasing a metric coefficient makes it easier for a variable to change when satisfying a constraint (A or B). That is, by decreasing $g_{\mathcal{I}}$ we can increase $\Delta_{\mathcal{I}}$, and vice versa.

Since only the relative values of the metric coefficients and gaps matter, the first metric update rule we might write down is[4]

$$g_{\mathcal{I}} \rightarrow g_{\mathcal{I}} + \gamma \left(\Delta_{\mathcal{I}}^2 / \Delta^2 - 1 \right) g_{\mathcal{I}} , \qquad (*)$$

where

$$\Delta^2 = \frac{1}{|\mathcal{P}|} \sum_{\mathcal{I} \in \mathcal{P}} \Delta_{\mathcal{I}}^2$$

is the average of the gaps. When using this update it comes with the understanding that $\gamma > 0$ is a small parameter, since our local analysis of RRR assumed a static metric. In this limit, we can think of $(*)$ as the system of differential equations

$$\frac{dg_{\mathcal{I}}}{dt} = \left(\langle \Delta_{\mathcal{I}}^2 \rangle / \langle \Delta^2 \rangle - 1 \right) g_{\mathcal{I}} ,$$

where $dt = \gamma$ is the metric-update time step and $\langle\ \rangle$ is an average over several iterations. The logic behind this slow, continuous evolution of the metric is

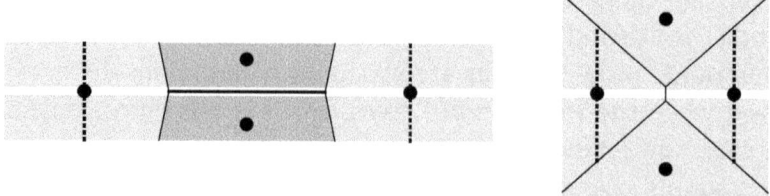

Figure 6.10 *Left*: Two points of A (black) lie on the line B (white); the associated fixed-point sets are the dotted lines. The two points off the line create a trap. Starting anywhere in the dark gray region, iterates end up hopping from one side of the black line segment to the other. *Right*: Horizontal contraction and vertical expansion, corresponding to a change in the metric, eliminates the region of trapping.

that variable-groups $\mathcal{I}$ with an above-average gap, or $\langle \Delta_{\mathcal{I}}^2 \rangle > \langle \Delta^2 \rangle$, should be made less compliant by giving their coefficient $g_{\mathcal{I}}$ a *positive growth rate*. Conversely, variables with a below-average gap are enabled to play a larger role in satisfying constraints by having their metric coefficients decay.

As long as $\gamma < 1$, rule ($*$) poses no risk to the sign of the metric coefficients. However, there is a risk of unbounded growth if the feedback on the metric does not have the intended effect. A slightly different rule is free of this problem:

$$g_{\mathcal{I}} \to g_{\mathcal{I}} + \gamma \left(\Delta_{\mathcal{I}}^2 / \Delta^2 - g_{\mathcal{I}} \right) . \tag{6.2}$$

Rather than setting growth rates, this rule sets target values for a steady-state metric:

$$g_{\mathcal{I}} = \frac{\langle \Delta_{\mathcal{I}}^2 \rangle}{\langle \Delta^2 \rangle} . \tag{6.3}$$

The inverse of γ is the number of iterations for the metric to relax to the non-isotropic form (6.3). This cures the unbounded growth problem because the gaps are bounded. Rule (6.2) also has the nice feature that the average of the metric coefficients relaxes to the value 1. In particular, if we set $g_{\mathcal{I}} = 1$ for all $\mathcal{I} \in \mathcal{P}$ initially, the average of the metric coefficients is fixed at this value.

Figure 6.10 shows how the auto-tuning heuristic (6.2) extricates RRR from a trap in two dimensions. Set A comprises four points and B is a line that passes through two of them. If the geometry is as shown in the left panel, and the iterate (x, y) starts somewhere in the dark gray region, then there is trapping on a limit cycle. The point-line flow in this region is convergent on a line segment piece of B. Because of transverse contraction (to the axis

$\gamma = 0.02$	0.01	10^{-3}	10^{-4}	10^{-5}	10^{-6}
success (%)	85	99	99	98	75	7
iterations	3 300	2 300	2 100	2 400	6 900	220

Table 6.4 Results of metric auto-tuning when solving `coly013`.

defined by the points in the dark gray region), v_x approaches zero while v_y alternates between $\pm y_0$ in the limit, where y_0 is the distance between the line and the two points. Since $\Delta_x^2 = 0$, $\Delta_y^2 = y_0^2$ in the limit, with average $\Delta^2 = y_0^2/2$, the two metric coefficients evolve as

$$g_x(t) = g_x(0)e^{-t} \, ,$$
$$g_y(t) = 2 + \big(g_y(0) - 2\big)e^{-t} \, ,$$

where $t = \gamma \times$ iterations. The right panel of Figure 6.10 is the rescaling of A and B corresponding to the changed metric (compression in x, expansion in y). The Voronoi regions of the four points have changed, eliminating the trap. Now all initial points iterate to one of the two fixed-point sets (dotted lines).

In one of the exercises you can work out the details of a trap that RRR faces when trying to solve the `coly013` puzzle, and it is different from the example just described. Still, a general feature does carry over. Traps are local entities, involving just two of the four points in Figure 6.10. The two other points are distant and effectively out of the picture. Metric evolution is complicated, in general, but there's a good chance that the more distant parts of a constraint are brought closer and made relevant. Traps are defeated when new projection-options arise.

As explained at the end of Section 4.6.2, each of the $9 \times 9 \times 9$ variables in our suduko formulation is easily given its own metric coefficient. When we do this, initializing them all at the value 1, we have an additional parameter for solving sudoku: γ. The effect of this parameter in solving `coly013` is summarized in Table 6.4. Since $\beta = 0.2$ works well when there is no trapping, this parameter was unchanged. The results show that trapping is effectively eliminated when the metric is evolved with rule (6.2) and $\gamma \sim 10^{-2} - 10^{-4}$. At smaller γ, the metric evolves too slowly for the iterates to escape traps in the 10^4 iterations allowed for each trial. Performance also degrades for $\gamma > 10^{-2}$, apparently because the metric evolution is no longer adiabatic at this rate.

Because sudoku puzzles are made unique by their clues, it seems reasonable that there is a uniquely best metric for RRR to use when solving any puzzle.

Expression (6.3), from that perspective, is a proposal for that best metric. If we apply this logic to *blank puzzles* (sudokus with no clues), we should gain nothing by using rule (6.2) on them. That's because only the isotropic metric is singled out by the enormous symmetry of the blank puzzle. However, as we learn in Section 6.3.3, rule (6.2) is a powerful heuristic even in the presence of symmetry!

6.3.3 Cooperativity

Filling out a blank sudoku puzzle is harder than you think. Much of the time you cannot rely on inference: choices must be taken. Unfortunately, some choices have bad, causally opaque consequences.

Systems of many variables that exhibit the freedom of a much smaller number of variables are said in physics to be *cooperative*. The hidden free variables are the true decision-makers, and the fewer their number the easier it is, in principle, to find a combination of values that solves the problem. This picture suggests a very different rationale for tuning the metric, unrelated to symmetry: to increase the cooperativity of the RRR system of variables.

Let's suppose the number of hidden free variables is n. More formally, though still abstractly, we model the RRR gap as the sum of n independent gap-random-variables Δ_i:

$$\Delta = \sum_{i=1}^{n} \Delta_i .$$

Using $\langle \ \rangle$ to denote the average over RRR iterations, we can define the following averages:

$$\langle \Delta \rangle = \sum_{i=1}^{n} \langle \Delta_i \rangle$$

$$= n\mu ,$$

where

$$\mu = \frac{1}{n} \sum_{i=1}^{n} \langle \Delta_i \rangle$$

is an n-independent constant. Using independence of the Δ_i,

$$\langle (\Delta - \langle \Delta \rangle)^2 \rangle = n\sigma^2 ,$$

$$\sigma^2 = \frac{1}{n} \sum_{i=1}^{n} \left(\langle \Delta_i^2 \rangle - \langle \Delta_i \rangle^2 \right) ,$$

	$\gamma = 0.02$	0.01	0.005	0.002	0.001
iterations/solution	–	3 600	680	1 900	6 700
χ	0.086	0.158	0.197	0.143	0.109

Table 6.5 Results of metric auto-tuning when filling out a blank sudoku.

where σ is another n-independent constant. The cooperativity χ is defined as the scale-independent ratio

$$\chi^2 = \frac{\langle(\Delta - \langle\Delta\rangle)^2\rangle}{\langle\Delta\rangle^2}$$
$$= \frac{1}{n}\left(\frac{\sigma}{\mu}\right)^2.$$

If the model applies, we can assess cooperativity from gap statistics. An intervention that increases χ corresponds to fewer variables (smaller n) and an easier problem.

When RRR arrives at a solution, all the gap-random-variables are significantly smaller than their average in the search, or

$$\Delta_i < \langle\Delta_i\rangle\delta$$

for some suitably small, n-independent δ. If $\rho_i(0)$ is the probability density of Δ_i at zero (during the search), then

$$\prod_{i=1}^{n}\left(\rho_i(0)\langle\Delta_i\rangle\delta\right) = (\omega\,\delta)^n$$

is the solution probability, where ω is the n-independent geometric mean of the numbers $\rho_i(0)\langle\Delta_i\rangle$. Our model predicts the probability that RRR stumbles upon a solution improves exponentially with an increase of cooperativity (decrease in n).

Table 6.5 shows the effect of metric auto-tuning on the problem of filling out a blank sudoku puzzle. A total of 10^5 RRR iterations with $\beta = 0.2$ and various relaxation constants γ were performed, the variables reinitialized to random values and the metric made uniform each time a solution was found. The number of iterations per solution is lowest when the cooperativity χ is highest. While the effect increases with γ, it breaks down as expected when adiabaticity is compromised (no solutions at all for $\gamma = 0.02$). The best solution time $0.2 \times 680 = 136$, for $\gamma = 0.005$, matches that of escargot without auto-tuning ($\gamma = 0$).

In problems with symmetry we can think of the metric update heuristic

	$\gamma = 10^{-3}$	10^{-4}	10^{-5}	10^{-6}
success (%)	45	96	85	19
iterations	100 000	61 000	98 000	170 000
χ	0.456	0.750	0.707	0.223

Table 6.6 Results of metric auto-tuning on a hard Turán set-cover instance. The iteration limit was 2×10^5.

(6.2) no differently from how it explains extrication from traps (Figure 6.10). The metric (6.3) is not customized for the problem instance at hand, but for the situation at hand. This metric is transient, a target defined by the current iterate. By making variables with large gaps metrically less compliant, and low-gap variables more compliant, the gap becomes more equally distributed over the variables. As this promotes coordination in the distance minimization for a larger group of variables, the system's cooperativity is enhanced.

One of the most spectacular examples of cooperativity by metric auto-tuning is the set-cover instance for the Turán number $T^*(18, 4, 3)$. There are 12 240 edges/variables and solutions correspond to very special subsets of size 330 of the 816 broadcast nodes. Table 6.6 echoes the relationship between iterations and cooperativity we saw for blank sudoku. The χ values are averages over entire runs in 200 trials. But as the plot in Figure 6.11 of the gap makes clear, χ evolves significantly and gradually, also on the time scale of $1/\gamma$. Initially, when the gap's fluctuations are a small fraction of its average, χ is very small. Eventually, when the gap fluctuations are comparable to the average, χ reaches the value 1. The detail of the gap at late times suggests a coordinated switching behavior. We know that many variables are switching in unison because the switches take the gap to large values between the multiple near-solutions.

One detail about the set-cover experiments should not go unreported as a cautionary tale when developing general principles. The results presented are for $\beta = 0.8$; performance degrades significantly not just at larger values, but also in the flow limit. Apparently the flow limit conjecture does not hold for the formulation of set-cover that uses B_{sum} for the B constraint! Fortunately the metric auto-tuning heuristic does quite well all by itself in this application of RRR.

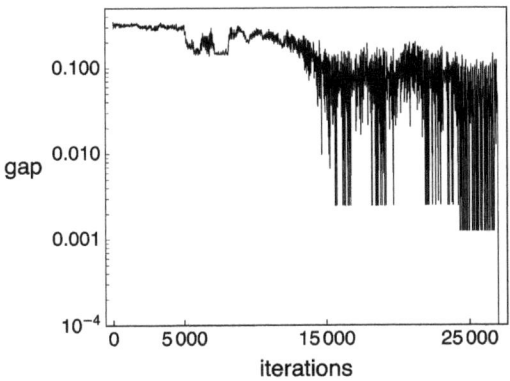 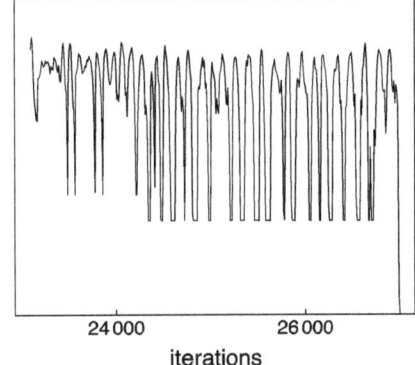

Figure 6.11 Growth of cooperativity (for the case $\gamma = 10^{-4}$) revealed in the behavior of the gap in a solution of the set-cover instance for the Turán number $T^*(18, 4, 3)$. As the late-time detail on the right shows, the solution is found through the highly coordinated dynamics of the system's 12 240 variables.

6.4 Good Practice

We conclude this chapter with a short list of recommendations to improve your experience with bipartisan problem solving, the experience of the computers that do the actual work, and the experience of the audience that receives your results.

Choose your scales with care

The similarity invariance of the RRR algorithm is not a license to be cavalier about scales. Define your variables so the projections change their values by the same $O(1)$ amount, for all sizes of your problem. This is understood when working with indicator variables, but should also apply when packing spheres, folding origami, or training networks. Sphere diameters and the triangles edges in the origami paper should be $O(1)$ regardless of how many spheres you are packing or how complex the fold. The only sensible normalization of a neuron's inputs, outputs, and weights is where each one typically fluctuates by $O(1)$ in the course of training, independent of the structure and complexity of the network.

Normalize the gap with care

If the gap decreases by a large amount, you've probably solved the problem. The absolute value of the gap can also be revealing, especially when the problem is resisting solution. Say you are trying to pack spheres of diameter 1. The gap normalization

$$\Delta^2 = \frac{1}{n} \sum_{i=1}^{n} \|x_i^A - x_i^B\|^2$$

is the best choice you can make, even if the sphere-count n includes divide-and-concur replicas (which is almost always the case). Now if you see $\Delta = 0.5$, it means there is typically one sphere radius between where constraints A and B want a sphere to be. Dividing the sum of squares by the total number of terms in the sum is a good default. By using that rule separately on distinct variable types, say neuron inputs and weights, the type-specific gaps will let you assess whether the different types are participating equally.

Avoid empty success statistics

The human cognitive system is too complex to attempt any kind of modeling. If `arachnizine` reduces spider anxiety in 20% of cases, the best you can do is report that one number. Hopefully this book has taught you ways to characterize and diagnose RRR behavior that allows you to paint a more complete picture of how the algorithm is doing on your problem. When it fails, is it because of traps, dismal cooperativity, or the problem is just too hard? The last possibility can be assessed by experiments with problem ensembles, tuned from easy to hard by size or some other parameter. Cooperativity can be gleaned from gap statistics, but there are more direct ways. Consider plotting the time series for a few variables. Is the algorithm searching well, or just spinning in circles?

Render the gap with respect

If the gap is the main source of information about your algorithm's behavior, consider how best to convey its features. A logarithmic vertical scale is great for checking the linear convergence promised by correctly implemented projections, but is also useful when RRR confronts a hierarchy of subproblems. A horizontal logarithmic scale is a good way to juxtapose the gaps of easy and hard versions of the problem to spot trends. The sphere packing application in Section 7.2 uses this device to define something analogous to a thermodynamic transition. Apart from the choice of axes, are your gap samples fine enough to convey the time scale of the dynamics? If a sufficiently fine sampling is not feasible for rendering, consider plotting a short-time detail.

Do not increment thousands of variables by zero

In some problems with selector (1-hot) vectors, the number of vector components can be quite large. Do not subject your computer to the indignity of setting many variables to zero, only to follow up by incrementing many other variables by those zero values. Keeping track of the position of the nonzero element and its value saves both time and energy. This recommendation can be ignored while you're in the constraint-design stage and experimenting with small problems.

Do not set thousands of variables all to the same value

The number of replicated variables in divide-and-concur formulations can be quite large. Instead of reserving memory and then writing identical values to all those memory locations, it saves time and energy (and some memory) to store just one copy of each concur projection. This recommendation may also be ignored, for the sake of program transparency, during the design stage.

Think twice about using Python

A nice feature of the bipartisan approach is that most of the heavy lifting is consigned to the projections P_A and P_B. If you are programming in an interpreted language like Python, and the two projections can be implemented using compiled functions (methods) in the language, the inefficiencies relative to completely compiled code are minor. But suppose the variables are not packaged in rectangular arrays, or compiled functions for computing the projections are not available. Consider the projection to the discrete "cover" constraint in set-cover. The computation, on variables that live on the edges of a mixed-degree graph, is very simple: round to 0 or 1, and if all edges incident to a node are rounded to 0, go back and round the largest edge to 1. Because there is no Python function for this task, you must implement it with Python loops and conditionals. On the Turán-number graphs considered in this chapter, Python code takes 60 to 150 times as long to compute the projection as a C program, depending on the size of the graph.

Engage the public with animations

Most of the math behind RRR will be lost on the general public. But that's no reason to give up on connecting with your audience, especially if the technique has helped you discover something interesting. Use distance minimization, at the core of the algorithm, to argue that

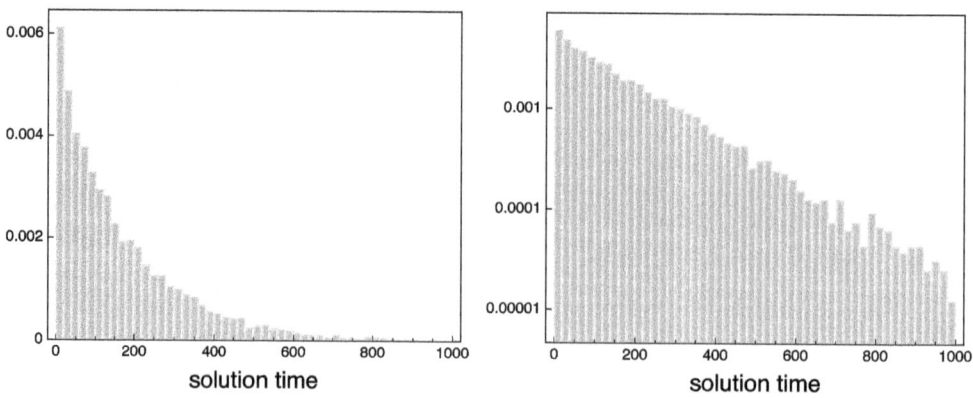

Figure 6.12 Distribution of solution times for the `escargot` sudoku puzzle, replotted with a logarithmic vertical scale on the right to show the exponential decay.

RRR is *deterministic*. That the deterministic "moves" are also *smart* can be conveyed through animations/movies. Spider-reconstruction movies are an obvious and fun choice, but fall short because Fourier magnitude is not a constraint that comes naturally to most people! Packing movies (of disks, spheres, ...) need no commentary because they resonate with an aptitude that runs deep in human nature. Discrete problems work well too, if constraint compliance is visually direct. The dance of tiles in an edge-matching puzzle, or the moves of a court of queens trying to dominate a chessboard, are good examples. The frame-to-frame changes are small and the very opposite of random.

A distribution's shape is more interesting than its mean

Earlier in this chapter we reported that the mean solution time of the notorious `escargot` sudoku puzzle was 140, topping newspaper puzzles by a large factor. What we should also have done is display the *distribution*. The omission is redressed in Figure 6.12, and indeed it is quite interesting. The shortest solution times are the most probable! The exponential shape is exactly what we expect if the RRR dynamics is ergodic with a short mixing time. After relatively few iterations, all memory of the initial point is lost and there is a time-independent probability for the algorithm to stumble on the solution – a Poisson process. It's very different from the nearly direct route RRR finds for easy puzzles, whose solution-time distributions have puzzle-specific structure.

Exercises

6.1 At the beginning of Section 6.3, it was pointed out that a diagonal modification of the metric in phase retrieval runs into the problem that the diffraction-intensity constraint cannot be implemented. However, a special kind of nondiagonal modification *can* be implemented and is often used when the other constraint imposes contrast-sparsity, as in crystallography. To keep things simple, in this exercise the crystal contrast x is one-dimensional, with period ℓ.

(a) Consider the following nondiagonal rescaling of the contrast:

$$\tilde{x}_p = \sum_k f_k \, x_{p-k} \ .$$

The index arithmetic is mod ℓ and sums always run over a complete period (from 1 to ℓ). You can interpret f as a convolutional filter. Show that the corresponding change in the Fourier transform is a *diagonal* rescaling:

$$\hat{\tilde{x}}_q = \sqrt{\ell} \hat{f}_q \, \hat{x}_q \ . \tag{6.4}$$

(b) Imposing the isotropic metric on $\hat{\tilde{x}}$ (and consequently also on $\tilde{x}$) is equivalent to a diagonal metric modification on $\hat{x}$ with coefficients

$$\hat{g}_q = \ell |\hat{f}_q|^2 \ .$$

What is the corresponding nondiagonal metric on x?

(c) Let f be the filter

$$f_k = \begin{cases} 1/(1-2s), & k = 0, \\ -s/(1-2s), & k = \pm 1, \\ 0, & \text{otherwise,} \end{cases}$$

where the parameter s satisfies $0 \le s < 1/2$. Work out $\hat{g}_q$ explicitly for this filter and show that when you are reconstructing the transformed contrast $\tilde{x}$, you will be using the transformed intensities

$$\tilde{I}_q = \hat{g}_q \, I_q$$

in the diffraction-intensity constraint.

(d) The filter f becomes the discrete Laplacian in the limit $s \to 1/2$. How might the quality of diffraction data at high spatial frequencies[5] compromise the reconstruction of $\tilde{x}$ in this limit? The motivation for using the filter despite this problem is that it makes the other constraint, on the contrast sparsity, stronger. As you can see in Figure 6.13, the four broad "atoms" in x are made sharper in $\tilde{x}$, and there is

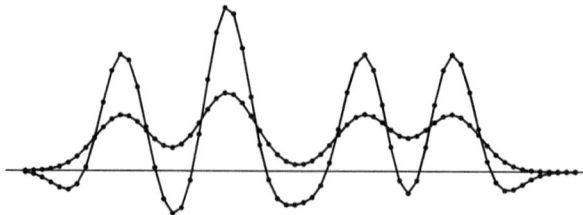

Figure 6.13 Contrast sharpening with a Laplacian filter ($s = 0.48$).

a sizeable reduction in the number of elements p where x_p is positive. As the filter creates negative constrast between atoms, how should the usual sparsity constraint be modified to better define "atomicity"?

6.2   In this exercise you supply the details that show the symmetry group of the graph $\mathcal{G}(n, k, r)$ acts transitively on the edges. The antenna nodes $\mathcal{A}$ of this bipartite graph are all the k-subsets of $\{1, \ldots, n\}$; the broadcast nodes $\mathcal{B}$ are all the r-subsets. Any permutation σ of $\{1, \ldots, n\}$ induces an action on $\mathcal{G}(n, k, r)$ and we use the same symbol, σ, for that action.

(a) Consider any two edges of $\mathcal{G}(n, k, r)$,

$$e_1 : b_1 \to a_1 \, ,$$
$$e_2 : b_2 \to a_2 \, ,$$

where b_1, b_2 are elements of $\mathcal{B}$, and a_1, a_2 are elements of $\mathcal{A}$. Show that these r-subsets and k-subsets can be expressed as

$$
\begin{aligned}
b_1 &= c_0 \sqcup c_1 \, , \\
b_2 &= c_0 \sqcup c_2 \, , \\
a_1 &= c_0 \sqcup c_1 \sqcup c_3 \sqcup c_4 \, , \\
a_2 &= c_0 \sqcup c_2 \sqcup c_3 \sqcup c_5 \, ,
\end{aligned}
\tag{6.5}
$$

where c_0, c_1, c_2, c_3, c_4, and c_5 are disjoint subsets of $\{1, \ldots, n\}$, and $|c_1| = |c_2|$, $|c_4| = |c_5|$.

(b) Use the decomposition (6.5) to show there exists a symmetry element σ such that $\sigma(e_1) = e_2$.

6.3   For $k > 1$ variables, consider the local constraint sets

$$
\begin{aligned}
A &= \{(1, 1, \ldots, 1)\} \, , \\
B &= \mathbb{R}^k_{k-1} \, .
\end{aligned}
$$

(a) Determine explicitly the coordinates of an RRR k-cycle by solving the equation

$$\mathcal{R}(x^*) = \mathcal{R}(x_1^*, x_2^*, \ldots, x_{k-1}^*, x_k^*)$$
$$= (x_2^*, x_3^*, \ldots, x_k^*, x_1^*),$$

where the coordinates of x^* in the positive orthant are strictly increasing, and β satisfies $|1 - \beta| < 1$ (the condition for local convergence when both A and B are affine). Because of the cyclic symmetry of this k-cycle, all the RRR gaps

$$\left\| \mathcal{R}^{i+1}(x^*) - \mathcal{R}^i(x^*) \right\| / \beta$$

for $i = 0, 1, 2, \ldots$ are equal.

(b) By explicitly calculating $\mathcal{R}^k(x^* + y)$ for an arbitrary perturbation $y \ll x^*$, show that the k-cycle you determined above is the limit of a *limit cycle*. The restriction $|1 - \beta| < 1$ is important here too.

6.4 Construct the flow in the plane for the special case $k = 2$ of the previous problem.[6] Show that for $\beta \to 0$ there is trapping at the origin for any initial point.

6.5 Figure 6.14 provides all the relevant details for analysing the 2-cycle trap that RRR encounters when trying to solve the `coly013` sudoku puzzle. Only the assignments of digits 6 and 9, in four blocks, and four rows and columns, change from one iteration to the next. By working out the coordinates of $\{a_1, a_2\} \subset A$ and $\{b_1, b_2\} \subset B$ – limiting yourself to the relatively few that change – show that this is an instance of the 2-cycle trap of Exercise 5.8.

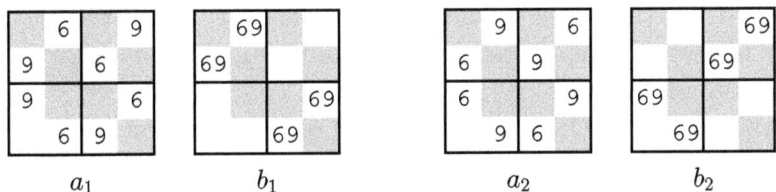

Figure 6.14 Local constraint sets $A = \{a_1, a_2\}$, $B = \{b_1, b_2\}$ in a 2-cycle trap that arises when solving the `coly013` sudoku puzzle. The gray cells contain givens for the other digits (and are off-limits for 6 and 9).

7

Divide and Concur

In abstract terms, bipartisan formulations are interesting because the constraints are imposed on two completely different factorizations of Euclidean space:

$$A = A_1 \times \cdots \times A_m \,,$$
$$B = B_1 \times \cdots \times B_n \,.$$

Thanks to the factorizations, the projections P_A and P_B can act efficiently and independently on different subproblems. Often the two factorizations are obvious, as when the variables live on the edges of a bipartite graph, with all the A constraints at the antenna nodes and the B constraints at the broadcast nodes (Section 2.4). Problems with the system-of-equations structure also fit this description. Instances of the same variable in different equations are made independent by replacing them with replicas. The equations, now independent, are the factors of the A constraint. The role of the B constraint is simply to force the replicated variables to be equal. Equations are broadly interpreted, and might be clauses in a logical conjunction.

Factorizations of B where each factor imposes equality on a set of variables, possibly with a side constraint, are called a *concur constraint*. However, the scope of the A constraint, for dividing up a problem, is considerably broader than the system-of-equations "textbook" example. The point of this chapter is to showcase some of these not-so-obvious ways of dividing up constraints. Before we start the tour, let's take a look at that textbook example.

7.1 Linear Programming

It may seem odd to see linear programming featured as an application of divide-and-concur. After all, linear programming is a firmly established optimization discipline for which software packages at many levels of sophistication abound. Why would you use divide-and-concur to solve a linear program when there is CPLEX, Gurobi, and a host of other packages? Besides the question of licenses and installation, one answer to this question is that writing a divide-and-concur LP (linear programming) solver is so easy it may take less time than learning how to format your problem for the software package!

With a tiny notational modification, linear programming problems are already "divided." There are n inequalities, labeled with an index i:

$$\sum_{j=1}^{m} c_{ij}\, x_j \leq d_i\,, \qquad i = 1, \ldots, n\,.$$

All n inequalities have to be satisfied. The unknowns are the variables $x_1, x_2, \ldots, x_m$. By replicating the variables with the equation-index i, the inequalities are made independent:

$$\sum_{j=1}^{m} c_{ij}\, x_{ij} \leq d_i\,, \qquad i = 1, \ldots, n\,.$$

We can think of x_{ij} as the replica of variable j in charge of satisfying inequality i. This inequality is constraint set A_i in the divide factorization

$$A = A_1 \times \cdots \times A_n\,.$$

The projection to A_i can be found in Section 4.2.3 and amounts to adding a suitable multiple of the vector $(c_{i1}, \ldots, c_{im})$ to the vector $(x_{i1}, \ldots, x_{im})$ to give the equality case, or doing nothing if the inequality is already satisfied.

The concur or B constraint is factored over the m variables and each projection simply replaces the n replicas by their average. Some LP problems are made harder by additionally constraining variables to be integers or indicators (0 or 1). One approach to this wrinkle is to add this as a side constraint on the concur value. For example, if x_j has to be an indicator, then after forming the average of its replicas, the concur value would be rounded to 0 or 1, whichever is nearest. However, doing that comes at a significant loss of information relative to the information held by a continuous concur value. To avoid this, it's better to define side-constraint replicas subject to additional divide constraints. In this example, for variable x_j, we would introduce a replica $x_{n+1\,j}$ and the divide constraint $x_{n+1\,j} \in \{0, 1\}$. This preserves a continuous concur value for x_j, with part of the average coming from an indicator. The contribution of the indicator to the average can be adjusted by tuning the metric.

So far our set of n inequalities just define a feasibility problem, whereas

LPs usually also have a linear objective function:

$$\sum_{j=1}^{m} e_j\, x_j = f \, . \tag{7.1}$$

The actual task is to *maximize* f for settings of the variables that also satisfy the n inequality constraints (and any indicator or integrality constraints). Suppose we replace the variables in (7.1) with objective-function replicas, and the equality with an inequality:

$$\sum_{j=1}^{m} e_j\, x_{0j} \geq f \, . \tag{7.2}$$

Now we have a feasibility problem as before, and the largest f that gives a feasible problem is the objective-function maximum.

RRR is very efficient at deciding whether the system of inequalities is feasible when there are no integrality or indicator side constraints on the variables. That's because A in that case is convex: the intersection of half-spaces. Set B, a hyperplane, is also convex. When these do not intersect, the iterates of $x_A = P_A(x)$ and $x_B = P_B(R_A(x))$ converge on a pair of proximal points on A and B and the gap converges to the distance between them.[1] Because weakening (7.2) by decreasing f decreases the gap, we can obtain a sequence of decreasing upper bounds on the objective function. If we've decreased f too much, and the gap goes to zero, we now have a lower bound on the objective.

In the convex case one can use the *alternating direction method of multipliers* (ADMM) [13], a method closely related to Douglas–Rachford. As discussed in Section 5.6, with ADMM one can maximize the objective function directly, instead of progressively tightening bounds. However, when one of ADMM's two objective functions is reserved for the LP objective, ADMM is no longer able to work with independent inequalities, as divide-and-concur does.

Besides the remarkably simple program design, the ease of introducing integrality and indicator constraints should make divide-and-concur a compelling alternative when writing a "mixed-integer" programming solver from scratch. As we've seen, these are introduced via additional divide constraints and replicas. Because of nonconvexity, one no longer has proofs of convergence. But whereas RRR is implemented no differently in these circumstances, simplex and interior-point algorithms have to be thoroughly revised (branching, integer-relaxation, etc.). To be fair, these revisions still result in exact algorithms and may be the only option for proving a solution (for a given f) does not exist.

Linear programs have dual formulations that may be advantageous when using divide-and-concur. In matrix notation and capitalizing the matrix, the "primal" LP above is usually written

$$\text{maximize:} \quad f = xe^{\mathsf{T}}$$
$$\text{subject to:} \quad x\,C \le d \;.$$

The corresponding dual LP is

$$\text{minimize:} \quad g = yd^{\mathsf{T}}$$
$$\text{subject to:} \quad y\,C^{\mathsf{T}} = e \;,\; y \ge 0 \;.$$

When both LPs are feasible, since

$$xe^{\mathsf{T}} = x(C\,y^{\mathsf{T}}) = (x\,C)y^{\mathsf{T}} \le dy^{\mathsf{T}} \;,$$

we see that

$$\max f \le \min g \;,$$

or that the optimum of one LP bounds the optimum of the other. Often one knows the primal problem is feasible and the solution is bounded. In that case, the strong duality theorem [74] asserts that the dual LP is also feasible and $\max f = \min g$, or that the two forms give us two handles on the same problem. One of the exercises will show you how different these LPs can be in the problem of finding maximum-weight bipartite matchings.

In the six remaining divide-and-concur applications, the structure of the divided constraints and their projections are much less obvious. Packing spheres in d dimensions is straightforward until one has to confront the periodicity of the space in which they live. Even the concur constraint is nonconvex in that setting. In "equidistant lines" we have a situation where concur is applied to orthogonal decompositions of space. The protein folding game SEQVENCE is the only discrete problem, but the indicator variables that make the most sense are not the most obvious. Landscape origami is about *designing* origami folds, where replicas that enforce topological constraints make an unexpected appearance. In the dimension-reduction algorithm eℓmo, the many (divided) inequality constraints on distances have no scale and so this task is taken up by concur. Finally, the neural network section is a blueprint for radically redesigning a core technology. Here, divide is responsible for individual neurons while concur defines their connectivity.

Figure 7.1 "Spheres" packed on the integer lattice in one, two, and three dimensions and interstitial holes, rendered as small spheres. In 20 dimensions (not shown) the space between spheres (in one unit cell) has grown so large it can accommodate over 20 000 spheres.

7.2 Periodicity and Sphere Packings

Geometric intuition in humans evolved for life in a three-dimensional world. Navigating the higher dimensions requires math, and any kind of open-ended exploration is impossible without a computer. One of the toughest challenges posed by high-dimensional geometry is the efficient packing of congruent spheres. In this section we show how divide-and-concur can be a useful tool for exploring that particular world.

Consider unit-diameter spheres in one, two, and three dimensions. In $d = 1$ these are better known as "intervals," and "disk" is the term used in $d = 2$. Figure 7.1 shows how these pack when arranged periodically on the integer lattice, $\mathbb{Z}^d$. In two and three dimensions, there are denser packings, but the integer lattice packings do not seem too bad. In these packings there are "holes" between the spheres that are unfortunately too small to accommodate additional spheres. This changes in the first dimension beyond our power to envisage, $d = 4$. In that dimension, spheres can be packed at the half-integer coordinates as well, because they have squared-distance $4 \times (1/2)^2 = 1$ from the integer-coordinate spheres. If doubling the number of spheres per unit cell is possible in $d = 4$, what can we expect in higher dimensions?

In $d = 10$ it is possible to pack 40 spheres per unit cell, with coordinates that are always combinations of integers and half-integers. This seems like a lot, but can be rationalized when the 40 coordinate sets are interpreted as length-10 binary codes with Hamming distance 4 [11]. As long as every pair differs by $1/2$ (mod 1) in at least four coordinates, the corresponding spheres will be disjoint.

While 40 spheres per unit hypercube is hard to fathom in $d = 10$, in

$d = 20$, it is known that there exists a distance-4 binary code of size at least 20 480 [11, 51]. Even more astounding, and perhaps a greater embarrassment for our low-dimensional intuition, is the fact that (still in $d = 20$) it's possible to arrange 4 000 spheres much more randomly, still disjoint, and with coordinates that range continuously between 0 and 1! In this section we show how divide-and-concur can construct such packings.

Some dimensions, like 2, 8, and 24, are exceptionally well suited for dense packings [22]. That's because lattices in these dimensions can take advantage of unusually large automorphism groups. However, from the fact that the final entry in the complete catalogue of finite simple groups was gleaned from the study of the 24-dimensional Leech lattice [25], one might suspect that lattices have already exhausted the benefits of interesting groups. Also, in some dimensions denser packings are obtained by relaxing the lattice restriction. The known examples of this phenomenon are still periodic packings, but have $k > 1$ spheres in the fundamental parallelotope (unit cell). The current density record holders in $d = 10, 11, 13, 18, 20$ have, respectively, $k = 40, 72, 72, 512, 4$ [23].

To explore how the packing density behaves in large dimensions, the unit cell geometry may not be all that important, provided k is very large. Since the unit cell volume and the number of spheres in that volume grow exponentially with d, to reach the largest dimensions, the unit cell should be as small as possible. For unit-diameter spheres, the smallest hypercubic cell is the unit hypercube, because spheres centered at the vertices must be disjoint, as in Figure 7.1.

In low dimensions, the integer-lattice periodicity severely restricts the tangencies available to any sphere. For dimensions $d < 4$, there are none besides the coordinate axis contacts, and as we have seen, just a discrete set can be added in $d = 4$. But our low-dimensional intuition is misleading, as shown in the plot of Figure 7.2. This gives the fraction of a sphere's surface area available for tangencies with other spheres in packings with the $\mathbb{Z}^d$ periodicity [39]. The inflection point at $d = 16$ means that the effects of periodicity (with the smallest possible unit cell) become increasingly less important above that dimension.

Before RRR can be deployed for the proposed experiments in high dimensions, it needs to be modified for work in periodic spaces. Projections are modified as well, with even the simple concur constraint becoming nonconvex. Fortunately these changes do not incur much extra computational cost.

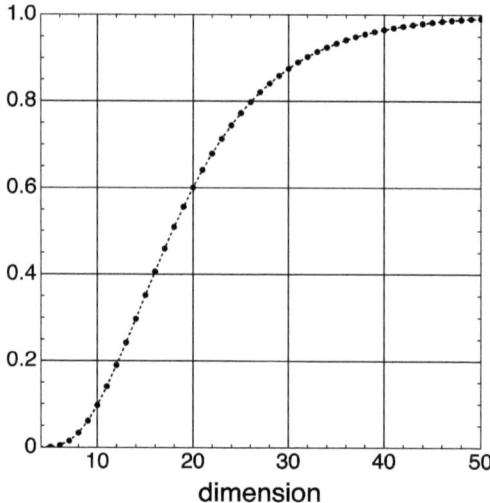

Figure 7.2 Fraction of a sphere's surface area available for tangencies with other spheres in a packing with $\mathbb{Z}^d$ periodicity [39].

7.2.1 RRR with Periodicity

Projections and the RRR algorithm, as defined so far, act on elements of ordinary Euclidean space, $\mathbb{R}^m$. The compact space

$$T^m(w) = \mathbb{R}^m/(w\,\mathbb{Z}^m) \,,$$

where points of $\mathbb{R}^m$ differing by an element of the integer lattice scaled by w are declared equivalent, is metrically the same as $\mathbb{R}^m$ on small scales. In computations one may use convenient coset representatives of $T^m(w)$. For us these are m-tuples of numbers with absolute value bounded by $w/2$. We'll refer to $T^m(w)$ as the m-torus of width w. There is nothing special about the widths being the same in all dimensions, except that this case is easier to analyze.

The distance between a pair of coset representatives x and x' of $T^m(w)$ is defined to be

$$\mathrm{dist}(x, x') = \| \, (x - x')_w \, \| \,, \tag{7.3}$$

where $\| \, \|$ is the standard Euclidean norm and $(\ )_w$ "reduces" the difference of m-tuples so all elements have absolute value bounded by $w/2$ (by adding an integer multiple of w). The maximum absolute value over all elements is usually written as the ∞-norm and in this case is the statement

$$\|(x - x')_w\|_\infty \leq w/2 \,.$$

The projection to a set (of cosets) $A \subset T^m(w)$ is defined with respect to the periodic variant of distance:

$$P_A(x) = \arg\min_{y \in A} \| (y - x)_w \| .$$

This definition is consistent with a reformulation in $\mathbb{R}^m$ with the periodic set $\tilde{A} = A + w \, \mathbb{Z}^m$ replacing A.

A good example of the effects of the torus geometry on projections is the concur constraint. Suppose we need two copies of a variable $x \in T^1(w)$ in a divide-and-concur formulation of a problem in the 1-torus (circle). This is the same as working with variables $(x_1, x_2) \in T^2(w)$ and the concur constraint

$$B = \left\{ (x_1, x_2) \in T^2(w) : x_1 = x_2 \right\} .$$

Reformulated in $\mathbb{R}^2$, B is replaced by a set of parallel lines $\tilde{B}$ of slope 1 and passing through all the points of $w \, \mathbb{Z}^2$. Because $\tilde{B}$ is nonconvex, some points – those exactly halfway between the lines – have nonunique projections:

$$P_{\tilde{B}}(w/2, 0) = \left\{ (w/4, w/4), (3w/4, -w/4) \right\} .$$

As long as projections are computed with distance (7.3), the RRR algorithm is unchanged because a constraint problem in T^m is equivalent to a constraint problem in $\mathbb{R}^m$ with periodic constraint sets. However, the update rule gets one minor modification:

$$x \mapsto \mathcal{R}(x) = x + \beta \Big(P_B(R_A(x)) - P_A(x) \Big)_w .$$

Without the reduction step, projections $x_{\tilde{A}}$ and $x_{\tilde{B}}$ to the periodic sets and differing by a nonzero element of $w \, \mathbb{Z}^m$ would not give a fixed point and be recognized as solutions!

7.2.2 Periodic Concur

When using divide-and-concur to pack k spheres periodically into a hypercubic unit cell of width w in d dimensions, each sphere center has $n = k - 1$ replicas that live in the product of n tori $T^d(w)$. The replicas make projecting to the disjointedness constraint (set A) tractable because it involves independent pairs of spheres. The concur constraint (set B) ensures that the n replicas are equal, but with the qualification that their coordinates can differ by an element of $w \, \mathbb{Z}^d$. Since the equality constraint, even with periodicity, is independent over the d dimensions, we will analyze the subproblem of projecting to the set

$$B = \{x \in T^n(w) : \; x_1 = \cdots = x_n \quad \mathrm{mod}\; w\} \, .$$

For example, the variables $x_1, \ldots, x_n$ above may correspond to the seventh of d coordinates for each of n replicated spheres.

Anticipating that replicas for pairings with nearby spheres should have larger metric coefficients than the replicas for distant spheres, we will need the projection P_B for the metric

$$\mathrm{dist}(x, x') = \sum_{j=1}^{n} g_j\big((x_j' - x_j)_w\big)^2 \tag{7.4}$$

in our subproblem. To keep formulas simple, we assume the coefficients satisfy

$$\sum_{j=1}^{n} g_j = 1 \, .$$

Because our packing experiments will have several thousand replicas, it is important the computation of P_B scales linearly with n, that is, no worse than concur without the periodicity wrinkle. Our theorem is stated with this consideration in mind:

Theorem 7.1 (Periodic concur) *The projection of a point $(x_1, \ldots, x_n)$ to set B with metric (7.4) is the point*

$$(\bar{x}_i, \ldots, \bar{x}_i) \, , \tag{7.5}$$

where $\bar{x}_i$ is computed as follows. First reduce the numbers $x_1, \ldots, x_n$ mod w (to occupy an interval of size w) and relabel them (and the corresponding metric coefficients) so that

$$x_1 < \cdots < x_n \, . \tag{7.6}$$

Next, compute the average

$$\bar{x}_1 = \sum_{j=1}^{n} g_j\, x_j \, , \tag{7.7}$$

and define shifted averages by the recursion

$$\bar{x}_{j+1} = \bar{x}_j + g_j w \, . \tag{7.8}$$

Then, starting with the squared distance

$$d_1^2 = \sum_{j=1}^{n} g_j (x_j - \bar{x}_1)^2 \, , \tag{7.9}$$

(of the unshifted average) and the recursion

$$d_{j+1}^2 - d_j^2 = 2g_j w(x_j - \bar{x}_j) + (g_j - g_j^2)w^2 \ , \qquad (7.10)$$

compute the differences

$$\Delta_j = d_j^2 - d_1^2 \qquad (7.11)$$

for $j = 1, \ldots, n$. The index i of the smallest difference Δ_i is the index of the shifted average $\bar{x}_i$ to be used in (7.5).

Proof Because the concurring coordinate x^* of the projection minimizes

$$d^2 = \sum_{j=1}^n g_j \big((x^* - x_j)_w\big)^2 \ ,$$

the squared distance may be rewritten

$$d^2 = \sum_{j=1}^n g_j (x^* - x_j)^2 \ ,$$

with the understanding that the numbers $x_1, \ldots, x_n$ have been reduced mod w so the differences $x^* - x_j$ lie between $-w/2$ and $w/2$, since otherwise a shift by an integer multiple of w decreases d^2. Moreover, the value of x^* has no effect on the cyclic order of $x_1, \ldots, x_n$ on the line of real numbers. We therefore have n cases to consider, depending on which of these numbers is smallest. Take one such case and relabel the numbers so the numbers are ordered as (7.6). Call this case 1, for which we already know by Proposition 4.6 for the non-periodic case that $x^* = \bar{x}_1$, the simple average in (7.7). The squared projection distance for this case is (7.9). All the other cases, where x_2, then x_3, and so on is the smallest number, involve adding w first to x_1, then both x_1 and x_2, and so on. The corresponding distance-minimizing concur values are given by recursion (7.8) and recursion (7.10) gives the corresponding change in the squared projection distance. Finally, by keeping track of the squared differences relative to case 1 by the differences (7.11), the case i with the smallest squared distance is identified and the corresponding shifted average $\bar{x}_i$ gives the concur value. $\qquad \square$

7.2.3 Periodic Disjoint Spheres

To modify the disjoint-spheres projection (Proposition 4.13) when there is periodicity, we only need to modify the more fundamental projection that imposes the norm $2r$ (twice the sphere radius) on the difference-of-centers vector u. Without periodicity, that projection is just the rescaling $u' = (2r/\|u\|)u$.

For vectors u defined by the difference of two points in $T^d(w)$, one may impose the constraint $\|u\|_\infty \leq w/2$. We will later see, when projecting to disjoint spheres, that this same constraint should be included in the usual norm constraint:

$$N(2r) = \left\{ u \in \mathbb{R}^d : \|u\|_\infty \leq w/2 , \|u\| = 2r \right\} .$$

The projection to this constraint has some similarities with the projection to the probability simplex (Section 4.3.2). The cases we had to consider there involved the number of nonzero probabilities. Here the cases correspond to the number of elements of u with absolute value $w/2$. The inequality $2r \leq w$ restricts that number of cases to five.

Proposition 7.1 (Projection to norm constraint with periodicity) *Relabel the components of $u \in \mathbb{R}^d$, $\|u\|_\infty \leq w/2$, so the absolute values are decreasing:*

$$|u_1| > \cdots > |u_d| .$$

The projection of u to set $N(2r)$ is the vector $u' = u'_\| + u'_\perp$ where $\|$ is the space defined by the first ℓ components of u, $\perp$ is the orthogonal complement,

$$u'_\| = (w/2)\, \mathrm{sgn}\, u_\| ,$$
$$u'_\perp = (\lambda/\|u_\perp\|)\, u_\perp , \tag{7.12}$$

$$\lambda = \sqrt{(2r)^2 - \ell(w/2)^2}$$

and $\ell \in \{0, 1, 2, 3, 4\}$ is the smallest for which $\|u'_\perp\|_\infty \leq w/2$.

Proof We make use of two elementary facts: The projection preserves the signs of the components, and by the order-preserving matching Lemma 4.5,

$$|u'_1| \geq \cdots \geq |u'_d| .$$

Because $2r \leq w$, at most four components can equal $w/2$ in absolute value. These are first in the ordering and their number ℓ satisfies $0 \leq \ell \leq 4$. The distance-minimizing change to the remaining components that satisfies the periodicity-unrestricted norm constraint is the rescaling (7.12). This is an admissible projection point of $N(2r)$ only if $\|u'_\perp\|_\infty \leq w/2$. Let ℓ^* be the smallest ℓ where this is true in the strict sense, where the first component of $u'_\perp$ has absolute value less than $w/2$. With probability 1, the ℓ to be used in the projection therefore satisfies $\ell \geq \ell^*$. But if some $\ell > \ell^*$ were used instead of $\ell = \ell^*$, then the vector $u'_\perp$, padded with $\ell - \ell^*$ suitably signed $w/2$ components (to also have $d - \ell^*$ components), disagrees with the true distance minimizing projection to the same norm (having components strictly less than $w/2$ in absolute value). We therefore know that $\ell = \ell^*$.   □

To pack k spheres with divide-and-concur, one uses $k(k-1)$ replicas x_{ij} for the sphere centers. Replica ij of sphere i is the one tasked with being disjoint from sphere j, and vice versa for replica ji. Here is the disjoint-sphere constraint for such a pair, modified for the torus geometry:

$$A_{ij} = \left\{ (x_{ij}, x_{ji}) \in T^d(w) \times T^d(w) : \; \|(x_{ij} - x_{ji})_w\| \geq 2r \right\} . \qquad (7.13)$$

As in the projection to the norm constraint, the sphere radius and torus width satisfy the inequality $r \leq w/2$. The projection to this constraint uses the projection to the periodicity-restricted norm constraint:

Theorem 7.2 (Periodic disjoint sphere projection) *Let* $u = (x_{ij} - x_{ji})_w$ *be the reduced difference vector of the projection input. If* $\|u\| \geq 2r$*, the projection output* (x'_{ij}, x'_{ji}) *is unchanged, otherwise,*

$$\begin{aligned} x'_{ij} &= x_{ij} + (u' - u)/2 \,, \\ x'_{ji} &= x_{ji} - (u' - u)/2 \,, \end{aligned} \qquad (7.14)$$

$$u' = P_{N(2r)}(u) \,.$$

Proof From equations (7.14) we see that

$$(x'_{ij} - x'_{ji})_w = u'$$

identifies u' as the reduced difference vector of the projection. These equations are also consistent with the elementary fact that the components of the difference vector maintain sign and only increase their magnitude when increasing the norm in a distance-minimizing way. Moreover, the change in the difference vector is shared equally and oppositely by each pair of coordinates. Since the periodicity restriction also applies to u', the distance-minimizing change $u' - u$ to a vector to have norm $2r$ is exactly how $P_{N(2r)}(u)$ was defined.   □

7.2.4 Packing Spheres Periodically with RRR

The most general metric compatible with the projections presented in the two previous sections is one which distinguishes replica pairs:

$$\operatorname{dist}(x', x) = \sum_{i=1}^{k-1} \sum_{j=i+1}^{k} g_{ij} \left(\|(x'_{ij} - x_{ij})_w\|^2 + \|(x'_{ji} - x_{ji})_w\|^2 \right) .$$

The coefficients g_{ij} have no effect on the projection to the disjoint-sphere constraint (7.13) and introduce replica weighting in the concur projection (7.7). Replica ij of sphere i receives weight g_{ij} if $i < j$, and weight g_{ji} otherwise. The auto-tuning heuristic for the weights (metric coefficients) is

$$g_{ij} \rightarrow g_{ij} + \gamma \left(\Delta_{ij}^2 / \Delta^2 - g_{ij} \right) ,$$

where

$$\Delta_{ij}^2 = \|(x_{ij}^A - x_{ij}^B)_w\|^2 + \|(x_{ji}^A - x_{ji}^B)_w\|^2$$

is the replica-pair-specific constraint discrepancy and Δ^2 is the average discrepancy.

For the initial RRR point, we made the choice that it exactly satisfied the concur constraint but was otherwise as random as possible. That is, for each of the k spheres being packed in $T^d(w)$, a center was generated with all d coordinates uniformly sampled between 0 and w. The corresponding $k-1$ replicas were initialized to the same coordinates. Interestingly, with this initialization and $\beta = 0.5$, the RRR gap first *increases* for several iterations (depending on d). This suggests that exact concurrence is not representative of the configurations that arise in a divide-and-concur packing search.

7.2.5 The Best Packing in 10 Dimensions

As far as anyone knows, 10 dimensions is the lowest where the densest packing is *not* achieved by a lattice arrangement of the spheres [23]. The champion nonlattice (though still periodic) packing was discovered by Marc Best [11] as a byproduct of finding Hamming-distance-4 binary codes with the greatest number of words, 40. As explained earlier, such codes give the coordinates of a packing of spheres in the tightest possible torus, $T^{10}(1)$, for spheres of diameter 1. From Figure 7.2 we see that only about 10% of a sphere's surface is available for tangencies with other spheres because of the periodicity restriction in 10 dimensions. Finding packings in these circumstances really puts the periodicity-restricted disjoint spheres projection to the test!

Figure 7.3 shows the RRR gap in a successful run with $\beta = 0.5$ and $\gamma = 0.01$. There is trapping on near-solutions when both of these parameters are smaller. The success rate is about 28% when the iteration limit is set at 10^5. Because each sphere is tangent to 22 of the other spheres in a solution, this instance of the sphere packing problem is nonregular because the constraints are not locally affine near solutions. Though fixed-point convergence may not be as good as in a regular formulation, we see from Figure 7.3 that this is not a problem. Shortly after the gap plummets, the coordinates are so

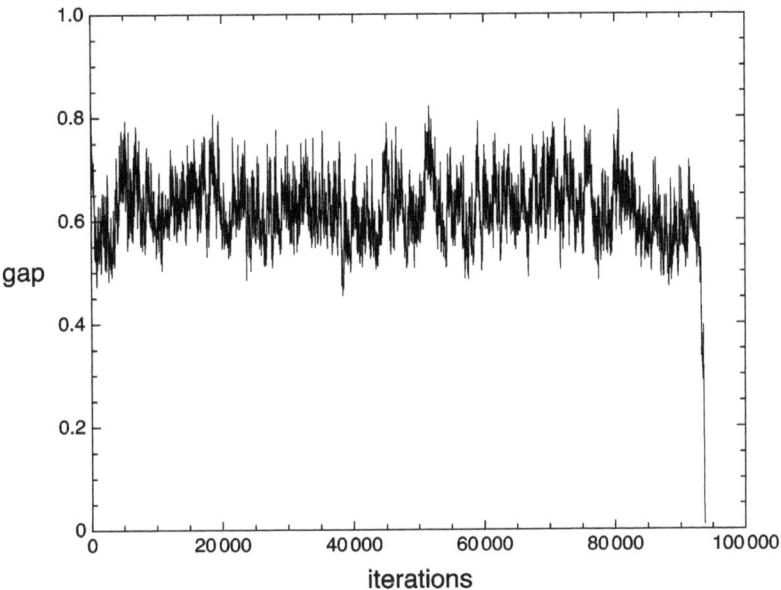

Figure 7.3 RRR gap when packing 40 spheres in a 10-dimensional unit-width hypercubic unit cell.

Figure 7.4 The 40 columns of this array are the sphere coordinates in Best's packing in 10 dimensions [11], where gray and black correspond to the numbers 0 and 1/2. All pairs of columns having the minimum of four contrasting elements are separated by Euclidean distance 1.

close to being binary that a solution's integrity can be confirmed by rounding and checking the Hamming distances. A solution, or Best's code, is rendered in Figure 7.4. Best also established the uniqueness, up to symmetries, of this code [11].

7.2.6 Rigorous and Empirical Density Lower Bounds

It's unlikely that packing spheres in the torus with RRR will break any density records. The Best packing in 10 dimensions was a happy coincidence, where an unusually dense packing aligned with the periodicity for which we have a simple divide-and-concur formulation. But while there is no doubt that finding the densest packing in any given dimension is a noble pursuit, there is another sphere packing question where RRR can at least provide interesting information.

Every explicit sphere packing construction in d dimensions serves as a lower bound on achievable densities in that dimension. But lower bounds obtained in this manner may not be very useful for general dimensions d, if the construction relies on features that are special to a particular d. In a *general-dimension lower bound* only the dimension d may appear as a parameter. The survey below will make clear that such bounds are still a wide open problem! Most striking is the fact that the bound derived from the obvious general-d lattice candidate quickly becomes uninteresting in high dimensions.

Because it is independent of the sphere-radius convention, we will use the fraction of space covered by spheres, the *density* Δ, to compare packings. For example, the density of the hypercubic packing, with spheres centered on the integer lattice, is

$$\Delta_{\text{cubic}} = v_d(1/2) \, ,$$

where

$$v_d(r) = \frac{\pi^{d/2}}{\Gamma(d/2 + 1)} \, r^d$$

is the volume of the d-dimensional ball of radius r. But as every grocer who has stacked oranges knows, there is a much better general-dimension lattice scheme called the *checkerboard*, where spheres are placed on just the "white" lattice points (those with even coordinate sum). In $d = 2$ this is just a rotated square lattice (no change from Δ_{cubic}), but starting with $d = 3$ where it is also known as *face-centered-cubic*, the density is increased:

$$\Delta_{\text{fcc}} = 2^{d-1} \, v_d(1/\sqrt{2})$$
$$= 2^{d/2-1} \, \Delta_{\text{cubic}} \, .$$

Thanks to a heroic effort this was proved optimal for $d = 3$ [52], and it is also conjectured to be optimal for the next two dimensions. However, if we mostly care about high dimensions, there is a much better and even simpler packing we overlooked.

A *saturated packing* is any packing of spheres in $\mathbb{R}^d$ with the property that no additional spheres can be added (without intersecting a sphere already in the packing). There's a simple proof that the density of such packings always exceeds this bound:

$$\Delta_{\text{sat}} = 2^{-d} \, .$$

Let the spheres have radius r. First note that any point not covered by a sphere will be within distance $2r$ of one of the sphere centers. Otherwise, a sphere could be placed at that point without intersecting another sphere and the packing would not be saturated. But if all the uncovered points are within distance $2r$ of a sphere center, then doubling all the sphere radii (from r to $2r$) would cover even these points. This radius-doubling increases each sphere volume by a factor 2^d and could not account for the complete covering of all of space unless the fraction covered before doubling was at least 2^{-d}.

Comparing saturated and checkerboard packings,

$$\Delta_{\text{sat}} / \Delta_{\text{fcc}} = 2 \, \frac{\Gamma(d/2+1)}{(2\pi)^{d/2}} \, ,$$

we find that saturated packings are denser starting at $d = 28$, and by a ratio that grows alarmingly as $d^{d/2}$. There are lattices with packing density much greater than Δ_{sat} in these dimensions [25], but unlike the checkerboard lattices, these do not fall into infinite families for general d.

Lattices become competitive again if we admit lattices whose existence is established much less directly than an explicit construction. The first step in that direction, by Hermann Minkowski and proved by Edmund Hlawka [54], was a lower bound for general dimension that only doubled Δ_{sat}. Keith Ball [5] obtained the improved bound

$$\Delta_{\text{Ball}} = 2(d-1)\Delta_{\text{sat}}$$

by using a lemma proved by Thøger Bang [6] of a conjecture by Alfred Tarski about covering convex bodies in d dimensions with "planks." Ball's lower bound beats the checkerboard lattice above $d = 16$.

Disordered packings, but with enhancements over the saturated construction, made a comeback in 2004 when Krivelevich, Litsyn, and Vardy [65] obtained a bound that matched Ball's but with a worse overall constant. The factor of d was gained using graph-theoretic techniques. The current state-of-the-art in this kind of bound is the result

$$\Delta_{\text{graph}} = \frac{d \log d}{2} \Delta_{\text{sat}}$$

by Campos et al. [19] that adds a $\log d$ factor. Because of the worse constant factor, this does not surpass Ball's bound until 50 dimensions.

Whether constructed explicitly or not, all the lower bounds above are

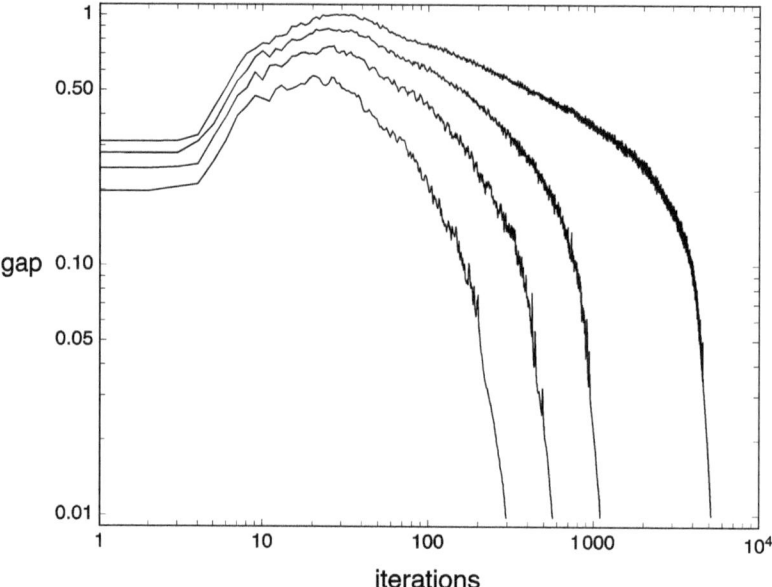

Figure 7.5 RRR gaps when packing spheres into the 17-dimensional unit-width torus to achieve densities 1, 1.5, 2, and 2.5 times Ball's lower bound.

rigorous mathematical results. On the other hand, arriving at reasonably definitive answers to questions such as,

Can spheres in 17 dimensions be packed at twice the density of Ball's bound?

is another matter. Figure 7.5 answers this question in the affirmative. It shows the RRR gap when packing 227, 341, 454, and 568 spheres into the 17-dimensional unit-width torus. The corresponding densities are respectively 1, 1.5, 2, and 2.5 times Δ_{Ball}. For the first few dozen iterations the gap increases, which we interpret as divide-and-concur finding more sensible replica configurations than the concurring replicas of the initial point. As in the search for Best's packing we used $\beta = 0.5$, but decreased the metric relaxation parameter to $\gamma = 0.001$.

Though the plots in Figure 7.5 do not show the gaps arriving at zero, it is easily checked that the final small value is due to several sphere pairs (concur projections) having separations very slightly smaller than 1, rather than a few spheres intersecting significantly. By slightly shrinking all the spheres in the incompletely converged solution, we get a true solution with only a slightly worse density. The downward curvature of the gaps also gives confidence that arbitrarily small gaps are within the scope of the algorithm. That the horizontal (iteration) axis is logarithmic does not change this conclusion. To demonstrate that a packing exists, we only care that the work involved is finite.

We will refer to lower bounds obtained by the method just described as *empirical* general-dimension bounds. They can never replace the rigorous general-dimension bounds, but neither do they have the narrow scope of bounds obtained by exhibiting an isolated, symmetry-inspired packing. The 568-sphere packing in 17 dimensions constructed by RRR is disordered and qualitatively similar to 569-sphere packings the algorithm will surely succeed in constructing as well. The equanimity of the construction of empirical bounds with RRR, bounds that surpass the rigorous bounds by large factors, can be interpreted in two ways. First, it may be the case that the computationally feasible dimensions for empirical bounds fall short of the "asymptotic regime," and empirical bounds have (for whatever reason) an unfair advantage in low dimensions. However, it's also possible the sacrifices made in formulating bounds for which proofs can be produced are sufficiently great that the shortfall relative to empirical bounds is not a low-dimension artifact at all. Understandably, in this book we take the second position.

7.2.7 Princess Packings

Formulating hypotheses that invoke *transitions* is second nature for physicists and not all that farfetched in the context of packing spheres with divide-and-concur. Though the dynamics is non-Newtonian, the "laws" behind each iteration are no less particle-like. Disordered fluid packings are qualitatively different from their denser, solid counterparts. Position correlations in fluids decay on the scale of a few sphere diameters, but in a solid extend throughout the entire packing – crystals being an extreme case. Because of the weaker correlations, fluid packings are easier to find and the work is less cooperative in nature than packing spheres in a solid. And if there is a true transition in the qualitative behavior of the algorithm as a function of density, maybe this can be used to define an empirical lower bound.

We have evidence of a transition when an unreasonably large change is caused by a small change in a parameter. In our case, for finding sphere packings in d dimensions with divide-and-conur, the evidence is a dramatic increase in the work required to find packings when the density exceeds some threshold, $\Delta_\circ$. Packings poised right at the hypothesized transition are called *princess packings*, a historical reference to the princess whose sleep-restfulness was sensitive to the presence of even a single sphere (in the form of an uncooked pea) [2]. The corresponding density, $\Delta_\circ$, is the *princess density* for that dimension. This density is de facto an empirical packing-density

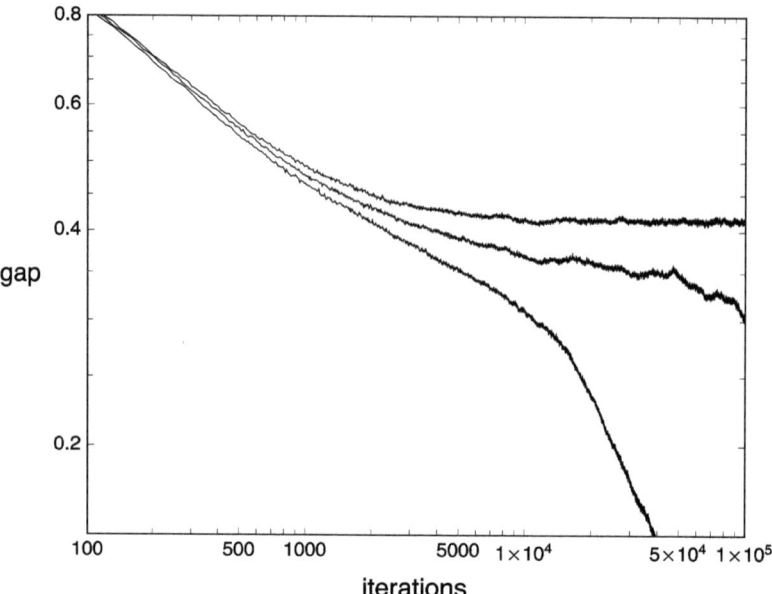

Figure 7.6 RRR gaps when packing 605, 610, and 615 spheres into the 17-dimensional unit-width torus. The 610-sphere packing, at the transition between easy and hard searches, would be a princess packing when completed.

lower bound since it comes with the understanding that packings below $\Delta_\circ$ exist because they are easy to find.

Figure 7.6 presents a striking contrast with Figure 7.5 in that the density differences are only 0.8%. The highest density experiment, with 615 spheres, is qualitatively different from the others in that the gap has leveled out. This is what we expect when the spheres are not just negotiating for space in their immediate neighborhood but have to work out a more solid-like arrangement that involves all the spheres in the torus. RRR searches for super-dense arrangements, like Best's packing in 10 dimensions, are well described by this kind of gap. We expect the gap in searches for disordered solid-like packings to be similar. Additional evidence of a transition is the enhanced gap fluctuations at the intermediate density (610 spheres). All three plots are, in fact, gaps averaged over 20 runs.

The results just described give the estimate $\Delta_\circ \approx 2.69 \, \Delta_{\text{Ball}}$ for the princess packing density in 17 dimensions. As explained above, this (or a slightly lower value where the completion of the search can be more convincingly demonstrated) serves as an empirical density lower bound for that dimension. The large disparity with the best rigorous, general-dimension bound is hard to ignore. In 18 dimensions it appears the transition occurs at approximately 1 125 spheres, with density $\Delta_\circ \approx 2.72\,\Delta_{\mathrm{Ball}}$. Before interpreting the trend, we have to remember that these disordered packings are still subject to a rather tight unit cell! One could study these effects by repeating the experiments with torus widths $w > 1$. But increasing w by even 10% in 18 dimensions brings a more than fivefold increase in the number of spheres, or a factor of 30 more time and memory because these scale with the number of replica pairs. Leveraging the computational advantages of divide-and-concur for gaining evidence that informs the density lower bound question is the subject of ongoing research [39].

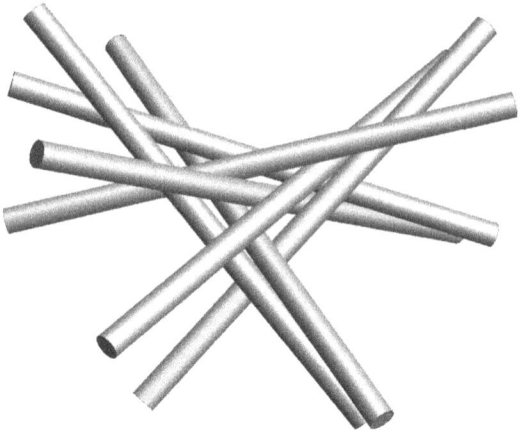

Figure 7.7 Seven equidistant lines in three dimensions rendered as mutually tangent, congruent cylinders. The seven lines are close to being in a common plane.

7.3 Equidistant Lines

In d dimensions there can be at most $d+1$ equidistant points, and the configurations are the regular simplices (equilateral triangles, regular tetrahedra, and so on). The corresponding problem for (infinite) lines is completely open, even for d as small as 3. The distance between a pair of lines is the minimum distance between all pairs of points, one on each line. We are free to set the (nonzero) common distance, between all pairs of lines, equal to 1. In two dimensions, equidistant lines must be parallel (for the distance to be nonzero) and there can be at most two lines.

The equidistant-lines problem quickly becomes complex in higher dimensions. In three dimensions it's possible to have seven equidistant lines, and Figure 7.7 shows such a configuration. The lines are rendered as unit-diameter cylinders. Because a pair of unit-distance lines is the same as a pair of tangent unit-diameter cylinders, the equidistant-lines problem is also known as the problem of finding arrangements of mutually tangent congruent (infinite) cylinders.

It's far from obvious that the configuration shown in Figure 7.7 is a solution for seven lines. There is no symmetry or known construction principle that can be used in a proof. The problem can be formulated as solving a system of 20 polynomial equations, and Bozóki et al. [14] found two numerical solutions of the system in a systematic, month-long search. After refining the solutions to 12 digits of precision, they were able to use Steve Smale's α-theory [102] to rigorously establish the solutions' existence.

In Exercise 3.10 you derived the formula

$$d^*(d, n) = 2(d - 1)n - d(d + 1)/2 - n(n - 1)/2 \qquad (7.15)$$

for the virtual dimension of the set of configurations of n equidistant lines in d dimensions. The values $d^*(3, 7) = 1$ and $d^*(3, 8) = -2$ are consistent with John Littlewood's problem 7 [70] :

Is it possible in 3-space for seven infinite circular cylinders of unit radius each to touch all the others? Seven is the number suggested by constants.

To get discrete solutions, instead of one-dimensional continua, Bozóki et al. additionally constrained one pair of lines to be orthogonal.

In this section we use divide-and-concur to construct equidistant-line configurations in three and four dimensions. After a chaotic search, the RRR algorithm's gap exhibits linear convergence and with enough work would reach the level of precision required by Smale's numerical solution certifier. In four dimensions we looked for configurations of 11 lines, since $d^*(4, 11) = 1$, but were unsuccessful. The 10-line configuration we found, with virtual dimension 5, resembles the 7-line configuration in that all the lines are close to a 2-plane.

7.3.1 The Proximal Pair Representation

In any number of dimensions, two points suffice for specifying a line. For configurations of n lines, $n - 1$ replicas of such point pairs for each line could be used in a divide-and-concur scheme to impose the distance-1 constraint between all pairs of lines. Exercise 4.13 even dealt with the projection to the distance constraint in this representation. Concur in this scheme just replaces the $n - 1$ point-pair replicas with their average.

We should have second thoughts about the point-pair representation when the number of contacts (cylinder tangencies) per line, $n - 1$, is much greater than two. Small shifts in the two points can translate to a large re-orientation of the line, especially when the distance between the points is small. Amplification of small changes is good from the perspective of chaos, but in this case

Figure 7.8 Three lines in three dimensions showing all three pairs of proximal points and the associated proximal vectors.

may be going too far if it takes many constraints out of the conversation! A more general criticism is that non-unique representations should be avoided because they may bias or confound the search in unproductive ways. The two-point representation of lines is nonunique and clearly dangerous when the two points are near each other.

A better representation of lines for the problem at hand is based on *proximal pairs*. Figure 7.8 explains the idea in three dimensions when there are three lines. In any number of dimensions $d > 2$ and a general number of lines $n > 1$, each line has $n - 1$ points that are proximal, or nearest to, one of the other $n - 1$ lines. Associated with each proximal point is a proximal vector. Let x_{ij} be the point on line i that is proximal to line j. The proximal vector v_{ij}, when added to x_{ij}, is the point x_{ji} on line j that x_{ij} is proximal to. We give the proximal vector associated with x_{ji} its own name, v_{ji}, even though we know it is just the negative of v_{ij}.

The proximal vectors, in addition to relating pairs of proximal points, have two properties that together express the distance-1 constraint between lines in any number of dimensions. The first is that v_{ij} is orthogonal to line i (the line on which the associated proximal point x_{ij} sits), and the second property is that it has norm 1.

If we can separate all these constraints into sets A and B that have efficient projections, we will be able to use RRR to find configurations of equidistant

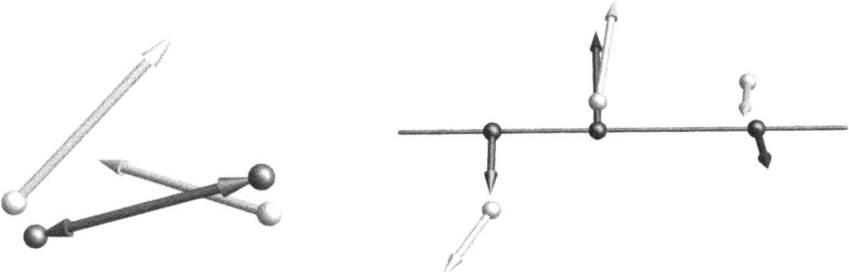

Figure 7.9 *Left*: Divide projection applied to a pair of proximal point-vector pairs. After the projection (dark points and vectors), the norms of the vectors and the distance between the points equals 1. *Right*: Concur projection applied to a triple of proximal point-vector pairs. After the projection (dark), the points are collinear and the vectors are orthogonal to the line.

lines. Divide-and-concur is our guide for how this is done. All the competing contact constraints acting on a single line are constraints we would like to divide into independent constraints. Forcing all $n - 1$ proximal points of a line to be collinear, and have that line also be orthogonal to all its proximal vectors, is concur, now generalized from point equality to subspace equality.

7.3.2 Divide Constraint and Projection

Configurations of $n(n - 1)$ proximal point-vector pairs in d dimensions live in the space $\mathbb{R}^{n(n-1) \times 2d}$. For the divide constraint, we use the following factorization:

$$A = \textstyle\prod_{i<j}^{\times} A_{ij} ,$$

$$A_{ij} = \left\{ (x_{ij}, v_{ij}, x_{ji}, v_{ji}) \in \mathbb{R}^{4d} \ : \ x_{ij} + v_{ij} = x_{ji} \, , \ v_{ij} + v_{ji} = 0 \, , \ \|v_{ij}\| = 1 \right\}.$$
$$(7.16)$$

Since proximal points are qualitatively different from proximal vectors, we apply a metric coefficient g to the squared changes of vectors relative to the squared changes of the proximal points. An example of the projection to this constraint for $g = 3$ is shown on the left of Figure 7.9. After the projection one proximal vector is the negative of the other, their norms are 1, and the proximal points have been moved so their separation is consistent with the vectors.

Below is the projection to constraint (7.16) with the 1-parameter metric.

Proposition 7.2 (Proximal point-vector pair projection)

$$P_{A_{ij}}(x_{ij}, v_{ij}, x_{ji}, v_{ji}) = (x'_{ij}, v'_{ij}, x'_{ji}, v'_{ji}),$$

$$x'_{ij} = \bar{x} - \hat{u}/2,$$
$$v'_{ij} = \hat{u},$$
$$x'_{ji} = \bar{x} + \hat{u}/2,$$
$$v'_{ji} = -\hat{u},$$

where

$$\bar{x} = (x_{ij} + x_{ji})/2,$$
$$\hat{u} = u/\|u\|,$$
$$u = x_{ji} - x_{ij} - 2g(v_{ji} - v_{ij}).$$

The proof is left as an exercise. Because this projection does not change the centroids of the proximal point pairs, that can only happen through the concur projection (Section 7.3.3). To keep the line configuration centered near the origin one can add the weak constraint that the norms of all the centroids are less than some r. All that this changes in the projection is to rescale $\bar{x}$ to have norm r when it exceeds that value.

7.3.3 Concur Constraint and Projection

In the concur constraint, the n sets of proximal point-vector pairs, on each of the n lines, are independent:

$$B = B_1 \times \cdots \times B_n,$$

$$B_i = \left\{ \prod_{j \neq i}^{\times}(x_{ij}, v_{ij}) \in \mathbb{R}^{2d(n-1)} : \forall_{j \neq i} \ (x_{ij} - \bar{x})p = (x_{ij} - \bar{x}), \ v_{ij}\, p = 0, \right.$$

$$\left. \text{for some } p \in P_1 \text{ and where } \bar{x} = \tfrac{1}{n-1}\textstyle\sum_{j \neq i} x_{ij} \right\}. \quad (7.17)$$

Here P_1 is the set of rank-1, $d \times d$ projection matrices. Constraint B_i says that the $n - 1$ proximal points lie on a line through their centroid as specified by some one-dimensional projector p, and the proximal vectors are in the orthogonal complement. An example of the projection to this constraint, for $n - 1 = 3$, is shown on the right in Figure 7.9.

The projection to constraint 7.17 is a special case of the construction we used in Section 5.5.7 to find the best least-squares affine approximation $\tilde{A}$ of a constraint set A, given samples $(x, P_A(x))$ of projections. The specialization is that the approximating affine set is one-dimensional.

Proposition 7.3 (Concur projection of proximal point vectors)

$$P_{B_i}\big(\ \ldots\ , (x_{ij}, v_{ij}) , \ldots \big) = \big(\ \ldots\ , (x'_{ij}, v'_{ij}) , \ldots \big) \, ,$$

$$x'_i = \bar{x} + (x_i - \bar{x})p \, ,$$
$$v'_i = v_i - v_i\, p \, ,$$

where x_i and v_i are $(n-1) \times d$ matrices comprising the $n-1$ proximal points and vectors and

$$p = P_{P_1}\big((x_i - \bar{x})^{\mathsf{T}}(x_i - \bar{x}) - g\, v_i^{\mathsf{T}} v_i \big)$$

is the projection to $d \times d$ rank-1 projectors.

The proof of the proposition closely follows that of Theorem 5.4. There the vector $P_A(x) - x$ of each sample tries to be orthogonal to the approximation $\tilde{A}$, just like each of the proximal vectors tries to be orthogonal to its line.

7.3.4 Equidistant Lines in Three and Four Dimensions

Even when blessed with a natural set of variables (proximal point-vector pairs), a tunable metric coefficient, and small β to combat nonconvexity, equidistant line configurations are hard to find!

The gap in most runs of RRR transitions from an initial chaotic epoch to something markedly less chaotic, though still exhibiting large amplitude fluctuations. The mechanism behind this behavior is revealed, in three dimensions where this is possible, by rendering the colinear points of the concur projection (as cylinders). The cylinder axes start out with a wide range of directions, thanks to the random initialization of the variable replicas. But in the less chaotic regime that nearly always ensues, a large fraction of the cylinders – often all of them – form a compact, nearly parallel bundle!

This behavior makes sense because a pair of near-parallel cylinders has the potential to be tangent over much more of their length than cylinders whose axes subtend a large angle. Axis alignment develops irreversibly because of the property that if ℓ_1 is near-parallel to ℓ_2, and ℓ_2 is near-parallel to ℓ_3, then ℓ_1 is automatically near-parallel also to ℓ_3. When the lines are exactly parallel, the equidistant-lines problem in d dimensions has degenerated into the equidistant *points* problem in $d-1$ dimensions. But that limits the number of lines (points) to d, which we know is well below what is possible.

Parallelitis could be addressed by introducing additional vector variables for imposing a minimum angle between all pairs of lines. A unit vector u_{ij} would specify the direction of line i according to line j, u_{ji} the direction of line

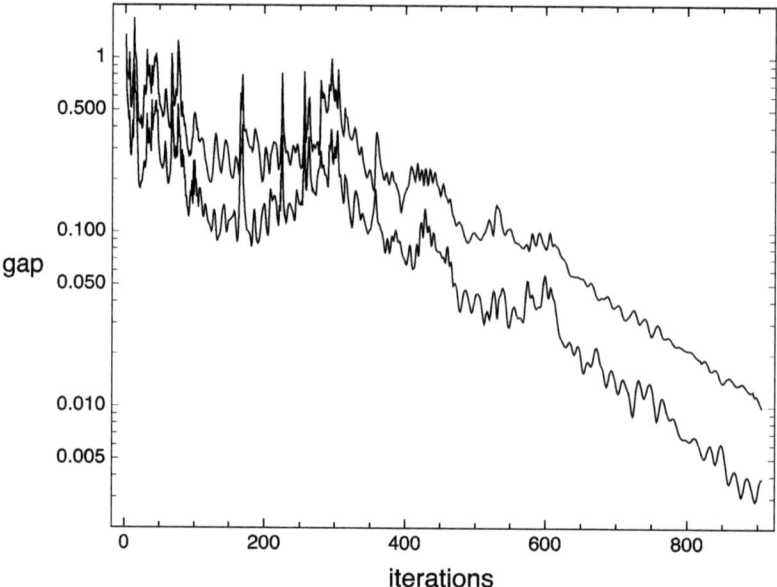

Figure 7.10 RRR gap in the discovery of a configuration of seven equidistant lines in three dimensions. The gap is broken down into a contribution from the proximal points and the proximal vectors. In this run, with metric coefficient $g = 3$, the vector contribution (lower curve) is suppressed relative to the point contribution.

j according to line i, and a divide constraint between these would impose their normalization and a minimum subtended angle. In the B constraint these variables would participate in the concur of the vectors defined by a line's proximal points. The new variables would have their own metric coefficient.

But there is a much simpler and lazier way to cure parallelitis: restart RRR as soon as the disease sets in! That's how the successful runs shown in Figures 7.10 and 7.11 were obtained. The restart criterion is based on the $n \times d$ matrix u of unit-normalized axis vectors produced by the concur projection. Since $\operatorname{Tr} u^{\mathsf{T}} u = n$, the squares of the singular values σ of u sum to n. Parallelitis is indicated if the square of the largest singular value is already close to n. In the search for seven lines in $d = 3$, RRR was restarted when $\sigma_{\max}^2 > 4.7$; for 10 lines in $d = 4$, the criterion was $\sigma_{\max}^2 > 6.5$.

Both searches used $\beta = 0.5$. Reducing the compliance of the proximal vectors relative to the proximal points, with $g = 3$, clearly helped in $d = 3$. In $d = 4$ this did not confer much of an advantage and the search used $g = 1$. Finding 10 equidistant lines in four dimensions proved to be much easier than seven lines in three dimensions, probably because the virtual

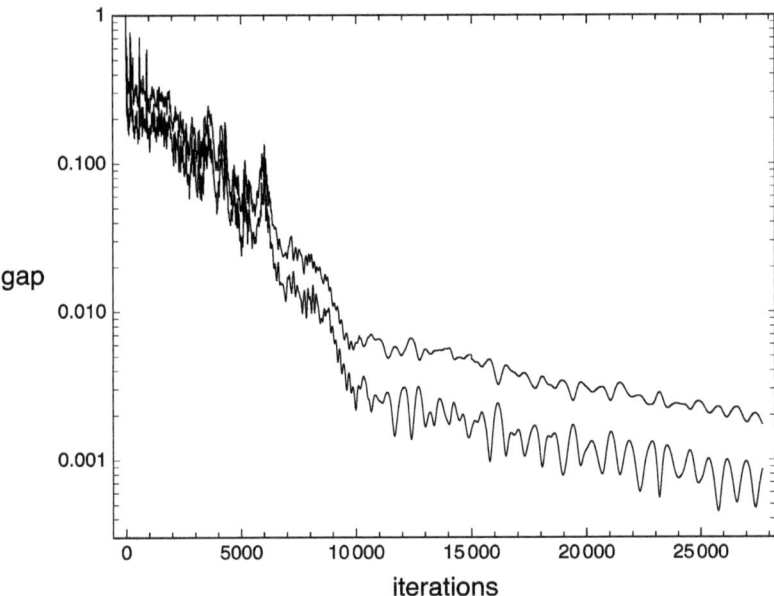

Figure 7.11 Same as Figure 7.10 but for 10 lines in four dimensions and with metric coefficient $g = 1$.

dimension $d^*(4, 10) = 5$ is larger. With the restart policy described earlier, the number of RRR iterations per solution was under 10^4 for the 10-line configurations and about 10^6 for seven lines in three dimensions.

Like the seven-line configurations in three dimensions that can be visualized, the 10 equidistant lines in four dimensions always had axes close to a two-dimensional plane. Here are the singular values of the matrix u in four solutions:

$$
\begin{array}{llll}
2.27603 & 2.11248 & 0.519686 & 0.294993 \\
2.27539 & 2.09839 & 0.532909 & 0.367907 \\
2.28142 & 2.14499 & 0.373781 & 0.233293 \\
2.23998 & 2.13144 & 0.571246 & 0.336407
\end{array}
$$

7.4 SEQVENCE

Put yourself in the position of the Creator, or more specifically, the subordinate in charge of living things. Your plan is to create organisms out of molecular building blocks, the number of which will be enormous. The problem is that you don't have the resources to build the correspondingly large number of factories. After all, your charge is to create life, not factories.

Luckily, you hit upon an idea so brilliant even your divine peers take notice. The trick is to use a handful of amino acids as the raw material and combine these into covalently bonded polymer chains. The polymers/building blocks differ only in how the amino acids are sequenced. Provided a universal factory can be built that is able to assemble any sequence it is given, the building-blocks of life can be made to order by a single factory!

There are many practical engineering details. Perhaps the most important is that all of life is wet. That is, the building blocks and their assembly happen in an aqueous environment, called *the solvent*. Whenever the universal factory spits out a polymer, the solvent forces its shape to assume a unique three-dimensional form and violà, a building block is formed! This works because the solvent forces are much weaker than the polymer's covalent bonds, thereby preserving the integrity of the building block's sequence even in its 3D shape.

By now you've realized the building blocks are *proteins*, the universal factory is the *ribosome*, and the instruction sequence is another polymer called *mRNA*. After the newly minted polymer settles down into a particular shape in the solvent environment, we say it is *folded*. You may also know that solvent aversion plays a big role in the folding process. Some amino acids are like oil and would condense into droplets (if allowed by the polymer chain) to avoid contact with the solvent. A key part of the protein sequence's design is to make it possible for the polymer to fold in such a way that all of the oily parts are buried and not exposed to solvent. This is the principle behind encoding 3D shapes by 1D sequences.

Because it would be a shame if the sequence-to-shape principle required a PhD in biochemistry, the puzzle SEQVENCE[2] was created to convey the idea in the simplest possible terms. A different toy model of protein folding, that additionally was meant to address some theoretical questions was proposed in 1985 by Ken Dill [28]. SEQVENCE shares the simplicity of Dill's model in that there are only two types of amino acid: *hydrophobic* (H) and *polar* (P). H-amino acids are oily and shun solvent, so the protein tries to fold into a shape where only P-amino acids are on the surface. In Dill's model, each fold has a numerical score, and the object is to find the fold with the

Figure 7.12 The three H-tiles (dark) and three P-tiles (light).

best score. SEQVENCE is more like a puzzle in that folds are either valid or invalid. Probably the most important difference between the models is that SEQVENCE is fun to play. Unlike Dill's (more realistic) geometry of polymers on a 3D cubic lattice, the polymers of SEQVENCE meander in 2D, over the sites of a triangular lattice. This not only makes the structure of folds more transparent, but the puzzle can be played by placing hexagonal tiles on a flat surface.

SEQVENCE was motivated by pedagogy and only accidentally became a vehicle for showcasing divide-and-concur. It turns out that divide-and-concur can capture some elements of how humans go about solving the puzzle. But before we get around to that you first need to learn the rules.

7.4.1 Rules

The puzzle is played with the six hexagonal tiles shown in Figure 7.12. Dark tiles are the H-amino-acids, lightly shaded tiles are P-amino-acids. The white lines on the tiles show the polymer chain. In a valid fold, the chain forms a single continuous path with two ends. The tiles force the chain to make only four kinds of turns: sharp or wide, and either to the right or left (the chain never goes straight). The instructions for a *hexein* (a puzzle protein) are just the H/P sequence along its chain. We can think of a hexein's solvent environment as tiled by white S-hexagons. What makes SEQVENCE challenging are the restrictions on the adjacency of the H, P, and S tile types.

Figure 7.13 shows three hexeins, all with sequence length $\ell = 30$ and 10 H-tiles. We see that in each example the H-tiles are completely shielded from solvent by a single layer of P-tiles. These characteristics are formalized by two adjacency rules:

> Each **H-tile** is adjacent to at least one H-tile and never to solvent.

> Each **P-tile** is adjacent to solvent, exactly two other P-tiles, and at least one H-tile.

These rules strike a nice balance between not being too restrictive, so many

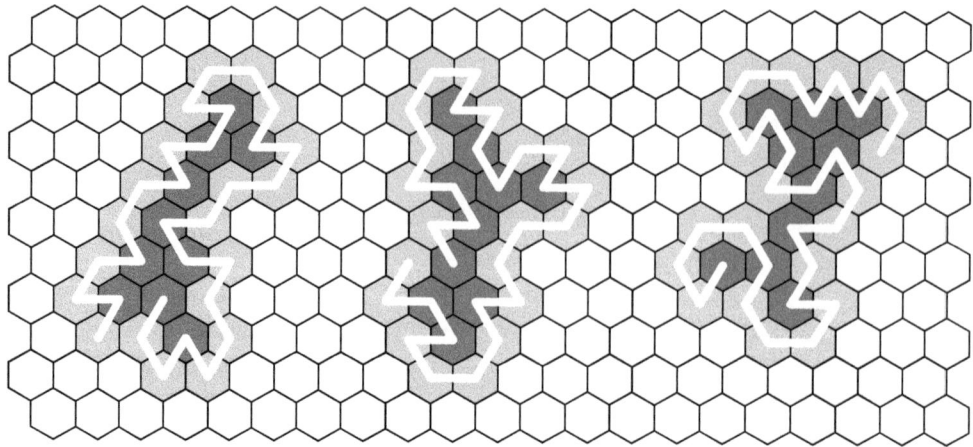

Figure 7.13 Three hexeins whose sequences have 10 H-tiles and 20 P-tiles. The surrounding white tiles are solvent (S). The hexein on the right is called `tripairin`.

valid folds exist, and being restrictive enough to confer uniqueness to many valid folds. About 11% of all 10/20 H/P-sequences, three million, have unique folds.

7.4.2 Litemotifs

Establishing that the hexeins shown in Figure 7.13 are the unique[3] folds for their sequences involved performing exhaustive depth-first searches – not the subject of this book. In brief, the chain is extended one tile (of the given type) at a time, backtracking (removing tiles and reextending by making a different turn) whenever it is impossible to extend the chain further in a way that respects the adjacency rules. If we count additions and removals of single tiles as moves, exhaustive searches for 10/20 sequences comprise about one million moves. Depth-first search is clearly not the method used by humans to solve these puzzles!

Human solvers tend not to be very obsessive about maintaining a single continuous chain. Instead, small pieces of chain are assembled quasi-independently and with due attention to the bigger picture of an H-core enclosed by a P-layer having the correct number of tiles of the two types. The tricky part is joining the bits of chain in a way that agrees with the hexein's sequence. Sadly, all of this is too vague to be the basis of an algorithm that can be executed on a computer.

Taking advantage of strong local constraints is the "divide" half of divide-

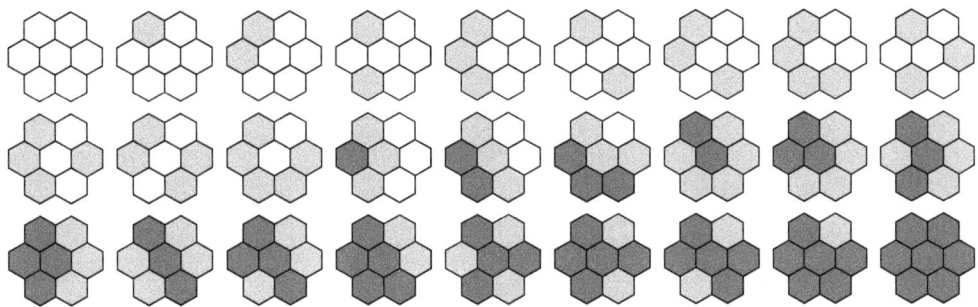

Figure 7.14 The 27 7-hexagon litemotifs allowed by the adjacency rules. This number grows to 144 when rotations and reflections are included.

and-concur. Consider the structure of a hexein when examined just over the six tiles adjacent to a given central tile. Without any adjacency restrictions, there are $3^7 = 2\,187$ possible arrangements of H, P, and S. Filtering these to be consistent with the adjacency rules, we end up with just the 27 *litemotifs* shown in Figure 7.14 and their rotations and reflections (giving a total of 144). Litemotifs are much smaller than the familiar secondary-structure motifs (α-helices, β-sheets) of real proteins. It's likely that evolution played as much of a role in the filtering of those big motifs as chemistry. The smaller, more primitive hexein litemotifs of Figure 7.14 call to mind the presumably wilder and more daring designs that prevailed in early life.

The litemotifs are a good start for a plan to use divide-and-concur to solve SEQVENCE puzzles. In some bounded region, one defines variables that declare hexagons to be H, P, or S. Because the 7-hexagon litemotifs overlap, replica variables need to be defined whose equality is imposed by the concur constraint. A side constraint on the concur values would ensure that exactly the right number of hexagons of the three types fill the region.

But this plan is massively incomplete because it makes no reference to the hexein sequence at all! Fortunately, this is remedied by switching to more finely resolving variables.

7.4.3 Sequence Variables

Instead of each hexagon being in one of three states (H, P, or S), we expand the set of states to $s \in \{0, 1, \ldots, \ell\}$, where ℓ is the length of the sequence. When the hexagon at position p is in state s, it means the hexein is folded so that element s of the sequence (an H or P) is located at p. The state $s = 0$ for that same hexagon means the fold never visits position p. In a valid fold, all of the S-hexagons will be in sequence-state $s = 0$.

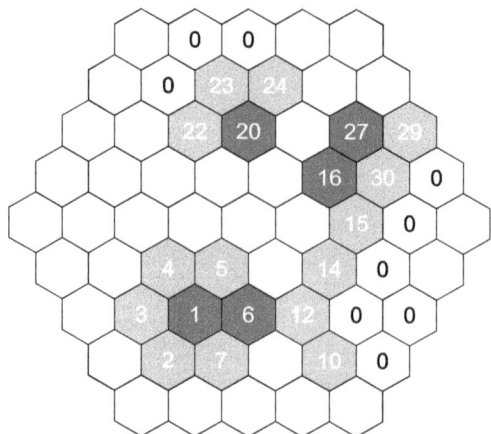

Figure 7.15 Four examples of 7-hexagon clusters (centers numbered 0, 1, 23, and 30) with sequence states that satisfy consecutivity. The s-values correspond to the `tripairin` fold shown in Figure 7.13. There are 37 (overlapping) 7-hexagon clusters in this region of 61 hexagons.

A very strong constraint that can be applied to sequence variables is "consecutivity." A hexagon in sequence-state s will always be adjacent to a pair of hexagons in states $s-1$ and $s+1$. There are two exceptions, $s=1$ and $s=\ell$, when only one adjacent hexagon will have a consecutive state. Solvent hexagons, with $s=0$, are a third exception because they have no consecutivity relationship with the hexagons used by the polymer.

Figure 7.15 shows a region of hexagons with four valid examples of sequence-states in 7-hexagon clusters (previously used to define litemotifs). The sequence values shown correspond to the `tripairin` fold of Figure 7.13. In one example, the central state is 0 and the surrounding six are 0 or distinct positive integers. The positive integers are not strongly constrained because the order in which the fold visits these positions is not required to be consecutive. This changes in the example where the central state is 1. Now one of the surrounding states must be 2. The remaining five hexagons are as unconstrained as in the first example (but happen to be consecutive in the solution). The example where the central hexagon is in state $\ell=30$ is similar to the previous case, except that now one of the surrounding states must be $\ell-1=29$. The fourth example, where the central hexagon is in state 23 (not 0, 1, or ℓ), is the most constrained because both 22 and 24 must be present among the surrounding hexagons, and not on opposite sides of the 23 (because the polymer always makes turns).

Like the states of number cells in sudoku, the correct way to represent the sequence state of a hexagon is with indicator (1-hot) variables. At each hexagon, there is a vector of $\ell + 1$ indicators, with all but one equal to zero in a solution, the exception having value 1.

7.4.4 Replicas

To "divide" the sequence constraints on 7-hexagon clusters, and make projections like the four shown in Figure 7.15 tractable, we introduce variable-replicas. Un-replicated sequence variables would be written x_{ps}, where

$$x_{ps} = \delta_{s\, s^*(p)}$$

means the hexagon at position p is in state $s^*(p)$. All hexagons except those on the boundary of the region belong to seven 7-hexagon clusters. We distinguish the corresponding seven replicas using the vector k that points to the replica's cluster center, and denote the replicated variables x_{pks}. Here is a useful way to think about these variables:

$$x_{pks} : \text{state-}s \text{ indicator of hexagon } p, \tag{7.18}$$
$$\text{according to the cluster centered at } p + k.$$

Hexagons at the boundary of the region have fewer than seven replicas. In general, hexagon p has replica k if $p+k$ is also in the region. For the hexagon-shaped region of Figure 7.15, the hexagons at the corners have three replicas, those on the edges have four. When the central hexagon of a 7-hexagon cluster lies on the boundary of the region, the hexagons outside the region are always assigned the state $s = 0$ (solvent).

7.4.5 Divide Projection

The divide projection is an independent computation at each hexagon p of the region and imposes consecutivity and litemotif constraints on the 7-hexagon cluster centered at p. As an example, consider the hexagon p with sequence state 23 in Figure 7.15. The sequence states s^* on the cluster centered at p (from the Figure) is

$$\left(s^*(p, k_0)\,,\, \ldots\,,\, s^*(p, k_6)\right) = (23\,,\, 0\,,\, 0\,,\, 0\,,\, 24\,,\, 20\,,\, 22)\,, \tag{7.19}$$

where $k_0 = 0$ and the other six vectors are ordered clockwise around p. Because $1 < s^*(p, 0) < \ell$, consecutivity in this instance means there are two nonopposite k's, in this case k_6 and k_4, such that $s^*(p, k_6) = s^*(p, 0) - 1$ and $s^*(p, k_4) = s^*(p, 0) + 1$. There are many valid s^* besides this one and all that satisfy consecutivity must be considered.

The 7-tuple s^* must also satisfy the litemotif constraint. Let $a(s)$ be the amino-acid type at sequence position s, that is, our hexein's H/P sequence. For example, from Figure 7.13 you can read off the `tripairin` sequence:

$$\big(a(1), \ \ldots \ , a(30)\big) = (\text{H}, \text{P}, \text{P}, \text{P}, \text{P}, \text{H}, \ \ldots \ , \text{P}) \ .$$

It will be convenient to define $a(0) = \text{S}$ (for all hexeins). The litemotif constraint on our 7-tuple s^* is that

$$\Big(a\big(s^*(p, k_0)\big) \ , \ \ldots \ , \ a\big(s^*(p, k_6)\big)\Big) \in \mathcal{M} \ , \tag{7.20}$$

where $\mathcal{M}$ is the set of 144 litemotifs. The litemotif that applies in example (7.19) is $(\text{P}, \text{S}, \text{S}, \text{S}, \text{P}, \text{H}, \text{P})$.

Now suppose we have determined the 7-tuples s^* at all hexagons p in the region that minimize the distance to the divide constraint, which we call A. We will get to the computation of the distance shortly. Using the notation $x^A = P_A(x)$ for the projection, our earlier definitions give us the following formula:

$$x^A_{p+k \ (-k) \ s} = \delta_{s \ s^*(p,k)} \ . \tag{7.21}$$

To understand the minus sign on the second index, consider the hexagon at $p' = p + k$. Recalling the interpretation (7.18), $x_{p' \ (-k)}$ are the indicators of hexagon p' according to the constraints coming from the 7-hexagon cluster centered at $p' - k = p$. But these are exactly the 1-hot values that $s^*(p, k)$ specifies.

When computing distance with indicator variables, it is convenient to express the distance relative to the projection where all indicators are 0 (even if this point is not in the constraint set). If y is a single indicator, its projection to 1 relative to 0 is

$$(y - 1)^2 - (y - 0)^2 = 1 - 2y \ .$$

Applying this to the seven indicators specified by $s^*(p, k)$, we obtain the following formula for the relative distance to constraint A for the point x:

$$\text{dist}\big(x, s^*(p)\big) = \sum_k \big(1 - 2\, x_{p+k \ (-k) \ s^*(p,k)}\big) \ . \tag{7.22}$$

We now have all the parts in place to describe the divide projection. For each hexagon p, find the 7-tuple $s^*(p)$ that

- satisfies consecutivity of the central hexagon,
- satisfies the litemotif constraint (7.20),
- minimizes the 7-term distance (7.22),

and then use the distance-minimizing $s^*(p)$ to assign the projection (7.21).

To see how the work to compute the projection scales with the sequence length ℓ, we start with the consecutivity constraint. There are altogether

$$1 + 6 + 6 + 6 \times 4 \times (\ell - 2)$$

cases to consider, broken down here by the choice made for the central hexagon (0, 1, ℓ, otherwise). Applying the litemotif filter and computing the distance are both independent of ℓ. The work to find $s^*(p)$ therefore scales as ℓ. Unless we are unlucky and the shape of the fold is not very compact, the set $\mathcal{P}$ of hexagon positions p can be compact and have a size that scales as ℓ. Computing P_A therefore scales as ℓ^2.

To make the projection scale well, we had to slightly relax the constraint that the positive sequence states are distinct. In the four cases for the central hexagon, there are respectively 6, 5, 5, and 4 surrounding hexagons that are not in a consecutive relationship with the choice made at the center. For these we just use the distance to select the optimal s^*, even if this results in the positive s^* values not being distinct. We can get away with this because the B constraint will ensure that each of the positive integers $1, \ldots, \ell$ is assigned to just one hexagon.

7.4.6 Concur Projection

As usual, the concur projection (set B) is much more straightforward than the divide projection. The B constraint is a standard case of "concur with a side constraint." Formally, the equations

$$\forall\, (p, k, s) : x^B_{p\,k\,s} = \bar{x}^B_{p\,s}$$

express equality of the replicas, where $\bar{x}^B$ are the equality (concur) values. There is a linear side constraint on the concur values:

$$\sum_{p \in \mathcal{P}} \bar{x}^B_{p\,s} = \begin{cases} |\mathcal{P}| - \ell\,, & s = 0\,, \\ 1\,, & s > 0\,. \end{cases}$$

This just ensures that each element of the hexein polymer is found at exactly one hexagon of the region $\mathcal{P}$, and the remaining $|\mathcal{P}| - \ell$ hexagons are not visited by the polymer at all (and are solvent). Projecting to side constraints was covered in Section 4.1.7. Since concur is imposed independently for each p and s, the work to compute P_B, assuming $|\mathcal{P}| \propto \ell$, also scales as ℓ^2.

β	0.70	0.75	0.80	0.85	0.90	0.95	1.00
solution rate	97%	99%	98%	99%	89%	16%	0%

Table 7.1 `tripairin` success rate in 100 trials and less than 10^5 iterations.

7.4.7 Folding with RRR

Since we do not know how compactly a hexein will fold, its sequence length only gives us a rough idea for the size of the region $\mathcal{P}$. There is no downside in using a generous region other than the extra time per iteration from the many selectors (sequence variables) that get projected to state $s = 0$ (solvent). As in other applications, divide-and-concur is indifferent about the initial point and we used the simplest, where all $\ell + 1$ components of every selector are drawn independently from the uniform distribution.

Our SEQVENCE formulation is similar to set-cover and Boolean satisfiability, in that one set is complicated and discrete, and the other is affine. We therefore expect RRR to be similarly sensitive to β. Table 7.1 shows the behavior of the solution success rate for `tripairin` with the iteration cutoff set at 10^5. Most sequences exhibit this behavior: a solution rate that plummets for $\beta > 0.9$.

With time-step $\beta = 0.8$ RRR reliably folds `tripairin` in a few thousand iterations. Figure 7.16 follows the progress of the concur projection x^B in a solution that took 700 iterations. To generate these images, the sequence state $s^*(p)$ that maximizes x_p^B was first found for each hexagon p in the region. The hexagons were then rendered by our 3-grayscale convention for $a(s^*(p))$. Some of the H and P hexagons appear in a lighter shade because the value of x_p^B, often less than 1, was used to move their grayscales in the direction of white. As a reference, the solution in the last frame has the darkest grayscales because all the x_p^B values are 1.

White lines are drawn between adjacent hexagons p and p' whenever $s^*(p)$ and $s^*(p')$ are consecutive positive integers. Though the pattern of white lines is chain-like over the whole solution process, a single continuous chain appears only in the last frame (solution). It is this characteristic of the RRR search that resembles a strategy often used by human solvers.

The time series of the gap that goes with Figure 7.16 is plotted in Figure 7.17 (left panel). Because the gap is quasi-decreasing, the 700-iteration solution had the luck of the initial point landing in a basin of attraction. When the initial point is not so lucky, the gap first fluctuates in a long steady state. The Poisson statistics of basin arrival times is reflected in the corresponding distribution of iteration counts, shown in the right panel.

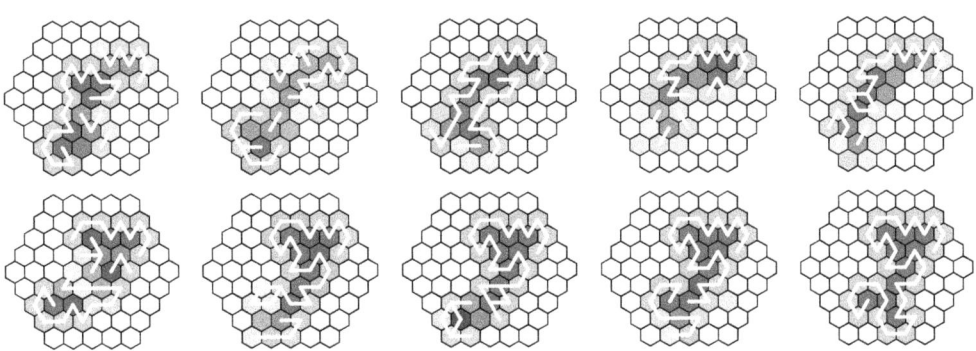

Figure 7.16 Progress of the concur projection in a 700-iteration solution of `tripairin`. The white lines indicate adjacency in the current best sequence assignment estimates (largest concur values). Except in the last frame, the polymer chain is fragmented.

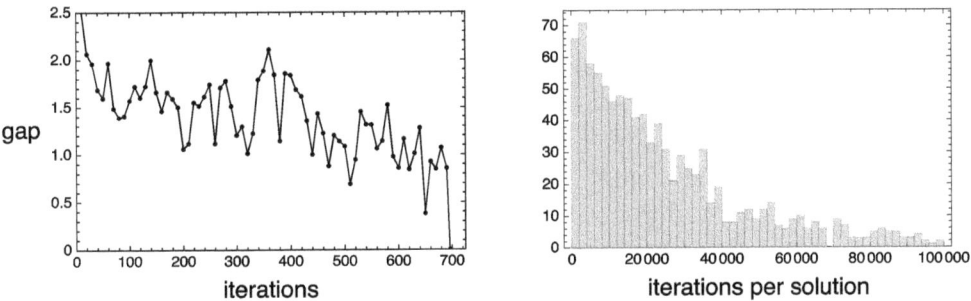

Figure 7.17 *Left*: Behavior of the gap for the search shown in Figure 7.16. *Right*: Distribution of iterations for folding `tripairin` in 10^3 trials.

7.4.8 Comparison with Backtracking

As a discrete problem, the SEQVENCE puzzle is amenable to exhaustive search. When that is done in a reasonably efficient manner, how does it compare with divide-and-concur? In a proper comparison, one would study the growth in the work with sequence length, and how that depends on solution uniqueness and other factors, such as the compactness of the fold. In this section we do nothing that comprehensive and instead compare the work in folding just one sequence at the limit of what is feasible. Figure 7.18 shows the fold of `prop49`, a designed length-49 sequence:

$$H\,(H\,P\,P\,H\,P\,H\,P\,H\,P\,P\,P\,H\,P\,P\,H\,H)_3\,.$$

The backtracking algorithm, which is described later, confirmed that the fold in Figure 7.18 is unique. That computation lasted eight days.

With $\beta = 0.8$ and the 91-hexagon working region of Figure 7.18, RRR

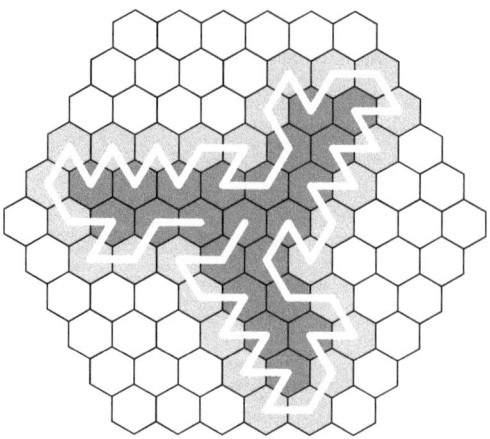

Figure 7.18 The unique fold of `prop49`.

averages about 8.5 hours to fold `prop49`. In trials terminated at one million iterations, the success rate was 40% and the iteration average of the successful attempts was 3.6×10^5. There were no signs of trapping.

To make the comparison as fair as possible, the backtracking search was informed by the same litemotif data used by divide-and-concur. Whenever the chain is extended by a hexagon, the algorithm is required to commit to a 7-hexagon litemotif centered on the added hexagon. We see in Figure 7.14 that there are 16 litemotifs with an H or P at the center (81 when symmetries are included). These litemotif commitments not only restrict the four possible turns at each link of the chain, but are used to force backtracking when a simple supply-and-demand criterion cannot be met. Suppose a total of n tiles have been committed to be P by choice of litemotifs (centered on visited tiles), but are still waiting to be visited by the chain. If there are less than n P's left in the sequence to be placed, no continuation of the current branch can produce a solution. With this early termination scheme (using H's as well as P's), the `prop49` search tree still has $1\,392\,762\,614\,109$ leaf nodes.

It's remarkable that a program doing floating-point arithmetic can outperform an integer-arithmetic, logic-based program by a large margin on a discrete task. No doubt the gap can be narrowed with clever schemes to further trim the size of the backtracking tree. But already the bookkeeping and two kinds of recursion (chain extension, litemotif selection) without any refinements are quite a challenge to implement. A large part of the relative simplicity of implementing divide-and-concur comes from the distributed nature of the computations. Exactly the same, independent computation is performed at all the 7-hexagon clusters in the A projection, and also for the seven replicas at each hexagon in the B projection. The solution time, which is dominated by these operations, would be divided by the number of hexagons when carried out in parallel. There is nothing analogous in a backtracking search.

We should not forget that for proving fold-uniqueness, there is no (known) substitute for backtracking. However, in the time required by backtracking to establish uniqueness, divide-and-concur can fold `prop49` about 31 times. If those folds are the same each time, the empirical evidence for uniqueness is quite strong.

7.4.9 Real Proteins

Though SEQVENCE nicely demonstrates how sequence can encode shape, the puzzle still falls short of a realistic model of proteins. Real proteins are *not* puzzles. Sequences with even mildly hard-to-find folds (by physical dynamics in an energy landscape) were eliminated by evolution in favor of sequences that realize more or less the same shape with a decision-free, funnel-like landscape. We cannot rule out the existence of hexein sequences that also fold easily, with a mostly decreasing RRR gap. Even so, we should not forget that RRR dynamics is very different from physical dynamics!

Do the litemotifs of SEQVENCE have a useful counterpart in real proteins? The answer to this question is probably "no," again as a result of evolution. Because the sequences of modern proteins were optimized by evolution, there are strong hereditary relationships even when the corresponding proteins do very different things. Instead of designing all the parts of a complex protein from scratch, it's much more likely that evolution repurposed parts of existing designs, making small edits as needed. The motifs most useful for folding real proteins are therefore anything but "lite." Moreover, they are on display in the Protein Data Bank (PDB) through the folds that were determined by X-ray crystallography. It would be a mistake to ignore this large-motif information when searching for the fold of a natural (evolved) sequence. When the very successful AlphaFold algorithm [96] constructs candidate folds, it works with close matches of length-64 subsequences it finds in the PDB.

Litemotifs may turn out to be a useful constraint in the inverse problem of *designing* a sequence that has a given folded shape. This is where the protein-artist creates a rough sketch of the fold and the algorithm is tasked with finding a sequence of amino acids that can realize the sketch with favorable energetics. Since the energetics is well captured by favored groupings of small numbers of atoms, the motifs would be small.

7.5 Landscape Origami

Origami is an art form, or craft, comprising diverse styles and techniques. All forms of origami have one thing in common: the folding of paper. Probably the best-known form of origami is where, between folds, the paper is flat, with different parts facing up or down. After all the folds have been completed, some slight unfolding nudges the paper into striking 3D shapes: birds, frogs, ... [32].

On the practical side, and before GPS navigation, origami was used for the compact storage of maps. Though the pattern of creases was simple, refolding a map for storage in the glove compartment was a formidable combinatorial task!

Though *flat-foldability* (of a pattern of creases) and counting the number of ways to fold a map have both been the subject of much mathematical research, this application of divide-and-concur is about a different form of origami. Called *landscape origami*, the rules are that one side of the paper is always facing up. The finished product is a paper landscape, with "mountain" and "valley" folds being exactly what their names imply.

The specific task we set for divide-and-concur is to design patterns of creases that can be landscape-folded very compactly, even as compactly as the flat-folded maps formerly used by motorists. Novel methods of compactly folding large flat objects were introduced to the world in the 1980s by astrophysicist Kōryō Miura, as a proposal for deploying solar panels in spacecraft [64]. The pattern of creases designed by Miura is exceptionally simple because it is periodic. We relax the periodicity constraint in our designs. When the compactly folded paper is incompletely flattened out, the result is an interesting landscape.

7.5.1 Triangulations and Rigidity

We model origami, physically, as a set of rigid polygons joined edge to edge. That is, whenever two polygons are joined along a common edge, the only allowed motion, apart from the usual rigid motions of an object in space, is a change in the dihedral angle between the two polygons. Physical paper is also subject to a non-intersection constraint. We can ignore this constraint in landscape origami because a landscape (single-valued function) never intersects itself.

Rigid triangles have an easy representation: three points with three pairwise distance constraints. And since any polygon can be represented as a

Figure 7.19 The $(3,5)$ triangulation of the square. There are $m = 3$ rows of up-down triangle pairs, and in each row there are $n = 5$ large triangles. A (m, n) triangulation has $T = m(4n + 2)$ triangles and $B = 4m + 2(n + 1)$ edges along the boundary of the square.

union of triangles, triangulations provide a universal representation of our physical model.

Figure 7.19 shows a triangulation of an origami square, the most popular shape for folding. All creases must be made along the triangle edges, though not all edges need to be utilized in this way. For example, the polygons in the Miura fold are parallelograms and correspond to adjacent coplanar triangles (no fold at the crease joining them). A valid landscape-fold of the pattern of creases in Figure 7.19 is any arrangement in space of the triangle vertices with the property that the 3D distance between every pair joined by an edge is the same as their 2D distance in the square.

We will use divide-and-concur to design the pattern of creases. That is, the *geometry* of the triangulation, like the many congruent triangles in Figure 7.19, is not fixed. However, we fix the topology of the creases, a mathematical graph, as this defines the variables used by RRR. The topology of the creases has interesting physical consequences by itself, to which we turn next.

Let T be the number of triangles in the triangulation and B the number of edges along the boundary of the square. For the $(m, n) = (3, 5)$ triangulation in Figure 7.19 (see caption for the definition of these numbers), $T = 66$ and $B = 24$. Remembering to count the exterior of the square as a face in the graph, there are

$$F = T + 1$$

faces altogether. The three edges of each triangle contribute half of a graph edge, and when combined with the half-edges from the edges on the boundary, all the graph edges are accounted for:

$$E = \frac{1}{2}(3T + B) \ .$$

Using Euler's theorem for planar graphs, we obtain

$$V = 2 + E - F$$
$$= \frac{1}{2}(T + B) + 1$$

for the number of graph vertices.

With the above information, we can compute the virtual dimension d^* of folded triangulations. Without additional effort, we will work this out when the paper is folded, more generally, in d dimensions. Folding two-dimensional paper in four dimensions, for example, is something like folding a one-dimensional polymer in three dimensions. In any case, our origami paper in d dimensions has a configuration space of dV dimensions (d coordinates for each vertex). Because each edge, independent of d, represents a single constrained distance, we obtain

$$d^* = dV - E$$
$$= \frac{1}{2}((d - 3)T + (d - 1)B) + d$$

for the virtual dimension. We see that for $d = 3$, $d^* = B + 3$, or that the configurations can be specified by a number of parameters that only grows as the number of vertices on the boundary. In particular, a single coordinate from each of the boundary vertices is enough to specify all the interior vertices!

In the map folding problem, there are constraints on the boundary of the paper as well. Let $S(w)$ be the boundary of the square after it has been scaled by a factor w. Also, let $\Gamma \subset V$ be the vertices on the boundary, so that $|\Gamma| = B$. The positions of the boundary vertices satisfy

$$\forall \, i \in \Gamma : \ x_{i\|} \in S(w) \ , \tag{7.23}$$

where the subscript $\|$ denotes coordinates in the plane of the origami square. There are no constraints on the out-of-plane coordinates, denoted $x_\perp$. Statement (7.23) is an oversimplification in that there are four corner vertices whose $x_{i\parallel}$ coordinates are completely fixed, and the remaining boundary vertices fall into four edge-sets, with constraints on just one coordinate. This breakdown of the constraints is understood when we use this notation. The total number of boundary constraints is $B + 4$, where the 4 comes from the fact that the corners are each subject to an additional constraint relative to the edges.

When boundary constraints are included, the virtual dimension is diminished by $B + 4$ and goes from $d^* = B + 3$ to $d^* = -1$. To admit solutions for even such a minor case of conspiracy, the pattern of creases must be designed. What this means for the map folding problem is quite interesting.

Suppose the creases are designed to admit compactification by a factor w^*. Let's fix those creases, and for compressions w in the range $w^* \leq w \leq 1$, use the relaxed constraint

$$\forall \, i \in \Gamma : \; x_{i\parallel} \in \bar{S}(w) \,,$$

where $\bar{S}(w)$ is the whole interior of the scaled square. In other words, during the compression process we only require that the shadow of the paper lie within some scaled square $\bar{S}(w)$, rather than the stronger constraint that all the boundary vertices lie on the boundary of some square. The virtual dimension is then diminished by a much smaller amount, a single constrained coordinate on each of the square's four edges: $d^* = B + 3 - 4 = B - 1$. This means that there are several independent continuous motions in play during compression. On the other hand, because of the way the creases were designed, these motions are able to terminate at a compression factor w^*, with the boundary vertices magically lined up, and the shadow of the compressed map a perfect square! An example of a map designed to compress to $w^* = 0.8$ is shown in Figure 7.20.

7.5.2 Triangle Variables

Triangulations of a square have an implicit topology. To see this, first consider the analogous construction in one lower dimension: the partition of an interval into subintervals. If we are partitioning the unit interval into n subintervals, we could introduce $n - 1$ variables that satisfy a chain of inequalities:

$$0 < x_1 < \cdots < x_{n-1} < 1 \,. \tag{7.24}$$

But systems of inequalities, on the coordinates of points, do not generalize well in higher dimensions.

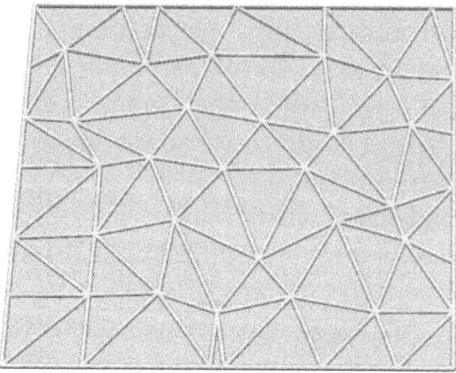 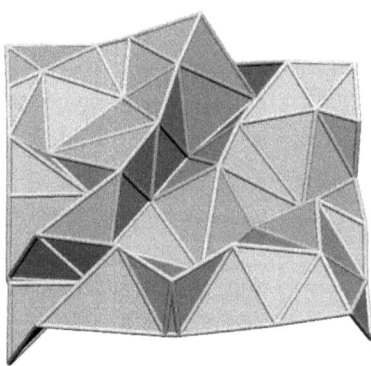

Figure 7.20 A map with the same crease topology as Figure 7.19 and designed by the RRR algorithm to compress by the factor $w = 0.8$. The folded map, on the right, casts a perfect square shadow.

A better approach is to let the subintervals themselves be the basis of a representation by variables. The pairs of points, (x_0, x_1), (x_1, x_2), ... (x_{n-1}, x_n) are 1-simplices and together comprise a *simplicial complex*. The inequalities (7.24) assert that all the simplices in the complex have a positive *orientation*, and that a part of the boundary of two of the simplices is known ($x_0 = 0$, $x_n = 1$).

Knowing that we have divide-and-concur at our disposal, we can keep the simplex variables independent. For example, the x_1 appearing in both (x_0, x_1) and (x_1, x_2) are independent replicas of the same point. Imposing equality of replicas, as always, is the job of the concur constraint.

We should think of each 1-simplex of the complex as a variable $x \in \mathbb{R}^{2 \times 1}$. The divide constraint on these variables is simply that their orientation (chirality) is positive:

$$\chi(x_1, x_2) = \text{sgn}\left(\det(x_2 - x_1) \right)$$
$$= \text{sgn}(x_2 - x_1)$$
$$= +1 .$$

This, together with the identification of replicas by the concur constraint, and the specification of the two boundary values (also in the concur constraint), completes the topological partitioning of the interval.

The symbols $t_{ijk} = (t_i, t_j, t_k)$ denote the analogous 2-simplices that comprise the triangulation of the origami square. Positive chirality corresponds

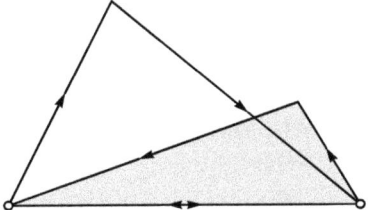 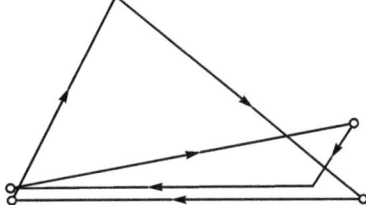

Figure 7.21 *Left*: Two incompatible triangles in the triangulation of the origami square, the shaded one having a negative orientation. *Right*: Orientation-reversing projection applied to the shaded triangle, with the added constraint that the triangle thickness (semi-minor axis of the smallest enclosing ellipse) is above a specified bound. The projection separates the triangles' shared vertices.

to a clockwise arrangement of the triangle vertices t_i, t_j, t_k in the plane, or

$$\chi(t_{ijk}) = \text{sgn}\left(\det(t_k - t_j, t_k - t_i)\right)$$
$$= +1 .$$

Divide-and-concur allows us to independently impose a chirality constraint on all the triangle variables because the concur constraint will impose equality of vertices that belong to multiple triangles. We use the symbols $\tilde{t}_{ijk} \in \mathbb{R}^{3\times3}$ for the variables that describe the corresponding triangles in space, that is, when the paper is folded.

Projecting a negative chirality triangle to the nearest positive-chirality triangle runs into the problem that this projection is the limiting case of a degenerate triangle, whose vertices are collinear (zero determinant). Degenerate or even very thin triangles are clearly unacceptable in an origami design! To avoid this we combine the projection to positive chirality with the property that the triangle's principal component magnitudes satisfy a lower bound ρ. The projection to this constraint is given in Theorem 4.20 of Section 4.5.3. The number $\sqrt{\rho}$ is the semi-minor axis of the smallest ellipse that encloses the triangle.

Projection to the nondegenerate positive chirality constraint is illustrated in Figure 7.21. Only two triangles of the simplicial complex are shown. The two triangles share one edge but have opposite chirality and therefore cannot both participate in a valid triangulation of the origami square. The projection (right panel) shows the distance-minimizing orientation reversal of the triangle with negative chirality, where the projection also satisfies the minimum thickness constraint. The fact that this ends up separating the two common vertices is not a concern, because vertex-replica equality is taken care of by the concur constraint.

Corresponding triangles $t \in \mathbb{R}^{3 \times 2}$ in the plane of the origami paper, and $\tilde{t} \in \mathbb{R}^{3 \times 3}$ after folding in space, are related by a rigid motion. Because the up-down orientation of the paper normal is preserved in a landscape fold, the in-plane projection of $\tilde{t}$, denoted $\tilde{t}_{\parallel}$, has the same (positive) chirality as t. We will also impose a nondegenerate positive chirality constraint on $\tilde{t}_{\parallel}$, though the thickness bound ρ may be smaller than the bound on the triangle t. When applied to $\tilde{t}_{\parallel}$, the in-plane shadow of $\tilde{t}$, the thickness bound ensures that the landscape does not have arbitrarily large gradients.

7.5.3 Edge Variables

Apart from labels (topology), the 2D triangle variables t and 3D triangle variables $\tilde{t}$ are completely independent. Their only mode of interaction is via their constituent points, and this happens through constraints on edge variables. For each graph edge $(i, j) \in \mathcal{E}$, there is an edge variable $e_{ij} = (e_i, e_j) \in \mathbb{R}^{2 \times 2}$ for each crease in the flat origami square. It has a counterpart $\tilde{e}_{ij} = (\tilde{e}_i, \tilde{e}_j) \in \mathbb{R}^{2 \times 3}$, when the paper is folded. Because e_{ij} and $\tilde{e}_{ij}$ are related by a rigid-body motion, the corresponding lengths must be equal:

$$\forall (i, j) \in \mathcal{E} : \|e_i - e_j\| = \|\tilde{e}_i - \tilde{e}_j\| .$$

The projection to this constraint is a minor modification of the projection to the known-distance constraint in Section 4.5.4 (the shorter edge is made longer, the longer made shorter, both symmetrically about their midpoints and along the line between their endpoints). These equal-distance constraints are independent for all the edges in $\mathcal{E}$. For example, the constraint between e_{ij} and $\tilde{e}_{ij}$ is independent of the constraint between e_{ik} and $\tilde{e}_{ik}$ even though the two edges have vertex i in common. Vertex-replica equality is imposed in the concur constraint.

7.5.4 Divide Projection

From the perspective of the divide projection, the variables for landscape origami have the following factors:

$$X = \left(\mathbb{R}^{3 \times 2}\right)^{|\mathcal{T}|} \times \left(\mathbb{R}^{3 \times 3}\right)^{|\mathcal{T}|} \times \left(\mathbb{R}^{2 \times 2} \times \mathbb{R}^{2 \times 3}\right)^{|\mathcal{E}|} .$$

For each triangle $(i, j, k) \in \mathcal{T}$, there is a two-dimensional triangle variable t_{ijk} and its three-dimensional counterpart $\tilde{t}_{ijk}$. And for each edge $(i, j) \in \mathcal{E}$ the projection acts on a 2D, 3D edge-variable pair $(e_{ij}, \tilde{e}_{ij})$. We write the corresponding factors of the divide constraint set $A \subset X$ as follows:

$$A = \left(\mathbb{R}^2 \times A_3\right)^{|\mathcal{T}|} \times \left(\mathbb{R}^2 \times \mathbb{R}^3 \times A_{\tilde{3}}\right)^{|\mathcal{T}|} \times \left(A_{2,\tilde{2}}\right)^{|\mathcal{E}|}. \tag{7.25}$$

The factor $\mathbb{R}^2$ within the first factor is a formal way of asserting that there is no constraint at all on the triangle's centroid. The only constraint, A_3, is on the triangle's chirality and shape (thickness):

$$A_3 = \left\{ t \in \mathbb{R}^{3\times 2}, \sum_{i=1}^{3} t_i = 0 \ : \ \chi(t) = 1, \ \mathrm{spec}\left(t^\mathsf{T} t\right) \in P_0 \right\}.$$

The distribution P_0 is the simple one that imposes a lower bound ρ_2 on the covariance eigenvalues, corresponding to a minimum semi-minor axis of $\sqrt{\rho_2}$ for the triangle-enclosing ellipse. The factors of $\mathbb{R}$ in the second factor of (7.25), on the 3D triangles, make manifest the fact that there are no constraints on a triangle's in-plane centroid or its three out-of-plane coordinates:

$$A_{\tilde{3}} = \left\{ \tilde{t}_\| \in \mathbb{R}^{3\times 2}, \sum_{i=1}^{3} \tilde{t}_{i\|} = 0 \ : \ \chi(\tilde{t}_\|) = 1, \ \mathrm{spec}\left(\tilde{t}_\|^\mathsf{T} \tilde{t}_\|\right) \in P_0 \right\}.$$

The lower bound on the thickness $\sqrt{\rho_3}$ of the $\|$-projected 3D triangles may be different from the bound $\sqrt{\rho_2}$ on the 2D triangles. Here is the constraint on the final factor of (7.25), on the 2D, 3D edge pairs:

$$A_{2,\tilde{2}} = \left\{ ((e_i, e_j), (\tilde{e}_i, \tilde{e}_j)) \in \mathbb{R}^{2\times 2} \times \mathbb{R}^{2\times 3} \ : \ \|e_i - e_j\| = \|\tilde{e}_i - \tilde{e}_j\| \right\}.$$

The projections to all three constraints can be found in Chapter 4.

7.5.5 Concur Projection

The concur projection imposes equality of the vertex constituents of the triangle and edge variables. There is also a side constraint on the vertices that lie on the boundary of the triangulation. To formally express the constraint, we define four sets. First, let $T_r(i)$ be the set of all replicas of 2D-triangle vertex variables that have the vertex label i. For example, in the triangulation shown in Figure 7.19, $|T_r(i)| \in \{1, 2, 3, 4\}$ when i is on the boundary of the square, while all the vertices in the interior have six incident triangles, or $|T_r(i)| = 6$. Similarly, $\tilde{T}_r(i)$ is the set of all replicas of the 3D-triangle vertex variables that have the vertex label i. Finally, $E_r(i)$ and $\tilde{E}_r(i)$ are the sets of all replicas of vertices in, respectively, the 2D and 3D edges that have vertex label i. Using these definitions, the concur constraint has the following factorization:

$$B = \prod_{i \in \mathcal{V}} B(i) \times \tilde{B}(i) \,,$$

$$B(i) = \left\{ (t, e) \in T_r(i) \times E_r(i) \; : \; t = e = \bar{x}, \; \bar{x} \in S(1) \text{ if } i \in \Gamma \right\} ,$$

$$\tilde{B}(i) = \left\{ (\tilde{t}, \tilde{e}) \in \tilde{T}_r(i) \times \tilde{E}_r(i) \; : \; \tilde{t} = \tilde{e} = \bar{x}, \; \bar{x}_{\parallel} \in S(w) \text{ if } i \in \Gamma \right\} .$$

The side constraint, on the concur values of the vertices on the boundary, is different for the 2D and 3D variables. The 2D concur values must lie on the boundary of the full-size square, while the scaled square constraint is imposed on the 3D values. The latter is a constraint just on the in-plane coordinates.

Projecting to set B – concur with some side constraints – is mostly a matter of bookkeeping. For example, if i is an interior vertex, then there are six 2D triangles and six 2D edges with that vertex as a constituent; the projection simply replaces the twelve variables by their average. When i is on the boundary, there are fewer constituents to be averaged, and one or both of the coordinate averages are replaced by a boundary value. Exactly the same averages are computed for the 3D variables, except that the boundary values apply to just the in-plane coordinates and correspond to the scaled square. If the four types of variables (2D/3D, triangle/edge) are given four different metric coefficients, the averages become weighted averages. Only the projection to the equal-edge-length constraint ($A_{2,\tilde{2}}$) is affected by such a metric, but the modification is minor.

7.5.6 Designing Map-Folds with RRR

If we plan to design origami landscapes over a range of triangulations, we should choose the scale of the triangles (or the origami square) so that results can be compared. Convergence is assessed with the RRR gap, so the scale should be chosen so that gaps for runs on coarse and fine triangulations are similar in magnitude. Since the gap is the root-mean-square change in a variable, and we expect triangle vertices to shift by an amount comparable to their size, the scale should be chosen to keep the triangle size fixed. For the (n, m) triangulations (Figure 7.19), we can do this by choosing the side of the square to equal n, the number of (large) triangles in each row. To also keep the triangles approximately equilateral, we set m to be $(n + 1/2)/\sqrt{3}$, after rounding to the nearest integer.

For the scale where the triangle edges have length 1 (in the starting geometry of Figure 7.19), the bounds $\sqrt{\rho_2}$ and $\sqrt{\rho_3}$ on the thicknesses of the
2D and 3D triangles (in projection) should be set at some modest fraction of
1. We used $\sqrt{\rho_2} = \sqrt{\rho_3} = 0.1$ for the $w = 0.8$ compression shown in Figure
7.20 and reduced both of these to 0.05 when compressing to $w = 0.4$.

As we saw in Section 7.5.1, the folding problem with *fixed creases* is very
mildly overconstrained, with a virtual dimension of -1. In *map design* we free
the geometry of the creases, with the result that the problem is now highly
underconstrained. In plain terms, this means that it should be relatively easy
to design highly compressible maps. Solutions will be far from unique and
strongly dependent on the initial point that starts the iterations.

Our initial points satisfy the concur constraint (B) and have the congruent-
triangle structure, like the one in Figure 7.19. This completely defines the
2D triangle and edge variables. The in-plane coordinates of the 3D variables
are the same, but scaled by w. For these variables we also have to specify
an out-of-plane coordinate, one per vertex. In a *random initialization*, these
are random numbers uniformly sampled between $-1/2$ and $1/2$. We contrast
these with *Miura initialization*, where the $\perp$ coordinates are $\pm 1/2$ and form
a $+ + - - + + - - \cdots$ pattern, left to right, on the vertices in Figure 7.19
(vertices in the same column have the same height above the square). The
only constraint not satisfied by either of these initializations is constraint
$A_{2,\tilde{2}}$, the equality of the 2D and 3D edge lengths.

Figure 7.22 shows the evolution of the gap for a run on a $(3, 5)$ triangulation, $w = 0.4$, and random initialization. Shown are the four contributions,
from 2D/3D and triangle/edge variables. For the first three thousand iterations, the 3D variables have the largest gap, with the triangles ($\tilde{t}$) exceeding
the edges ($\tilde{e}$). After this initial period, the gap discrepancies of the variable
types are mostly resolved and the four gaps settle into an oscillatory mode of
convergence. The period of the oscillations is about 100 iterations. We used
$\beta = 0.5$ in this run, so the length of the period cannot be explained by a
small time step. The long period is likely the result of the local, network-like
couplings of the variables across the triangulation. A constraint discrepancy
on one side of the origami paper needs $O(n)$ iterations just to propagate to
the other side, where n is the side length of the square measured in triangles.

Figure 7.23 shows the result of the run with random initialization. The
crease design is on the left, its compression to $w = 0.4$ on the right. The
compressed map casts a perfect square shadow.

The design that results with Miura initial conditions is shown in Figure
7.24. All the parameters, including the compression factor 0.4, were the same
as in the run that produced the random design. Miura's fold is simplest

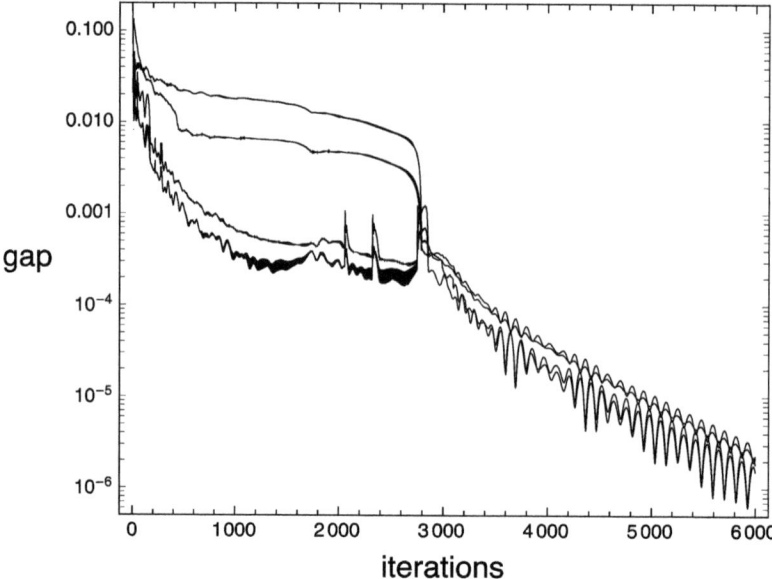

Figure 7.22 Behavior of the four contributions to the gap in a run with random initialization. In the first few thousand iterations, the 3D variables $(\tilde{t}, \tilde{e})$ have the largest gaps. This run produced the fold shown in Figure 7.23.

when the shape of the paper is compatible with the periodic pattern of the creases. Here RRR had to adapt Miura's idea (pattern of starting heights) to be consistent with the square shape of the paper. The pattern of creases in the random and Miura-initialized designs can be compared in Figure 7.25.

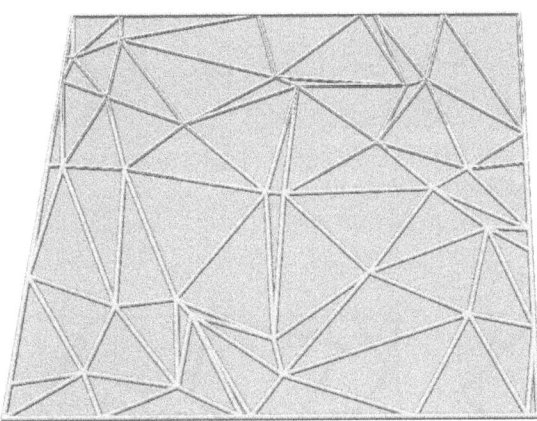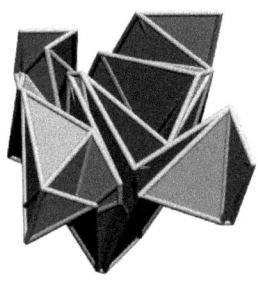

Figure 7.23 Design of a map that can be compressed by an in-plane scale factor $w = 0.4$. Random initialization of the out-of-pane coordinates was used to find this fold.

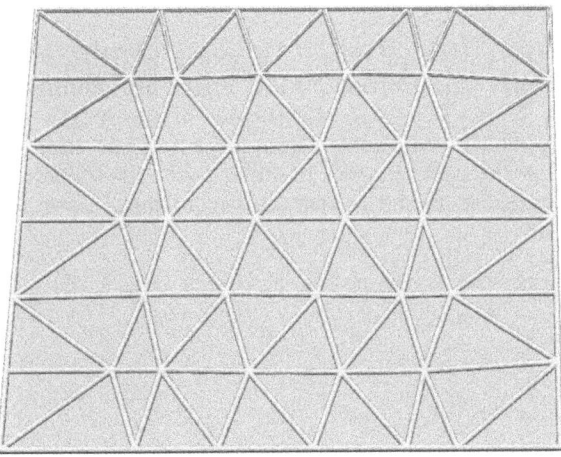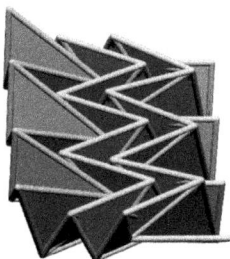

Figure 7.24 Same as Figure 7.23 except that the Miura initialization was used. Both folds cast a perfect square shadow.

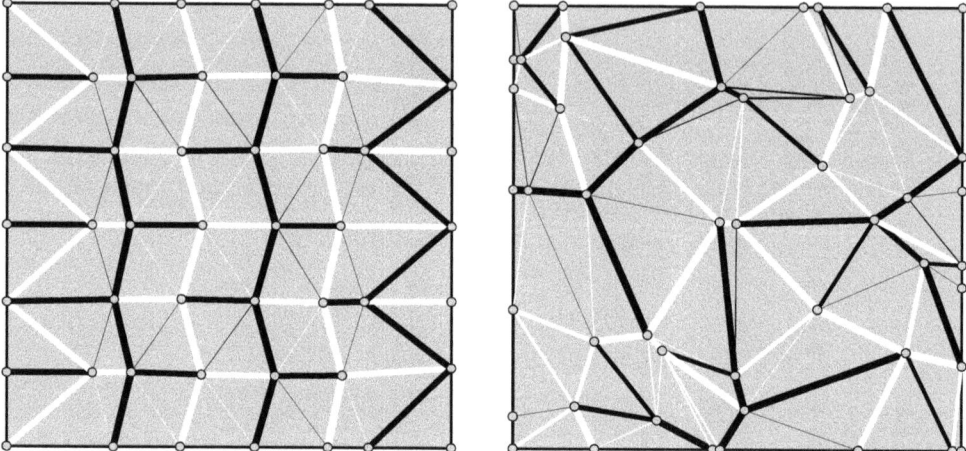

Figure 7.25 Comparison of the pattern of creases designed by RRR, with Miura (left) and random (right) initialization. White edges are mountain folds, black edges are valley folds. The edge thicknesses are proportional to the dihedral angles between the adjoining triangles when the map is compressed to $w = 0.4$. It is interesting that in both folds the dihedral angle is often very close to zero, so that the triangles form quadrilaterals and even pentagons.

7.6 Data Science with eℓmo

Humans have evolved the ability to spontaneously recognize patterns and geometrical structures ... when these inhabit two- or three-dimensional space. If we are counting on that same innate facility to make sense of the high-dimensional data we encounter in the modern world, we first have to find a way of rendering that data in the few dimensions accessible to our visual system.[4]

A very common form of data is simply a set of points. The attributes/features of a data item are its coordinates, of which there are many. Some form of dimension reduction is possible, in principle, when the points are all near some lower-dimensional set. When that set is a subspace, principal component analysis (PCA) will identify the best such approximating set. If we're lucky and a two- or three-dimensional subspace is already a good approximation, to render the data we simply project the points into that plane or 3D space.

Manifolds are also popular models for the geometry of the approximating set. When the data points are generated by a continuous process, the manifold model will usually apply even when the subspace model is too restrictive. For example, each grayscale image in the MNIST [67] data set was produced by the continuous motion of a pen on paper. Different instances of the same handwritten digit, executed by different writers, correspond to (mostly) smooth variations of that continuous trajectory. It would be a mistake *not* to build manifold structure into the dimension reduction if the data is known to possess that attribute. The uniform manifold approximation and projection (UMAP) algorithm [68] is one of the more sophisticated of such dimension-reduction schemes and is widely used to render data as two-dimensional images.

Some dimension reduction algorithms are agnostic on the geometry of the data and instead try to match the *distance distributions* of points near every point in the embedding to the corresponding distributions in the higher-dimensional data. When the long-tailed t-distribution is used to model the distances in the embedding, the algorithm is called t-distributed stochastic neighbor embedding (t-SNE) [109].

In this section we use divide-and-concur to reduce the dimension of data by a principle that is minimalist in the extreme. The name of the algorithm is eℓmo, which becomes an acronym when the first letter is moved to the end and a hyphen is added (without changing the pronunciation): ℓ-moe.

7.6.1 Preserving Only the Order of Distances

Let's start with the "moe" in ℓ-moe: *metric order-preserving embedding*. Clearly it is too much to ask that the embedding of points into our lower-dimensional Euclidean space is isometric, as in origami, where the distances within a polygon of the 2D paper are preserved when the paper is folded into a 3D frog. A greatly weakened criterion that still captures the relationships a data scientist cares about is that at least the *ordering* of distances is preserved.

Consider a point labeled 0 and its relationship to two others labeled 1 and 2. From the data we know d_{01} and d_{02}, the distances between 0 and the two other points. These might be Euclidean distances in a high-dimensional space, or some other kind of distance special to the data, such as is used when comparing genomes. In any case, we want the Euclidean distances of the embedded points to preserve the order of the data distances. Let m be the m-bedding dimension and x_0, x_1 and x_2 the corresponding points in $\mathbb{R}^m$. The constraint we impose on these points is then the following:

$$\|x_0 - x_1\| \leq \|x_0 - x_2\| \quad \text{if and only if} \quad d_{01} \leq d_{02} \ . \tag{7.26}$$

This is quite a strong constraint when applied to all the triples of points in our embedding. In fact, it is too strong when m is small.

To have a criterion that has a chance of being satisfied, we could relax the constraint (7.26) for triples where d_{01} and d_{02} are both greater than some cutoff $d_{\max}$. This local qualification is the ℓ in ℓ-moe. However, by using $d_{\max}$ as the locality parameter, we run the risk that the number of constraint-triples grows faster than the number of data. If eℓmo is going to be useful in data science, we need a different criterion for locality.

7.6.2 Locality, Triples, and the Divide Constraint

A much-used construct in data science is the k-nearest-neighbor graph. This graph has the N data items as its nodes, and its edges are the minimal set such that every node is joined to at least its k nearest neighbors. Let $\mathcal{T}_k$ be the set of triangles in the k-nearest-neighbor graph. Because the three nodes of a triangle $t \in \mathcal{T}_k$ cannot be in a strict cyclic k-nearest-neighbor relationship, at least one node of t has the property that the two other nodes are among its k-nearest neighbors. This implies that $|\mathcal{T}_k|$ is at most $k(k-1)/2$ times the number of nodes, N.

To set up the replicas in a divide-and-concur implementation of eℓmo, we compile the following set of triples:

$$\mathcal{T}_< = \left\{ (i_0, i_1, i_2) \in \mathcal{T}_k : d_{i_0\, i_1} < d_{i_0\, i_2} \right\} .$$

Associated with the triple $t = (i_0, i_1, i_2)$ is the triple of variables $x_t \in \mathbb{R}^{3 \times m}$ whose first element $x_{t\, i_0}$ is the replica of the embedded position of data i_0 according to triple (inequality) t, and similarly for the other two elements.

The divide constraint factorizes over triples

$$A = A_1 \times \cdots \times A_{|\mathcal{T}_<|} ,$$

where the constraint on triple t is

$$A_t = \left\{ (x_{t\, i_0}, x_{t\, i_1}, x_{t\, i_2}) \in \mathbb{R}^{3 \times m} : \| x_{t\, i_0} - x_{t\, i_1} \| \le \| x_{t\, i_0} - x_{t\, i_2} \| \right\} .$$

Aside from the replicas they apply to, all the triple constraints are the same, independent of t. The projection to one of them, called A instead of A_t, is given by the following:

Theorem 7.3 (Triple distance-inequality projection) *If the distance inequality*

$$\| x_0 - x_1 \| \le \| x_0 - x_2 \|$$

is already satisfied, then

$$P_A(x_0, x_1, x_2) = (x_0, x_1, x_2) ,$$

otherwise

$$P_A(x_0, x_1, x_2) = (x'_0, x'_1, x'_2) ,$$

where

$$x'_0 = \left((1 - q^2)x_0 + (q - q^2)x_1 - (q + q^2)x_2 \right)/(1 - 3q^2) ,$$
$$x'_1 = \left((+q - q^2)x_0 + (1 - q - q^2)x_1 - q^2 x_2 \right)/(1 - 3q^2) ,$$
$$x'_2 = \left((-q - q^2)x_0 + (1 + q - q^2)x_2 - q^2 x_1 \right)/(1 - 3q^2) ,$$

$$q = \frac{\sqrt{b^2 - 3a^2} - b}{3a} , \tag{7.27}$$

$$a = (2x_0 - x_1 - x_2) \cdot (x_1 - x_2) ,$$
$$b = \| x_0 - x_1 \|^2 + \| x_0 - x_2 \|^2 + \| x_1 - x_2 \|^2 .$$

The proof of the theorem is left as an exercise. When the input variables do not satisfy the distance inequality, the projection is to the equality case, an isosceles triangle. The barycentric-coordinate structure of the formula confirms the projection is similarity invariant.

The fact that some fraction of the point triples get projected to isosceles triangles does not imply that a similar fraction of triples has this property in the final embedding. The isosceles symmetry is "broken" by the projection to the B constraint (next section), and since this is a very different constraint, the isosceles triangles have a good chance of being resolved in a way that satisfies the A constraint, and are then unchanged by the projection to A in the next iteration.

7.6.3 Concur and Uniformizing the Embedding

Each embedded data point participates, through its replicas, in many triplet distance-inequality constraints. At a minimum, the concur or B constraint imposes equality of these replicas. But clearly something stronger is called for in the B constraint because the A constraint has no scale. In addition to the usual Euclidean motions (translations and rotations), solutions can be arbitrarily expanded or contracted!

Keeping with the rationale that the A constraint is only concerned with preserving the *local* ordering of distances, we can give the B constraint license to introduce a scale and impose global properties of our choosing. Without shame, we ask that the embedding also makes a pretty picture.

One of the simplest things we can do is to add a side constraint on the concur values, and the most natural for the problem at hand is a distribution constraint. To get a uniform distribution between 0 and 1 for the first coordinate, the latter would be projected to the set U_n comprising all coordinate permutations of

$$\left(1/n\,,\ 2/n\,,\ \ldots\,,\ n/n\right).$$

The same projection would be applied to the other coordinates. However, computing the projection (the nearest coordinate permutation) would be difficult because the order-preserving matching Lemma 4.5 does not apply when the metric is not the isotropic Euclidean metric. By Proposition 4.4, the metric for the concur values is

$$\sum_{i=1}^{n} r_i \left\| \bar{x}_i' - \bar{x}_i \right\|^2\,,$$

where r_i is the number of replicas of x_i, that is, the number of triples $t \in \mathcal{T}_<$ that contain data item i.

Divide-and-concur is not only useful for breaking up problems into independent subproblems, it also lets us avoid intractable projections. For the problem at hand, the first step is to introduce a new set of variables $y \in \mathbb{R}^{n \times m}$ to impose embedding uniformity and for which the metric is isotropic:

$$\sum_{i=1}^{n} \|y_i' - y_i\|^2 \ .$$

The A constraint now includes a constraint on the y variables which takes the form of m factors of U_n, one for each coordinate of the embedding. The y variables take care of what we first considered doing with a side constraint, but now with an easy projection computation. The B constraint is now purely a concur, imposing equality of the x-replicas and their corresponding y variable. If all of these have the same metric coefficient, then the concur value will be heavily dominated by the highly abundant x-replicas. In order to let embedding uniformity assert itself, the y variables should have a metric coefficient g that can be set to a large value relative to the x-replicas:

$$\sum_{t \in \mathcal{T}_<} \left(\|x_{t\,i_0}' - x_{t\,i_0}\|^2 + \|x_{t\,i_1}' - x_{t\,i_1}\|^2 + \|x_{t\,i_2}' - x_{t\,i_2}\|^2 \right) + g \sum_{i=1}^{n} \|y_i' - y_i\|^2 \ .$$

7.6.4 Parameter Selection

Parameters are unavoidable when working with real-world data. Keeping with eℓmo's minimalist premise, the only major parameters should be the m of the m-bedding dimension and the k of the k-nearest neighbor graph. We also accept that when m is small and k is large, the feasibility problem will only have near solutions. The most we can hope for is qualitatively similar solutions in runs differing only in the random initialization of the variables.

One way that eℓmo is able to work at the "edge of feasibility" is by taking advantage of the equality cases of the inequalities (7.26). That is, when a strict distance inequality cannot be realized in the embedding, then at least make the embedded distances equal. We therefore expect some fraction of the triples to form isosceles triangles. This presents a potential problem for the computation in Theorem 7.3 because the inner product a equals zero when x_0, x_1 and x_2 form an isosceles triangle. Though the $a \to 0$ limit of q in equation (7.27) is well defined, a convenient way to avoid division-by-zero complications is to slightly relax the constraint, so that the projection is performed only when

$$\|x_0 - x_1\| > \|x_0 - x_2\| + \epsilon_1$$

for some small positive ϵ_1. Since the coordinates are uniform between 0 and 1 thanks to the B constraint, by setting $\epsilon_1 = 10^{-3}$ the relaxation will have a negligible effect on the embedding.

In problems that have no true feasible points, one has to decide what will serve as the near-solution. The correct choice here is clearly the concur projection x^B, since we are interested in a single estimate for the embedded position of each data point. The degree to which this solution estimate aligns with eℓmo's preserving-the-order-of-distances objective can be assessed by the fraction of violated inequalities:

$$f_> = \frac{1}{|\mathcal{T}_<|} \left| \left\{ t \in \mathcal{T}_< : \|x^B_{t\,i_0} - x^B_{t\,i_1}\| > \|x^B_{t\,i_0} - x^B_{t\,i_2}\| + \epsilon_2 \right\} \right|.$$

The positive margin term ϵ_2 is related to the parameter ϵ_1 in the A constraint. Because some fraction of triples in a near solution will saturate the distance inequality with margin ϵ_1, we can avoid counting these as inequality violations by setting ϵ_2 comfortably larger than ϵ_1. By setting $\epsilon_2 = 10\epsilon_1$, then with $\epsilon_1 = 10^{-3}$, the margin ϵ_2 is still acceptably small for $f_>$ to count true distance-inequality violations.

Thanks to the locality provided by the k-nearest neighbor graph, finding eℓmo embeddings is a relatively easy problem and we should not be surprised to see a gap that is mostly decreasing. However, even these easy problems are slightly infeasible and the gap never arrives at zero. Because the gap is only indirectly related to the problem objective, setting parameters to minimize the final gap may not be a good strategy. We will instead set parameters by minimizing $f_>$.

Apart from the data scientist's parameters m and k, the algorithm for embedding the k-nearest neighbor graph in m dimensions (in an order-of-distances-preserving way) has only the RRR parameter β and the metric parameter g. Not surprisingly, $\beta = 1$ is the best choice for this relatively easy problem. We will set g by minimizing $f_>$. Recall that g is the weight given to the embedding-uniformity variables y in the concur projection, relative to the unit weights of the x-replicas. The average number of replicas per data item is $3|\mathcal{T}_<|/N$, where N is the number of data. The g that minimizes $f_>$ is somewhat smaller than this number.

7.6.5 Scaling with Data Size

It is important that the work involved to embed N data items scales linearly with N because of the enormity of N in typical applications. If we do not count the construction of the k-nearest neighbor graph, as that is a common

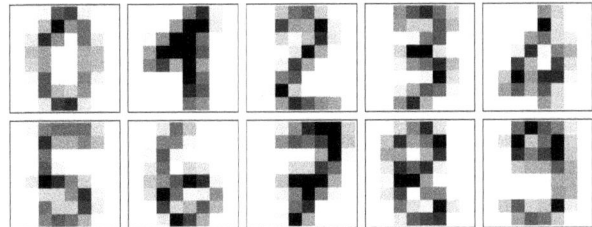

Figure 7.26 Samples of the reduced MNIST handwritten-digits data [44].

first step in embedding algorithms, then the work to produce an eℓmo embedding is the product of the work per RRR iteration and the number of iterations. The latter is surely highly dependent on the nature of the data and there is not much more to be said than what we learn, empirically, from experiments.

With k and m fixed, the work per RRR iteration in eℓmo scales linearly with N. Both projection computations are linear in N because the number of triples $|\mathcal{T}_<|$ is less than the number of triangles $|\mathcal{T}_k|$ in the k-nearest neighbor graph, and the latter is $O(k^2 N)$. The part of P_A that acts on the x variables comprises $|\mathcal{T}_<|$ independent computations that can be performed in parallel. The part that acts on the y variables is a sort of m sets of N numbers and is quasi-linear in N. The computation of P_B is the weighted average of a partition of $3|\mathcal{T}_<| + N$ m-tuples into N sets (one for each data item). Since the partitions are disjoint, the N averages can be computed in parallel.

7.6.6 Visualizing MNIST with eℓmo

A standard test case for dimension reduction is the MNIST dataset [67] of handwritten digits.[5] The original dataset comprises 60 000, 28×28, 8-bit (grayscale) images. We will test eℓmo on a popular reduced version [44] of $N = 3\,823$ images compressed to 8×8, 4-bit pixels. Samples of the digits 0 - 9 are shown in Figure 7.26. These data are points with 64 integer coordinates that range from 0 to 16.

Embeddings in $m = 2$ dimensions first start showing structure at $k = 6$. The clustering of the 10 classes of data improves with increasing k, but of course this comes with an increase in $f_>$. Results for three values of k are given in Table 7.2 and the corresponding embeddings can be compared in Figures 7.27 to 7.29.

Divide and Concur

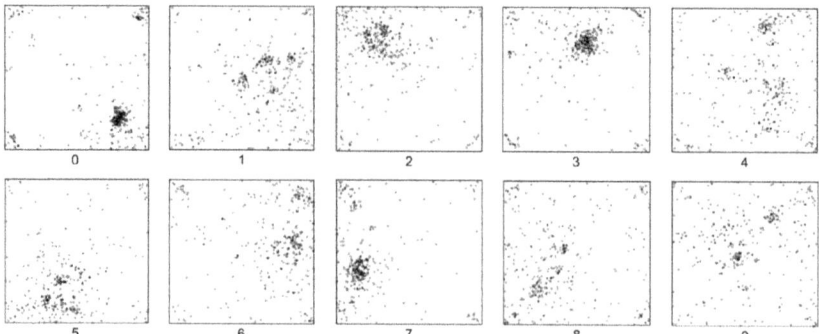

Figure 7.27 An eℓmo embedding of the MNIST data with nearest neighbor parameter $k = 6$.

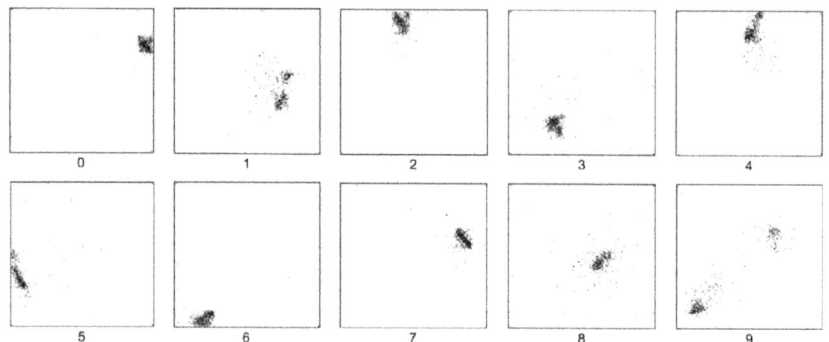

Figure 7.28 Same as Figure 7.27 but with $k = 10$.

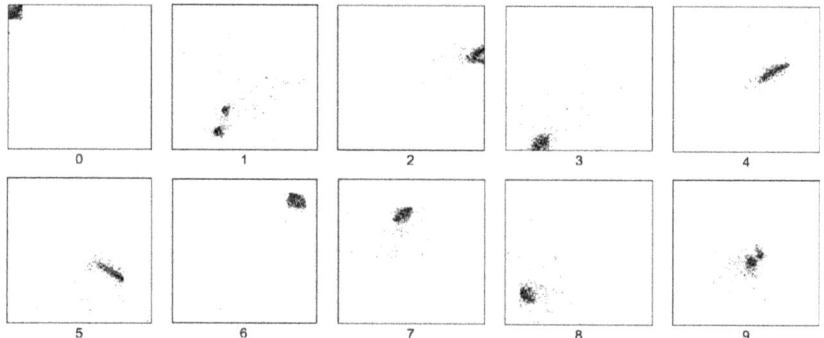

Figure 7.29 Same as Figure 7.27 but with $k = 20$.

k	$\lvert\mathcal{T}_<\rvert$	$3\lvert\mathcal{T}_<\rvert/N$	g	$f_>$
6	40 854	32.1	10	3.1%
10	127 010	99.7	50	9.9%
20	578 042	453.6	100	12.9%

Table 7.2 Distance-inequality violation fractions $f_>$ in two-dimensional MNIST embeddings for three values of the nearest-neighbor parameter k.

We initialize RRR as randomly as possible, with each coordinate of the x-replicas and the y variables sampled uniformly between 0 and 1. Some local features in the embeddings are reproducible, but the global layout of the clusters, for the most part, is not. In the $k = 10$ and $k = 20$ embeddings, the 1's and 9's clusters, unlike the others, have two components. On the other hand, the 6 and 9 clusters are adjacent in one embedding (Figure 7.28) and distant in the other (Figure 7.29).

The plot of the gaps for the two kinds of variables in Figure 7.30 makes the cluster-layout variability plausible. The gap for the y variables (dashed curve), which impose uniformity of the layout, is effectively zero after less than 100 iterations and *before* the x-variable gap undergoes a similar, much lengthier evolution. The layout of the clusters is determined first and quickly, while their concentration and refinement is a lengthier process. It is unlikely that the cluster arrangements are special (optimized in some way) if so few iterations go into their construction. The separation of the RRR time scales

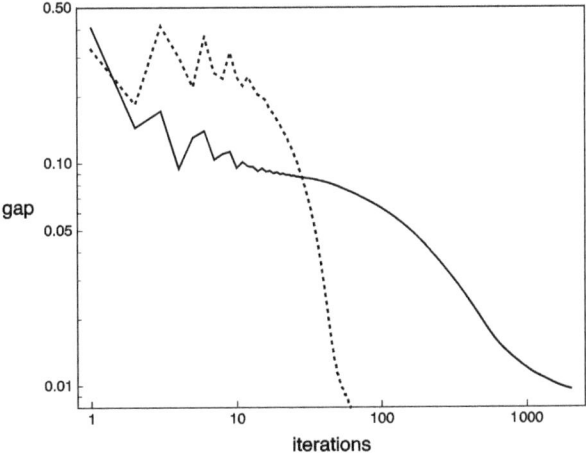

Figure 7.30 RRR gap for the x-replicas (solid) and y variables (dashed) when embedding MNIST in two dimensions with $k = 20$.

for the two kinds of variables is in hindsight not surprising, given the strong class structure of the MNIST data. It would be interesting to see how this behavior of eℓmo changes when class structure is absent.

Allowing embeddings to have variability, as demonstrated by eℓmo's arrangement of MNIST clusters, is preferable to the artificial reproducibility achieved when the initial, unoptimized embedding is not random (e.g., a PCA embedding). The final gap and $f_>$ metric in multiple (randomly initialized) runs of eℓmo are so similar, it would be a mistake to single out any one as being the "correct" embedding.

7.7 Neural Networks

Feed-forward artificial neural networks are the backbone of all implementations of artificial intelligence. These networks apply linear transformations to vectors of inputs, and also nonlinear functions between layers of the network, to produce output vectors for various tasks. When the input vectors are large and the output vector small, the network may be classifying images. When the sizes are reversed, the network is generating images when given a prompt.

The nonlinear functions are essential because composing just the layers of linear transformations results in a single linear transformation, defeating the whole idea of a layered architecture.

While their necessity is undisputed, the interpretation of the nonlinearities, called *activation functions*, has shifted over the years. The earliest activation functions were meant to mimic neurons and switched between two values as a function of a (single) continuous input. For example, a feature might be deemed absent or present, or a proposition declared true or false. A popular choice was the *sigmoid*,

$$\sigma(x) = \frac{1}{1 + e^{-x}} \, ,$$

which switches smoothly between 0 and 1.

The differentiability of activation functions would appear to be incidental to the operation of neural networks. In fact, this technical detail is at the center of the single greatest challenge to the successful deployment of neural networks in artificial intelligence: *training*. To properly capture the difference between a cassoulet from the Carcassone from one prepared in Montaubon, or the styles of sea shanties (capstan, windlass, etc.), networks require billions of parameters to define the mapping between inputs and outputs. Finding those parameters involves training the network on huge volumes of data, and in the current technology, this requires differentiable activation functions.

At the core of all training algorithms is an idea called *error back-propagation* [91]. Network performance during training is assessed by the discrepancy between the correct training-data outputs and the network's actual outputs. In quantified form, the discrepancy is the network's *loss*, and the idea is to minimize the loss over all the training data. When the activation functions are differentiable, the loss is a differentiable function of the network's many parameters. Error back-propagation is the industry term for the computations involved when moving the parameters along the downhill gradient direction of the loss.

A large part of the success of neural networks should be attributed to the efficiency of the error back-propagation algorithm. Loss is evaluated at the network outputs and its dependence on the network parameters, via the chain rule of calculus, descends layer-by-layer all the way to the network's input layer. Computations are local on the network and scale with the number of wires (edges) in the network.

The sigmoid has the drawback that its derivative vanishes rapidly for large inputs x of either sign. During error back-propagation, this means that some "neurons" and the wires that feed into them are left out of the parameter search. That is, sigmoids are blind to their role in fixing an incorrect network output when they operate in their saturated ranges. When this was realized, early in the development of the technology, the *rectified linear unit*,

$$\text{ReLU}(x) = \left\{ \begin{array}{ll} 0, & x < 0, \\ x, & x \geq 0, \end{array} \right.$$

took over as the activation function of choice. In having a strong gradient over half of its range, it reduced the "vanishing gradient problem" that beset sigmoid networks to some extent. The computational simplicity of the derivative was another advantage.

The adoption of the ReLU brought about a shift from activation functions as switches, to minimalist enablers of *universal approximation* [55, 99]. It was shown that a broad class of activation functions, when combined with linear transformations, could approximate arbitrarily complex data (input-output pairs). With that as an established mathematical fact, why not then use the computationally most expedient function? The downside is that non-switch-like activation comes with a loss of interpretability of the neural network's operation.

If network performance – sea shanty quality – is the only thing that matters, one might not care about how the network generated its output. Networks could be useful as black boxes. On the other hand, there are applications of neural networks where logical consistency, or the avoidance of bias, is highly valued. Should network interpretability in these cases be sacrificed, and only because of a technical obstruction involving differentiable functions?

This section of the divide-and-concur chapter could serve as a "white paper" for an entirely different approach to neural network training. We will abandon loss completely, and instead look at the matching of network outputs with training data as a constraint problem. And because switch-like activation is no longer off the table, the new approach holds the promise of bringing transparency to the black box.

Divide-and-concur plays a key role in the reformulation of the neural network training problem. RRR iterations roughly correspond to gradient steps, with a computational cost not that different from error back-propagation. Constraints are divided on the level of neurons, with the result that updates can be applied synchronously over the entire network. Because the error is similarly divided, there is no more waiting for the error to arrive via back-propagation.

We start with a review of a third kind of activation. Known as *Boolean threshold functions* (BTFs), these are a class of nicely parameterized "gates" in the broader study of Boolean functions. Because BTFs have discontinuous outputs and their inputs are integrated into their design, they come closer to realizing networks of "neuronal units" than do the combinations of linear transformations and activation functions in current technology.

BTFs operate on digital or symbolic data, and the two training applications featured are in that domain. However, we show how small analog-modifications can be made in the input and output layers so the new approach also works with images and other forms of analog data.

7.7.1 Boolean Threshold Functions

A BTF takes a vector of Boolean inputs, $x \in \{-1, 1\}^m$, and produces a Boolean output $y \in \{-1, 1\}$:

$$y = \text{sgn}(w \cdot x) \ .$$

The numbers $w \in \mathbb{R}^m$ are the BTF's *weight parameters*. As written, our BTF is *self-dual*. A Boolean function f is self-dual when negating all its inputs negates its output, or $f(-x) = -f(x)$. The self-dual property is relaxed if one of the inputs is held constant. For example, the two-input functions (for $m = 3$, $x_0 = 1$)

$$f(x_1, x_2) = \text{sgn}(w_0 + w_1 x_1 + w_2 x_2)$$

are non-self-dual and realize AND and OR gates for particular weights. Weights associated with constant inputs are also called *bias parameters*. Since y is unchanged when w is rescaled by a positive constant, we are free to apply a weight normalization

$$\|w\|^2 = m$$

that matches the normalization $\|x\|^2 = m$ of the inputs. If we think of the weights as assigned to the wires feeding the neuron, then flipping the sign of a weight on a wire is the same as the action of a NOT on that wire.

Returning to the m-input self-dual case, we see there are problematic

m	1	2	3	4	5	6	7	8
number	2	4	14	104	1882	94572	15028134	8378070864

Table 7.3 The number of BTFs on m inputs [84], also twice the number of m-cube bisections.

weights where for some $x \in \{-1, 1\}^m$ the argument of sgn() is zero. For weights without this problem, the hyperplane through the origin

$$X_0(w) = \{x \in \mathbb{R}^m : w \cdot x = 0\}$$

bisects the hypercube $\{-1, 1\}^m$ into halves X_+ and $-X_+$, where X_+ is the half on which the BTF output is $y = +1$. Twice the number of m-cube bisections gives the number of inequivalent Boolean functions that can be expressed as a BTF with m inputs (since either half can be the set X_+). Table 7.3 lists the number of BTFs up to $m = 8$ [84].

For any BTF, there are special weights w^* such that there exists a maximal subset $X^* \subset X_+$ of *extreme points* where

$$\forall x \in X^* : w^* \cdot x = d^*$$

for some positive d^*, and $w^* \cdot x > d^*$ for the remaining $x \in X_+ \setminus X^*$. Since $\|w^*\| = \sqrt{m}$, the number $d^*/\sqrt{m}$ is the distance between the hyperplane $X_0(w^*)$ and the extreme points X^*.

We can think of bisections with extreme parameters w^* as cuts of the m-cube with a knife of thickness $\theta = 2d^*/\sqrt{m}$, one side making contact with the extreme points X^*, the other side with the points $-X^*$. Figure 7.31 shows the situation in $m = 3$, where two knife thicknesses arise. If $X_0(w^*)$ is one of the coordinate planes, then $\theta = 2$, whereas $\theta = 2/\sqrt{3}$ when $X_0(w^*)$ is orthogonal to one of the cube's body diagonals.

Since d^* depends on the nature of the BTF (the kind of bisection), it is a de facto parameter for these activation functions. To implement this, we break up the neuron constraint into two statements:

$$y = \text{sgn}(w \cdot x), \tag{7.28a}$$

$$\delta \le |w \cdot x|. \tag{7.28b}$$

The first is a constraint relating the Boolean inputs and output, while the second constraint (together with $\|w\|^2 = m$) restricts the admissible weights. The parameter δ is called the *margin*. As an example, recall the possibilities for $m = 3$. At the most restrictive setting, $\delta = \sqrt{3}$, the only possible (normalized) weights are $(\pm\sqrt{3}, 0, 0)$ and its cyclic permutations. This kind of BTF simply transmits the value (or negated value) of one of its three inputs.

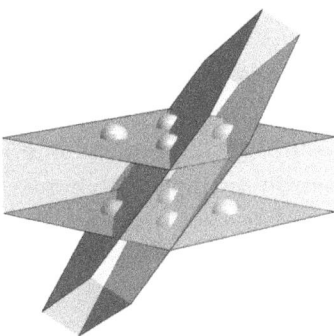

Figure 7.31 The two kinds of bisection of the 3-cube. A knife making contact with extreme points of the 3-cube has thickness 2 in one case, $2/\sqrt{3}$ in the other.

During training it only needs to learn which of the inputs to transmit, and if negation should be applied. To include majority gates, AND, and OR in a BTF's repertoire, the margin has to be set at the smaller value $\delta = 1$. Because a BTF's margin is set rather than learned, it is called a *hyperparameter*.

To access the full capacity of a BTF network, the margins should be set to the value that admits all possible hypercube bisections. We define $d^*_{\min}(m)$ to be the maximum value of this margin for a BTF with m inputs. A knife of half-thickness $d^*_{\min}(m)$, and no greater, can bisect the m-cube in all possible ways. Since the role of each BTF in a network is usually not known prior to training, it is safest to set $\delta = d^*_{\min}(m)$. If solutions persist with larger margins, with more BTFs tuned to extreme weights, the interpretability of the network is improved. Table 7.4 lists values up to $m = 7$.

m	1	2	3	4	5	6	7
$d^*_{\min}(m)$	1	$\sqrt{2}$	1	$\sqrt{4/7}$	$\sqrt{5/19}$	$\sqrt{6/59}$	$\sqrt{1/29}$

Table 7.4 The maximum margin that admits all BTFs in a neuron with m inputs. The values for $m \geq 6$ are conjectured.

7.7.2 Network Nomenclature and Notation

Even though layers are an important part of neural network design, the notation for variables that live on a network is simpler when this detail is left out. After all, constant inputs (to implement bias) reach into all layers, and there are designs where neuron outputs leap across layers. The only essential detail the notation should convey is an orientation for the network edges. We use Latin labels i, j, and so on for the nodes (neurons) of the network, and denote edges as $i \rightarrow j$, if say the output of neuron i is conveyed as an input to neuron j. A network of directed edges is a *directed graph*.

Feed-forward neural networks also have the *acyclic* property, in which directed cycles are absent. A directed acyclic graph has input nodes, with no incoming edges, and output nodes, with no outgoing edges. All directed paths starting from an input node end at an output node. The acyclic property is relaxed in some network models, such as recurrent neural networks. This property is conspicuously absent in the divide-and-concur training algorithm, in which variables are updated synchronously over the entire network. Not only is there no back-propagation, there is no time ordering at all. The same training algorithm, without modification, may be applied to networks with cycles. However, because of the simplicity of how data is introduced in acyclic networks, at the input and output nodes, we limit our two demonstrations of the method to acyclic networks.

In acyclic networks, the input and output nodes are denoted respectively $\mathcal{I}$ and $\mathcal{O}$. All other nodes, or those in the interior of the network, are denoted $\mathcal{N}$. The set of directed edges is $\mathcal{E}$. The subsets of incoming and outgoing edges at node i are denoted, respectively, $\rightarrow i \subset \mathcal{E}$ and $i \rightarrow \subset \mathcal{E}$.

7.7.3 Isolating and Dividing the Neuron

Figure 7.32 shows how one neuron in a network, with index j, is isolated from all the other neurons in the network. Its inputs are given variables like $x_{i \rightarrow j}$ that live on the incoming edges and are not shared by any other neuron. Similarly, its output is placed in the variable y_j which is distinct from variables like $x_{j \rightarrow \ell}$ that other neurons use for the input from neuron j. In the next section we will go over the obvious concur constraints that keep neuron inputs x, on edges, consistent with neuron outputs y that live on nodes. In this section we focus on the constraints among just the set of x variables and single y variable within one neuron.

The variables associated with neuron i are its output y_i, a vector of inputs x_i, and a vector of weights w_i. The elements of x_i and w_i live on the neuron's

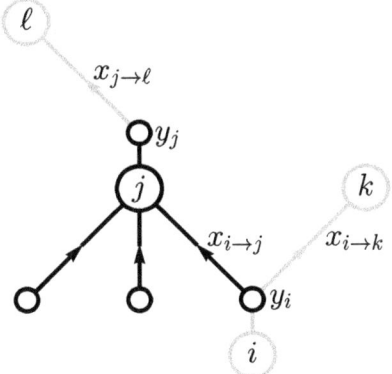

Figure 7.32 Neuron j is isolated from neuron i by having its input variable $x_{i \to j}$ distinct from neuron i's output variable y_i. Likewise, its output variable y_j is distinct from the input $x_{j \to \ell}$ received by neuron ℓ. The output neuron i sends to neuron k is also given its own variable, $x_{i \to k}$.

incoming edges, labeled $k \to i$ for all the neurons k that send their outputs to neuron i. Neuron i also has a margin hyperparameter δ_i.

The constraint on the variables of neuron i, written A_i, is the union of two sets,

$$A_i = A_{i+} \cup A_{i-} ,$$

where $A_{i\pm}$ correspond to the ± 1 cases of the output y_i. In the following, let the upper sign denote the value of $y_i = \operatorname{sgn}(w_i \cdot x_i)$ and let m be the number of inputs.

The margin constraint (7.28b) can be rewritten as

$$\pm\, w_i \cdot x_i \geq \delta_i ,$$
$$\|w_i\|^2 + \|x_i\|^2 - \|w_i \mp x_i\|^2 \geq 2\delta_i ,$$
$$2m - 2\delta_i \geq \|w_i \mp x_i\|^2 ,$$

where in the last line we imposed the normalization constraint $\|w_i\|^2 = m$ and the Boolean constraint $x_i \in \{-1, 1\}^m$. Because it is easier to impose the normalization constraints in set B, we define the divide constraint on neuron i as follows:

$$A_{i\pm} = \left\{ (w_i, x_i, y_i) \in \mathbb{R}^{m+m+1} \ : \ y_i = \pm 1, \ \|w_i \mp x_i\|^2 \leq 2m - 2\delta_i \right\} .$$

Projecting to this constraint is intuitive and easily computed. In the case of the upper sign, the points w_i and x_i (in $\mathbb{R}^m$) must be sufficient near each other, that is, the signs of the weights are encouraged to align with the Boolean signs. The distance-minimizing change, if the inequality is not satisfied, moves both points toward each other along the line that passes through them. The opposite happens, w and x are pushed toward the other's negation, when the lower-sign case $y_i = -1$ is considered. To project to A_i, the union of A_{i+} and A_{i-}, the $\pm$ option with the smallest projection distance is selected.

7.7.4 Concur and Data Constraints

The concur constraints that connect all the neurons in the network are the following constraints at the internal nodes $i \in \mathcal{N}$,

$$B_i = \left\{ (y_i, x_{i\rightarrow}) \in \mathbb{R}^{1+|i\rightarrow|} \: : \: \forall\, i\rightarrow j \, \in \mathcal{E} : \: y_i = x_{i\rightarrow j} \right\}, \qquad (7.29)$$

where $|i \rightarrow|$ is the number of outgoing edges at node i. Projecting to this constraint, the equality of $1 + |i \rightarrow|$ variables, is easy because no Boolean (± 1) constraints are imposed. The Boolean constraints are taken care of in the A constraint. Whereas constraint A allows the x's to take continuous values, the neuron outputs y_i are constrained to be ± 1 in that constraint. But that's enough, because constraint B_i forces the concurring $x_{i\rightarrow j}$ to be Boolean as well.

At the input nodes $i \in \mathcal{I}$, the data Y_i fix the values of the y variables. We can still use the concur constraint (7.29) at these nodes provided the A constraint is augmented to include, for $i \in \mathcal{I}$,

$$A_i = \left\{ y_i \in \mathbb{R} \: : \: y_i = Y_i \right\}. \qquad (7.30)$$

This data constraint, in set A, can be thought of as a replacement of the neuron constraint for those nodes that have no neurons.

The roles of A and B for imposing the data are reversed at the output nodes. Because there already are neuron constraints A_i that constrain y_i at nodes $i \in \mathcal{O}$, the data constraint must be in set B for these nodes:

$$B_i = \left\{ y_i \in \mathbb{R} \: : \: y_i = Y_i \right\}.$$

7.7.5 Weight Sharing and Normalization

In addition to normalization, the weights of the network are subject to a concur constraint that lies at the heart of the learning task itself. Called *weight sharing*, we will describe these constraints in a unified way.

Weight sharing is best known in the context of the convolutional filters that are learned by networks trained on image data. Different neurons process

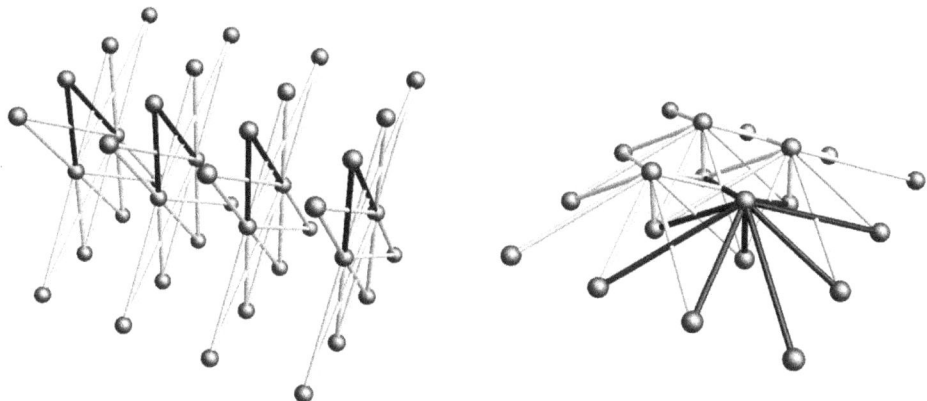

Figure 7.33 *Left*: Four instantiations of an autoencoder network, where all four instances of the same neuron share the same weights (dark edges). *Right*: A convolutional layer in one network, where each neuron receives inputs from a 3×3 set of nodes (dark edges). In this example all neurons share the same set of nine weights.

different parts of an image but are required to use identical weights in their filters. The premise behind the sharing principle in this instance is that the structures of elementary features (edges, etc.) are independent of where they occur in an image.

In a network with convolutional filters, there is weight sharing within the same network. A network without convolutional filters, but trained on multiple sets of data, also shares weights in the sense that the same weights are required to apply to every data item. This instance of weight sharing can be made to look like the convolutional one if we aggregate the multiple instantiations of the network, differing only by the imposed data, into a single network. The neurons in such an aggregate network do not interact across instantiations except for the fact that they share weights. But this too is a strong constraint and the foundation of generalization in machine learning.

Both types of weight sharing arise in our two featured applications and are contrasted in Figure 7.33. The left panel shows 2^n instantiations of an $2^n \rightarrow n \rightarrow 2^n$ autoencoder network, for the case $n = 2$. The 2^n networks differ only in the 2^n data constraints applied to their input and output nodes. If we view the aggregated networks as a single network subject to a single, aggregated-data constraint, then weights are being shared within the same network. In many training protocols, the data is partitioned into *batches*.

In that setting the aggregate network comprises one instantiation for every data item in the batch.

The right panel of Figure 7.33 shows the structure of a network tasked with learning a cellular automaton rule. The rule is defined by a BTF that takes its input from a 3×3 field of nodes in the layer below. Reconstructing the automaton rule is equivalent to learning BTF weights that correctly transform one automaton pattern into another. A key part of that learning process is the equality of the BTF rule across the network, that is, the sharing of the nine BTF weights.

Regardless of whether there is weight sharing because of data aggregation, or the model has an underlying symmetry (e.g., translation across layers, translation between layers), the neurons of the network can be partitioned into sets that share weights. If there are K weight-sharing sets, we can write

$$\mathcal{N} \cup \mathcal{O} = \mathcal{S}_1 \sqcup \cdots \sqcup \mathcal{S}_K$$

since there are no neurons associated with the input nodes. All the neurons in any set $k \in \{1, \ldots, K\}$ have the same number of weights on incoming edges, $\deg(k)$. The weight sharing constraint on neurons in the set $\mathcal{S}_k$, combined with normalization, is the following part of the B constraint:

$$B_k = \left\{ \left(\Pi_{i \in \mathcal{S}_k}^{\times} w_i \right) \in \mathbb{R}^{|\mathcal{S}_k| \deg(k)} \ : \ \forall i \in \mathcal{S}_k \ : \ w_i = \bar{w}, \ \|\bar{w}\|^2 = \deg(k) \right\}.$$

Projecting to B_k is a simple case of concur with a normalization side constraint: The shared weight vectors are replaced by their average after this average has been rescaled to have norm $\deg(k)$.

7.7.6 Analog Inputs and Outputs, Generative Models

With a relatively minor modification, our networks can work with analog inputs. When the m inputs to a neuron i are also network inputs, for which we have analog data $Y_{\to i}$, constraint $A_{i\pm}$ is replaced by

$$A_{i\pm} = \left\{ (y_i, w_i) \in \mathbb{R}^{1+m} \ : \ y_i = \pm 1, \ \pm w_i \cdot Y_{\to i} \geq \delta_i \right\}. \tag{7.31}$$

Because the data vector $Y_{\to i}$ is analog, there is no lower bound on the (positive) margin parameter δ_i. The projection for neuron i is almost the same as it was for the Boolean case, where both signs of the output y_i are considered and the one with the smallest projection distance is chosen. Rounding y_i to ± 1 has some cost, but the change to w_i, if it does not satisfy the inequality, may incur additional cost.

The simplest way to have the network produce analog outputs Y_i, $i \in \mathcal{O}$ is without any activation at all:

$$w_i \cdot x_i = Y_i \ . \tag{7.32}$$

Keeping the same normalization of the weight vector and Boolean ± 1 values for x_i, the data Y_i must be scaled because of the bound $|w_i \cdot x_i| \leq m$. However, this extreme case (realized with equal magnitude weights) should not guide our thinking about the correct scaling of the data. Instead, we observe that the average of the square satisfies

$$\langle (w_i \cdot x_i)^2 \rangle = m$$

when w_i is uniformly sampled on the sphere $\|w\| = \sqrt{m}$ and x_i on the m-cube. This suggests the data-rescaling rule

$$\langle Y^2 \rangle = \frac{1}{|\mathcal{O}|} \sum_{i \in \mathcal{O}} Y_i^2 = m$$

for the case where all the output nodes have m inputs.

Using the normalization constraint on w_i and Boolean constraint on x_i, we can rewrite (7.32) as a distance constraint:

$$\|w_i - x_i\|^2 = 2m - 2Y_i \ .$$

The replacement of constraint A_i, when there is analog data at output node i, is therefore

$$A_i = \left\{ (w_i, x_i) \in \mathbb{R}^{m+m} \ : \ \|w_i - x_i\|^2 = 2m - 2Y_i \right\} \ .$$

The input constraints in a *generative model* go a step further than the analog replacement (7.31) by having no input constraints at all. The idea is that even seemingly rich data sets are functions of a manageable number of *latent variables*. If the data is symbolic, like sea shanties, those variables might even be Boolean (merchant vessel? North Atlantic? etc.). The model attaches no meanings to these latent variables, asserting only that they exist. A generative network, trained with lots of data constraining its outputs, is tasked with generating that data given complete freedom in the settings of the latent variables. The independence of the latent variables is important. In a sea shanty generator, for example, a pair of latent variables with meanings military victory? and French? would not be independent. Models trained to generate new data that matches the characteristics of actual data would not be very useful if their latent variables had to be sampled in some unknown, correlated way.

To implement a generative network, one only needs to replace the data constraints (7.30) at input nodes $i \in \mathcal{I}$ by

$$A_i = \left\{ y_i \in \mathbb{R} \; : \; y_i = \pm 1 \right\}. \tag{7.33}$$

The only other constraint on y_i is the concur constraint B_i which allows y_i to take continuous values. The only role of the A constraint at input nodes in a generative is therefore to ensure those values are Boolean.

7.7.7 Initialization and Metric

Though initialization should not matter very much in hard constraint problems, in particular those with unique solutions, it is helpful to supply these details to others trying to reproduce results. For neural networks we thought it would be interesting to have the initial point satisfy *all* the constraints except those at the neurons.

Each neuron in a weight-sharing set is initialized with the same vector of weights w. The components of w are uniformly sampled between -1 and 1 and normalized. In defiance of the feed-forward operation of the network (once trained), the initial neuron outputs y_i are independently sampled from the uniform continuous distribution between -1 and 1. The corresponding variables $x_{i\rightarrow}$ on the outgoing edges are then initialized with the same value.

When training a network on multiple batches of data, the initialization would be done differently, since otherwise the weights learned would not carry over from one batch to the next. This is addressed by using the final, shared, and normalized weights from the projection to the B constraint in one batch as the initialization for the next. When the batch size is large, and the corresponding constraint problem has been solved, one can hope the network is already able to generalize. The neuron inputs and outputs in that case should not be initialized randomly, but initialized by feed-forward propagation from the data, if available, at the network inputs. The only constraint discrepancy is now at the network's output nodes. This mode of "educated" neuron-variable initialization is not possible in a generative network. Instead, those variables could be initialized by hill-climbing on the latent inputs with the cost being the discrepancy between the network's outputs and the data it was meant to generate.

Batch size is often deliberately set at a small value in gradient-based training because it introduces stochasticity in the loss landscape, a strategy that keeps the gradient from vanishing because of too many competing constraints. Since the RRR algorithm does not suffer from this problem, the batch size should be as large as possible for the available resources.

Because the three variable types – w, x, y – play very different roles in the constraint problem, they should be allowed to have different metric coefficients. Each neuron i, perhaps depending on the layer in which it resides, should also be allowed to fine-tune the metric coefficients of its (w_i, x_i, y_i) triple relative to those at other neurons. The following metric addresses both kinds of diversity:

$$(\text{dist})^2 = \sum_{i \in \mathcal{N} \cup \mathcal{O}} g_i \left(\|w_i' - w_i\|^2 + \xi \|x_i' - x_i\|^2 + \eta (y_i' - y_i)^2 \right) + \sum_{i \in \mathcal{I}} g_i \eta (y_i' - y_i)^2 . \tag{7.34}$$

The second sum is over the nodes not associated with neurons but network inputs. The metric coefficients of the x and y variables, relative to the weights w, are respectively ξ and η. For example, making neuron weights more compliant than neuron inputs in satisfying the neuron (BTF) constraints is accomplished by setting $\xi > 1$.

Because there are too many nodes, the node-specific coefficients g_i cannot be set by hand and we use the local-gap heuristic with update rule

$$g_i \to g_i + \gamma \left(\Delta_i^2 / \Delta^2 - g_i \right)$$

in each RRR iteration. When there is a neuron at node i, the local squared-gap is

$$\Delta_i^2 = \left\| w_i^A - w_i^B \right\|^2 + \left\| x_i^A - x_i^B \right\|^2 + \left(y_i^A - y_i^B \right)^2 ,$$

where the superscripts refer to variables after projection to the A and B constraints. At input nodes only the y variable contributes to the local gap. The average of Δ_i^2 over nodes is the squared RRR gap, Δ^2. By giving the variables associated with an above-average local gap an above-average metric coefficient, the burden of closing the gap is shifted from the local variables to other variables. The relaxation parameter γ controls the rate at which the equilibrium between gaps and metric coefficients is established. Initially $g_i = 1$ at all nodes.

The metric can be more finely ramified than (7.34), with different coefficients on all of a neuron's inputs. The restricted form has the advantage that the projection computations are only mildly affected.

7.7.8 The Binary Encoding/Decoding Problem

This neural network task was already described in Chapter 1, as motivation to study this book. The problem of encoding 2^n vectors into 2^n binary codes on n internal nodes, and then decoding them back to the original vectors, was used by Rumelhart, Hinton, and Williams (RHW) [91, 92] to demonstrate their landmark error back-propagation training algorithm. Though the trained *autoencoder* was a success – inputs exactly matched outputs – the sigmoid outputs at the internal nodes were not binary codes.

By using BTFs, the Boolean node values have a direct binary interpretation. Our implementation follows the definitions of the previous sections. We start with a *completely connected* two-layer network. In the encoder layer, each of the 2^n input nodes is connected to each of the n code nodes. The analogous all-to-all connectivity is used in the decoder layer. To allow for bias in the BTFs, the network is given an additional input node that sends a constant value $(+1)$ to all the neurons in the network. The code neurons therefore have $n + 1$ inputs and the output neurons have $2^n + 1$ inputs. We process all 2^n data in a single batch by aggregating 2^n weight-sharing autoencoders into a single network, as shown in the left panel of Figure 7.33 for $n = 2$. The aggregate network has 2^n shared copies of $n(2^n + 1) + 2^n(n + 1)$ weight parameters altogether.

As in RHW [91, 92], the 2^n data vectors have the 1-hot structure, with value $+1$ on one node, -1 on the other $2^n - 1$ nodes. Exercise 7.8 is about single-layer augmentations of the encoder and decoder that extend their functionality to an arbitrary set of 2^n Boolean data vectors.

In principle, the code- and output-node neurons should get different margin hyperparameters δ. Selecting good margins is tricky without knowing in advance the type of gate a neuron will implement. The safe values $d^*_{\min}$ (Table 7.4) will be very small when a neuron has many inputs and the resulting solution weights will not be as interpretable as a network trained with larger margins. To keep this demonstration simple, we set a global δ equal to 80% of the maximum possible margin. Part of Exercise 7.8 is showing that the decoding neurons (at the output nodes) have the smaller maximum margin, with value

$$\delta_{\max}(n) = \sqrt{\frac{n + 1}{(n - 1)^2 + n}} \, . \tag{7.35}$$

Provided β is sufficiently small, it is not difficult to find values for the other hyperparameters such that training is nearly always successful. We used $\beta = 0.2$ and kept the metric uniform across nodes by setting $\gamma = 0$. That leaves only the metric coefficients ξ and η, which control the compliance of neuron inputs and outputs relative to the weights. By our choice of normalization, all three sets of variables should have comparable fluctuations if there is cooperativity in the solution process. To diagnose if that is true, one can

n	2	3	4	5	6
parameters	22	59	148	357	838
bits of data	16	64	256	1 024	4 096
δ	0.800	0.605	0.496	0.428	0.380
ξ	1.0	1.0	1.5	2.0	2.5
η	2.5	3.0	4.5	6.0	7.5
iterations	580	2 100	7 700	34 000	330 000
success rate	100%	100%	99%	96%	92%

Table 7.5 Summary of binary autoencoder training experiments. The number of iterations is the average over the successes in 100 trials.

output the root-mean-square averages (over nodes)

$$(\delta w)^2 = \langle \, \|w^A - w^B\|^2 \, \rangle \,,$$
$$(\delta x)^2 = \langle \, \|x^A - x^B\|^2 \, \rangle \,,$$
$$(\delta y)^2 = \langle \, (y^A - y^B)^2 \, \rangle \,.$$

We should be worried, for example, if one of these is much smaller or larger than the others. To make the weight variables more (less) active in the search for solutions, one would increase (decrease) both ξ and η. Similarly, if too many neuron inputs are flipping signs relative to the sign flips of outputs, or $\delta x \gg \delta y$, one might want to increase ξ relative to η. However, these rules are just guides for how to explore the space of hyperparameters because in the end what counts is minimizing the number of RRR iterations.

Table 7.5 summarizes the binary autoencoder training experiments up to the $n = 6$ instance. Even with the loose margin, the pattern of the signed weights in a solution is easy to interpret. Solutions are nonunique, but only because of symmetry. Any of the n code nodes can take the role of the 1's bit, the 2's bit, the 4's bit, and so on. And there is a binary choice in how that bit is represented (as -1 or 1). Finally, there are $(2^n)!$ ways to assign the integers $0, 1, \dots, 2^n - 1$ to the 1-hot data vectors. Altogether this amounts to $(2^n n!)(2^n)!$ symmetry-related solutions.

The iteration statistics in Table 7.5 give an incomplete picture of the training complexity. We are not surprised to see that the work appears to grow exponentially with n. But comparisons of the time series of the gap, in Figure 7.34, show that the growing complexity is not about blindly finding needles in exponentially growing haystacks, but something more systematic. To train the n-bit autoencoder, it appears, the gap is first reduced to the level that was sufficient for the autoencoder with only $n - 1$ bits. In the

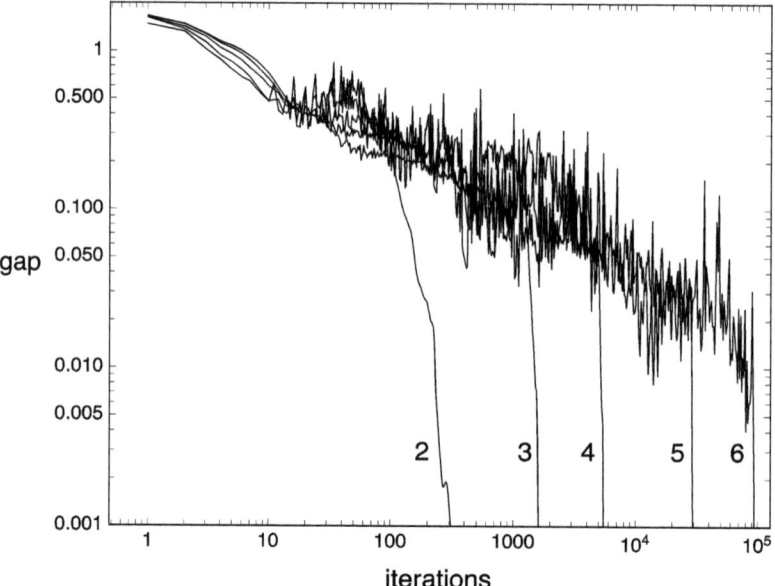

Figure 7.34 Overlay of typical gap time series in binary autoencoder training, spanning the range $n = 2$ to $n = 6$. After fluctuating on exponentially decreasing scales (over exponentially increasing times), training is complete when the gap plummets.

subsequent training, the gap for n bits evolves on just the smaller scale, until it too eventually plummets to zero. Since a small gap corresponds to fewer weights flipping signs, a better picture of the training process is one of progressive refinement.

The discovery of simple rules for assigning binary codes to 2^n data items, though remarkable, should not be seen as an achievement in *general* artificial intelligence. Learning the rule for $n = 5$ does not implicate the rule for $n = 6$. True learning of the meta-rule would involve a second stage of training that examines the weight patterns found for specific instances of n. However, we can be sure the success of meta-rule learning is helped by interpretable patterns to work with from the first stage!

7.7.9 Reconstructing Cellular Automata Rules

A standard artificial intelligence task is prediction, and the version of this task that best tests generalization is learning a rule for predicting arbitrarily far into the future. This is possible in principle if there exists a single rule that applies at all times – like a law of physics. The simplest worlds that fit this description, where the rules are also local in space and the same at every point, are *cellular automata*.

The most famous cellular automaton is John Conway's *Game of Life* (GoL). Space in this world is the infinite square grid, occupied by "cells" that are alive or dead. The rule that determines the alive/dead state of a cell at the next time is determined by the states of the surrounding eight cells. A live cell remains alive when surrounded by three or four live cells (otherwise it dies of loneliness or overcrowding). When the cell is dead, it is revived – becomes alive – when surrounded by exactly three live cells.

When presented with pairs of GoL patterns, showing t applications of the rule, how hard is it to infer the rule? All we are told is that the same rule was applied across the grid, t times, and that this rule is a binary function on the states of a 3×3 set of cells at the previous time. The case $t = 1$ is easy because a look-up table for the rule is constructed just by noticing what happened at various 3×3 parts of the grid. The rule is determined completely if the initial pattern contains all possible 2^9 patterns in one of its 3×3 neighborhoods. If the initial pattern is large and the states of the cells are randomly sampled, the odds of that are very good. For $t > 1$, the rule reconstruction problem is much harder because the states of unseen cells at intermediate times have to be reconstructed as well. Since either of two states, alive or dead, is a possible function value for each of the 2^9 states in the neighborhood, the number of rules, $2^{2^9} \approx 10^{154}$, is too large to try them out in turn.

Since the GoL rule and cellular automata rules more generally have the same structure as convolutional neural networks, a related problem is the representation of rules by network architectures and activation functions [45]. This and the training of such networks is already interesting for $t = 1$. With two exceptions, training for $t > 1$ has not received much scrutiny, probably because that problem is computationally much harder, whether attempted with neural networks or not. Jacob Springer and Garrett Kenyon [104] found that gradient-descent trained convolutional networks (with mostly ReLU activation) failed at learning the GoL rule from $t > 2$ data, and even with $t = 2$ data needed to have more parameters than the minimum set that could represent the rule. The only other attempt to train on $t > 1$ data was based on the RRR algorithm [38]. Though more successful, the networks in that study did not use activation functions in the usual sense.

Just as multiple layers of ReLU-activated networks are required to express the GoL rule [45], a single BTF is also insufficient. To keep our demonstration of rule reconstruction simple, and avoid the separate problem of architecture design, the rule we will try to reconstruct is one that is directly defined by

Divide and Concur

Figure 7.35 The extreme inputs x of `rule4379`, for which $\widetilde{w} \cdot x = 1$.

a single BTF acting convolutionally on a 3×3 field of inputs. This is still quite a hard problem because there are over 10^{13} 9-input BTFs [84] !

If the training data does not also serve as a source of stochasticity, then there is no reason why the network cannot be trained with a single pair of sufficiently large automaton patterns. For rules defined by a 9-input BTF, and data patterns related by one automaton time step ($t = 1$), we can approach the size-sufficiency question as follows. An m-input BTF is defined by its extreme inputs X^*, that is, the m-cube points touched by the thickest knife that can be used for the bisection. In the worst case, there are only m extreme points for which the BTF evaluates to $+1$ (and another m where it evaluates to -1). Figure 7.35 shows an example for $m = 9$. Unless each of the nine patterns or its negative appears in the input data, there is insufficient information to determine the rule. Let's say the input pattern has area A and contains that same number of 3×3 neuron inputs (on a periodic grid). In an unbiased random input pattern, the expected number of occurrences of each extreme neuron input, or its negative, is $\mu = A/256$. By Poisson statistics, it will be present with probability $p = 1 - e^{-\mu}$. Since all nine must be present, the probability of having sufficient information is p^9. Our rule reconstruction experiments use 32×32 periodic grids, so $\mu = 4$ and $p^9 \approx 0.85$. Those are acceptable odds for a single data pair to have sufficient information to uniquely reconstruct the rule. In their experiments to reconstruct the GoL rule, Springer and Kenyon [104] also used 32×32 pattern pairs, but their networks were trained on one million such data (instead of just one pattern-pair).

If data from $t \geq 2$ applications of the rule to random patterns also uniquely defines the rule, then information sufficiency is improved in the $t > 1$ reconstruction problems (relative to our estimate for $t = 1$). That's because the BTFs in the unseen layers might be confronted with inputs that were absent in the data layer. Even with sufficient data, reconstructions for even values of t have a two-fold nonuniqueness because $w \to -w$ flips the signs of the cells in alternating layers, leaving the network output unchanged.

The maximum margin $\delta_{\max}$ that can be imposed on a BTF varies by orders of magnitude, even with as few as nine inputs. Not surprisingly, the difficulty of rule reconstruction is a strong function of $\delta_{\max}$. At one extreme are the *dictator gates*, with a single nonzero weight equal to $\pm\sqrt{m}$. For these the neuron output is a function of just one input, all the points of X_+ are extreme, and $\delta_{\max} = \sqrt{m}$. Moving in the direction of greater difficulty and democracy, for odd m, are the *majority gates*. These correspond to all components of the weight vector equal to ± 1. The number of extreme points has shrunk to $\binom{m}{\lfloor m/2 \rfloor}$ and the maximum margin is correspondingly smaller, $\delta_{\max} = 1$. Finally, the most difficult BTFs, or those with $\delta = d^*_{\min}(m)$, have only m extreme points.

To observe the effects of rule difficulty on training, we performed experiments with the following two sets of BTF weights:

$$
\begin{array}{ccc}
-1 & 1 & 1 \\
1 & -1 & -1 \\
1 & -1 & 1
\end{array}
\qquad\qquad
\begin{array}{ccc}
33 & -7 & 10 \\
-6 & 36 & 28 \\
16 & -12 & 25
\end{array}
$$

$$\texttt{rule9} \qquad\qquad\qquad \texttt{rule4379}$$

These integer weights $\widetilde{w}$ have the property $\widetilde{w} \cdot x = 1$ when x is an extreme point of X_+. The largest possible margin is therefore $\delta = \sqrt{9}/\|\widetilde{w}\|$ once the weight vector is normalized. The majority gate on the left, called `rule9`, has $\delta = 1$, while `rule4379` is a strong contender for realizing $d^*_{\min}(9)$ with

$$\delta = \sqrt{9/4379} \approx 0.0453350 . \tag{7.36}$$

The nine extreme points of `rule4379` are rendered in Figure 7.35. Of course, any permutation and signs applied to these weight patterns keeps the margin unchanged. The choices shown were selected because the resulting automaton patterns are interesting – an aesthetic judgment.

Figures 7.36 and 7.37 show the evolution of `rule9` and `rule4379` from random initial patterns. Because the initial patterns have an unbiased distribution of ± 1 values, the self-dual property of the BTF preserves that distribution. A physicist would describe `rule4379` as a model of "coarsening" into three kinds of domain: black, white, and checkerboard.

The networks used to reconstruct the automaton rules are exactly the same networks that generated the time evolution. Data for t time steps of evolution is reconstructed on networks with $t+1$ layers of nodes, or t layers of neurons. The convolutional structure of the neuron inputs is shown in the right panel of Figure 7.33. To eliminate the effects of boundaries, the nodes in each layer have a periodic connectivity (their square-grid coordinates are integers mod 32).

Figure 7.36 An initial random pattern (left) of size 64×64 followed by one and four applications of `rule9`.

As in the binary autoencoder training, the initial point is a uniform random sample on constraint set B, that is, completely concurring and weight-sharing variables that do not satisfy the neuron constraints in set A. To include `rule9` among the admissible BTF weights, the margin has to be set at a value less than 1 when training on `rule9` data. We used $\delta = 0.9$. To learn `rule4379`, the margin must be set below the value in equation (7.36). We used $\delta = 0.045$ in those experiments. Unlike the binary autoencoder network, the neurons in the rule reconstruction network were *not* given a constant 10th input (and corresponding weight) as a way to implement bias.

Not surprisingly, reconstructing rules from $t = 1$ data is very easy. Using the Douglas–Rachford time step $\beta = 1$, metric parameters $\xi = \eta = 1$, and no metric auto-tuning ($\gamma = 0$), the `rule9` BTF is reconstructed in under 200 RRR iterations, and `rule4379` in 4 700 iterations. Besides the smaller margin setting, `rule4379` also requires a smaller stop-gap for terminating iterations, $\Delta_{\text{stop}} = 0.0001$, where 0.01 suffices for `rule9`. The RRR gap decreases nearly

Figure 7.37 Same as Figure 7.36 but for the `rule4379` automaton.

t	2	3	4
iterations	1 600	12 000	21 000
success rate	100%	32%	44%

Table 7.6 Summary of `rule9` reconstruction attempts from $t > 1$ data.

monotonically for both rules. Information sufficiency of the 32×32 data patterns was confirmed for both rules by evaluating the reconstructed BTFs on all 2^9 inputs.

Though rule reconstruction for $t > 1$ is significantly harder than for $t = 1$, because unseen cells at intermediate times must be reconstructed as well, this is also where the two rules differ markedly in difficulty. Results for `rule9` up to $t = 4$ are given in Table 7.6. In all of these experiments, there was no metric tuning of the three variable types, that is $\xi = \eta = 1$, but neuron-specific auto-tuning was enabled with $\gamma = 10^{-3}$. The most important change from the $t = 1$ settings was the RRR time-step $\beta = 0.2$. This is expected because RRR has to explore many more discrete configurations when there are unseen cells. Conspicuous in the results is the less than near perfect success rate for $t = 3$ and $t = 4$. From the iteration counts of the successful runs, this cannot be blamed on the generous 2×10^5 iteration limit. A more refined metric tuning might resolve this problem.

The increased difficulty of `rule4379` is conveyed by a typical gap time-series for the $t = 2$ reconstruction problem, in Figure 7.38. For this reconstruction, the x and y variables had to be made less compliant than the weights with $\xi = 2.0$ and $\eta = 1.6$. Also, a smaller time step, $\beta = 0.05$ and some neuron-specific metric tuning with $\gamma = 10^{-4}$ were used.

In Figure 7.38 we see that most of the search takes place when the gap is already quite small. On the other hand, when plotted on two logarithmic scales, there is evidence that the search is at least making incremental progress. To assess this progress, we count the number of wrong BTF values for reconstructed weights w,

$$\#(w) = \left| \{ x \in X_+ : w \cdot x < 0 \} \right|,$$

and average this for several runs that are terminated at a given stop-gap, Δ_{stop}. Statistics for $t = 2$ reconstructions with four geometrically decreasing values of Δ_{stop} are given in Table 7.7, each based on 200 solution attempts. From the average of the number of wrong function values, $\langle \# \rangle$, we see that a good approximation of the automaton rule can be learned with a modest amount of work ($\Delta_{\text{stop}} = 0.1$). In this respect, and at least for this learn-

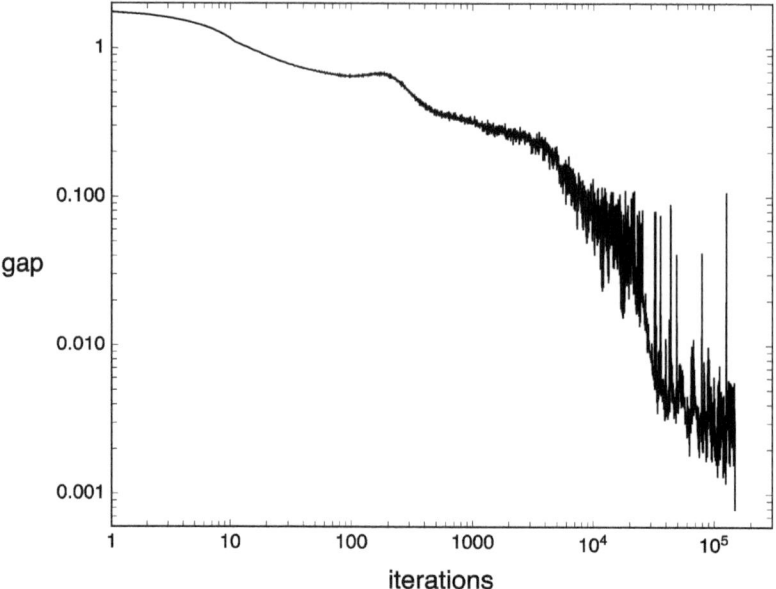

Figure 7.38 Log-log plot of the gap in a successful run when reconstructing `rule4379` from $t = 2$ data. To learn the rule in complete detail, not only does the gap have to decrease by orders of magnitude, but the iterations required also span orders of magnitude.

ing task, the RRR gap is not that different from the loss in gradient-based training. But even approximate rule learning quickly becomes difficult with increasing t for `rule4379`. When terminating training at $\Delta_{\text{stop}} = 0.1$, the success rate for $t = 3$ drops to 11%.

Δ_{stop}	$\langle \# \rangle$	iterations	success rate
0.1000	11.8	7 400	56%
0.0200	7.6	24 000	44%
0.0040	3.6	160 000	42%
0.0008	0.3	290 000	5%

Table 7.7 Summary of `rule4379` reconstruction attempts from $t = 2$ data.

7.7.10 Bipartisan AI?

Much of the ongoing revolution in artificial intelligence can be traced to developments in the high-level structure of networks. By contrast, the technology that underlies the individual components and their optimization has been relatively static. The transformer networks used for large language models still rely on feed-forward nets, ReLU activation, and back-propagation of errors for training. The divide-and-concur scheme for deconstructing networks, and the new approach to training this enables, is a compelling alternative for that core technology.

The bipartisan formulation of training, with constraint projections replacing gradient steps and the gap between sets A and B serving as a substitute for loss, cannot be dismissed by the usual engineering criteria. One RRR iteration scales with network size no differently than the computation of one gradient step. The machine operations that implement projections and the RRR update are local on the network and are dominated by simple computations (addition, subtraction, squaring, and vector multiplication by scalars). There are no computations of ad hoc transcendental functions. Also, unlike gradient-based training, the machine operations at the core may be carried out synchronously over all neurons of the network.

A number of headaches of the current technology – vanishing gradients, normalization questions, over-parameterization – go away with the new approach. Gradients can't vanish when none are computed, and there is no question about normalization when the neuron inputs and outputs are simply ± 1. Weights are tied to neurons and their normalization follows directly from the normalization of the inputs with which they are paired. And our two demonstrations (binary encoding/decoding, automaton rule learning) showed that networks can be trained in a strict teacher–student mode, where the teacher network generating the data has exactly the same structure and parameter set as the student network being trained. This is the very opposite of over-parameterization and the highest standard of model interpretability.

The metric parameters that control the projections have no obvious counterparts in the current technology. These may play a role similar to the additional variables, such as momentum, used by gradient-based optimizers. However, the exploitation of metric parameters is not so much an enhancement as an acknowledgment of the inherent inequivalence of the three variable types (w, x, and y). While the RRR algorithm's fixed-point property is unchanged by the metric, the *search* for fixed points is sensitive to the relative fluctuations of the neuron weights, inputs, and outputs. By their position in the network, neurons can also benefit from having their metric parameters tuned. The general-purpose RRR auto-tuning heuristic helped in both of the applications, but there may be better heuristics for constraints that live on networks.

Two extensions of network models based on the Boolean threshold activation function have yet to be explored. The first is to exercise more control over the margin hyperparameter. Instead of a global value of this parameter, it makes sense to have its setting depend on the type of neuron (number of inputs, position in the network). As we saw in the automaton application, the margin limits the complexity of a neuron's function. For example, $\delta = 1$ eliminates highly complex BTFs, like `rule4379`, but allows majority gates, such as `rule9`. Even larger margins drive neurons in the direction of totalitarianism, where outputs are a function of just a subset of the inputs. But there is a more direct way to achieve the same goal. This is the second extension, where the sparsity of a neuron's weights is specified through a side constraint (the same one that already imposes normalization). Sparsely connected networks have greater interpretability, and their architecture could be learned by imposing a sparsity constraint on a network with fully connected layers.

New technologies are not launched by abstract arguments, but by one exceptionally successful and exciting application. Binary encoding/decoding and automaton rule reconstruction do not quite rise to that level. Whereas current technology completely fails at these tasks, a more compelling demonstration of the bipartisan alternative is an open problem.

Exercises

7.1 Generalize divide-and-concur for systems of equations, when some number k of the n equations are allowed to be unsatisfied. Do this by adding a side constraint on the n independent divide constraints. Describe the divide projections and the projection to the side constraint for systems of the following kinds of "equations" (with examples):

real-linear equations	$3x_1 - x_2 + 2x_3 = 4$,
linear equations mod 2	$x_1 + x_2 = 0$,
linear inequalities	$3x_1 - x_2 + 2x_3 \leq 4$,
Boolean clauses	$x_1 \vee \neg x_2 \vee x_3$.

7.2 Let the elements w_{ij} of an $n \times n$ matrix represent edge weights between the two sides of a bipartite graph. As we saw in Section 4.6.3, the problem of finding the maximum weight bipartite matching in the graph is equivalent to solving the following LP:

$$\text{maximize:} \quad g = \sum_{i=1}^{n} \sum_{j=1}^{n} w_{ij} \, x_{ij} \tag{7.37}$$

$$\text{subject to:} \quad \sum_{k=1}^{n} x_{kj} = 1 \, , \ \sum_{k=1}^{n} x_{ik} = 1 \, , \tag{7.38}$$

$$\text{and} \quad x_{ij} \geq 0 \, , \ \forall \, i, j \, . \tag{7.39}$$

(a) Write a divide-and-concur solver of this LP using three replicas of the $n \times n$ matrix x. The replica for (7.37), after replacing the equality with a lower-bound inequality, is the objective replica and is subject to a single linear constraint (Section 4.2.3). Another replica can take care of all the row and column sums (7.38) because we know a very efficient projection to this symmetrical constraint (see Section 4.2.4). Finally, a third replica will impose all the nonnegativity constraints (7.39).

(b) As written, the LP above is what we have called the dual form. Show that the corresponding primal LP is expressed in terms of a pair of vectors $(y, z) \in \mathbb{R}^{2 \times n}$:

$$\text{minimize:} \quad f = \sum_{i=1}^{n} y_i + \sum_{j=1}^{n} z_j \, , \tag{7.40}$$

$$\text{subject to:} \quad y_i + z_j \geq w_{ij} \, , \ \forall \, i, j \, . \tag{7.41}$$

(c) Write a divide-and-concur solver for the primal LP with n replicas for each of the two vectors, giving two $n \times n$ matrices y and z. The concur constraint equates the columns of y and the rows of z. By combining this with a side constraint on the sums of the concurred values, you can satisfy (7.40) (when turned into an upper-bound inequality). The $n \times n$ inequalities (7.41) are extremely simple single-sum constraints on the elements of the two replicas.

(d) Compare the two methods. Both break down into very simple operations on $n \times n$ matrices (three in the dual, two in the primal). The replicas in the dual version have very different roles, while those in the primal just correspond to interchanging rows and columns (the sides of the bipartite graph). The dual-version replicas should therefore have different metric coefficients, unlike the primal replicas. In the dual version, the optimizing matching is the permutation matrix that x converges to (at the maximum g). How does one identify the matching in the primal version?

7.3 The norm constraint with periodicity can be generalized to lattices other than the scaled integer lattice, $\Lambda = w \, \mathbb{Z}^d$. The constraint

$$\|u\|_\infty \leq w/2$$

could have been written

$$u \in V_\Lambda(0) \, ,$$

where the Voronoi region of Λ at the origin, $V_\Lambda(0)$, is the set of points in $\mathbb{R}^d$ which have the origin as their closest element of Λ. The constraint set, for vectors of norm $2r$, has the general form

$$N(2r) = \left\{ u \in V_\Lambda(0) : \|u\| = 2r \right\} \, .$$

Consider this constraint for the scaled checkerboard lattice $\Lambda = w \, D_d$, where

$$D_d = \left\{ x \in \mathbb{Z}^d : \textstyle\sum_{i=1}^d x_i = 0 \quad \mathrm{mod}\ 2 \right\} \, .$$

(a) If this Λ is the periodicity of a sphere packing, of spheres of diameter $2r$, what is the minimum value of w?

(b) Work out the projection to $N(2r)$, assuming w exceeds the minimum value.

7.4 Prove Proposition 7.2, the projection to the divide constraint for equidistant lines.

7.5 Design a simple, infinite family of H/P sequences for the protein folding game SEQVENCE that have unique folds (up to symmetries).

7.6 Edge-matching tiles (Section 3.6) has a divide-and-concur formulation quite different from the one described in Section 4.6.2 (which does not use replicas). This exercise explores a formulation similar to the one used for SEQVENCE, with the constraint that every tile is placed and rotated so that it finds itself within known rectangular litemotifs of edge-matched tiles. To avoid boundary complications, the n^2 tiles are placed on an $n \times n$ board with the torus topology.

(a) Show that the number of matched edges per tile within a rectangular litemotif is maximized by minimizing the length of the litemotif's boundary. We therefore use square, $m \times m$ litemotifs for some integer $m > 1$. Also assume m is not too large, so that the complete litemotif library $\mathcal{L}$, for the tiles in the tile set, can be constructed. Each item in the library gives the identity and rotation of the tile at each of the litemotif's m^2 positions. Of course, the tiles in a litemotif have distinct identities.

(b) In SEQVENCE the sequence-selector variables had seven replicas because the litemotifs comprised seven hexagons. In the divide-and-concur formulation of edge-matching tiles, there are m^2 replicas for each tile variable. The tile variables are selectors with $4 \times n^2$ options giving the rotation and identity of the tile placed at a particular position on the board. Interpret the roles of the m^2 replicas and explain why one more replica is needed for the constraint that all n^2 tiles of the tile set are being used.

(c) Describe the computation of the projection to the divide constraint. Assume $\mathcal{L}$ is a precomputed resource.

(d) Explain how concur can address the fact that the litemotif-library-membership constraint is qualitatively different from the tile-set-usage constraint.

7.7 Prove Theorem 7.3, the isosceles-triangle (equality case) projection used by the dimension reduction algorithm eℓmo. It will help to know the parameter q in the projection formula is a Lagrange multiplier.

7.8   Here are some questions about the weights in the binary autoencoder of Section 7.7.8.

   (a) In the two-layer network trained on 1-hot data vectors, what is the pattern of weights in the encoder, including the weight for the constant input (bias)? What is the largest margin that can be imposed on the encoder neurons?

   (b) Same as the previous problem, except for the decoder stage. Confirm that (7.35) is the largest decoder-neuron margin and is smaller than the possible encoder-neuron margins.

   (c) By augmenting the two-layer autoencoder with fully connected $2^n \to 2^n$ layers on the input and output sides, the network can encode and decode arbitrary binary vectors. Show that this is possible with a design where the outer layers encode-to and decode-from 1-hot vectors. What are the maximum margins for the 1-hot-encoding and 1-hot-decoding neurons?

7.9   The figure below will help you work out the compression properties of the *infinite* Miura fold (avoiding the complications at the edges of the finite origami square). The left shows the pattern of mountain and valley folds when the paper is flat. The design is parameterized by the geometry of the repeating parallelogram. On the right we see the paper in a state of partial compression, and in projection. To understand what is going on in the third dimension, you only need to know that all the mountain folds that zig-zag vertically form a ridge at constant elevation, and similarly for the valley formed by their valley-fold counterparts.

   In the completely folded state, and neglecting the thickness of the paper, the infinite Miura-fold compresses to a horizontal line. Show that the compression in the horizontal dimension equals the cosine of the parallelogram's acute angle. This approaches zero as the parallelogram becomes a rectangle.

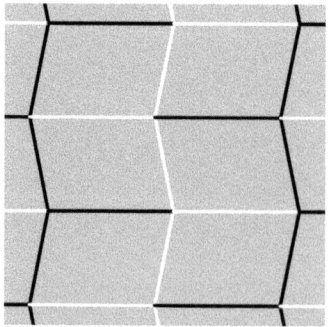
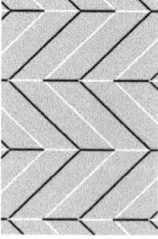

7.10 Show that the Game of Life rule can be implemented with three BTFs that take input from the 3×3 automaton cells and send their outputs to a fourth, 3-input, BTF. All your BTFs may have a bias parameter.

	4	1 1	2	0	2	1 1	1 1
2 2							
1 1 1							
1 1 1							
2 2							

7.11 *Nonograms* are named after Non Ishida, one of the two inventors who independently came up with puzzles such as the one shown above. In the unsolved puzzle, all cells in the 4×7 grid are gray. The solution has black cells that form contiguous horizontal and vertical sets of the lengths given in the margin. For example, in the top row there should be two sets, each of length 2 and separated by any positive number of gray cells, in this case 3. The numbers in the top margin give the same kind of information about the columns.

There is an obvious bipartisan formulation for nonograms where the variables are gray-black indicators for the cells. In the A constraint, each row must be consistent with its margin data, and B imposes the same, but on the columns. For example, there are six ways to arrange the two black dominoes in the top row, and 10 ways to have three isolated black cells in the second row.

Jon Borwein was fond of the bipartisan, row-column division of the constraints in nonograms, but lamented the fact that to compute the projections one had to precompute all the possible rows and columns,

There's a nice formulation which this margin is too narrow to contain, in detail. In outline, there are two sets of position variables p for each of the black cells. One set is row-wise, with fixed (integer) vertical coordinates and continuous horizontal coordinates. In the column-wise variables, these coordinate properties are reversed. There are spacing variables s for the row-wise p's and another set for the column-wise p's. These specify the spaces between the sets of contiguous black cells, and the space at the edges. The A constraint is the linear relationship between the p's and the s's. The B constraint imposes inequalities on the individual s's. It also imposes a matching of the row-wise p's and column-wise p's. This second part of the B constraint is a kind of "concur." -Numo

whose number grows exponentially in the size of the puzzle. Can you come up with a divide-and-concur formulation, where both projections have efficient computations?

8
Your Turn

If the author had an outsized role in the first chapter, that changes here. This chapter is all about you. Though the narrative has fictional elements, everything of a technical nature is true.

8.1 Four-Replica Sudoku

Your former college roommate, now the CEO of Pseudo-Co, calls you about plans for a new world headquarters comprising 16×16 cubicles on 16 floors – more than a fivefold expansion of the existing building (see Section 4.6.1). Everything, including floors and ceilings, would still be glass, and the same personal-space rules restrict cubicle occupation by the 16×16 employees. If you can come up with a valid occupation plan in 24 hours, you will be handsomely compensated.

The task at hand is to *design* a 16×16 sudoku solution with no givens – the solution of a blank puzzle. You could try a backtracking search, but debugging that code would take time. Your preference is simple operations performed on simple data structures. In the "glass building" representation you would be working with $16 \times 16 \times 16$ cubes of numbers. RRR works well with such data structures, where projections update array elements until a solution appears in the form of a fixed point. Unfortunately, the only bipartisan scheme you know involves projecting to 16×16 permutation matrices (e.g., on slices of the cube), and you don't have Hungarian-algorithm code for doing that. Is there another way to work with the simple data structures that avoids the complicated projection?

Pseudo-Co has four personal-space constraints (Section 4.6.1), and the permutation-projection-scheme combines these into just a pair of constraints, for A and B. But the divide-and-concur chapter has taught you there is another approach, where the four constraints are dealt with individually.

These collectively are your constraint A, and instead of computing with a single $16 \times 16 \times 16$ cube, your program works with four replicas subject to independent constraints:

$$A = A_1 \times A_2 \times A_3 \times A_4 .$$

For example, the "vertical" constraint A_1 on the first replica selects a single cubicle in each of the 16×16, 16-story vertical stacks to be a 1 (occupied), setting the remaining 15 to zero (unoccupied). Projecting to these selectors just involves finding the position (story) of the largest array element. Two of the other factors are similar: finding the index of the largest element for all (fixed) combinations of the other two indices. Only the "office" constraint (A_4) involves slightly more work because of the index arithmetic when selecting among the 4×4 cubicles within an office (for all 16×16 combinations of offices and floors).

Your concur constraint B imposes equality of the four replicas. The projection replaces each one by their average. It's easy to modify this when there are givens (the Pseudo-Co CEO sends a follow-up email with a list of cubicle assignments for his top management). Just override the concur-average with 1's at the specified occupations, and 0's at the vacancies in the associated stacks (of three kinds) and office.

Convinced you have a sound and easy-to-implement formulation of sudoku, you get to work. Because everything depends simply on the order $n = 4$ and size $n^2 \times n^2 = 16 \times 16$ of the puzzle, your code accepts n as a parameter. After debugging with $n = 2$ and $n = 3$ (blank puzzles), you try $n = 4$ and are not disappointed. Heeding the advice in *The User's Guide* to test your algorithm over a range of problem sizes, you compile averaged iterations for different β's and record the results in Table 8.1. For each n all 1 000 trials are successful, and the iteration count is growing only by a small factor upon increasing n. Surely, with a few thousand iterations, solutions for blank 36×36 sudokus can be constructed as well.

n	2	3	4	5
β	0.90	0.85	0.85	0.90
iterations	26	86	254	970

Table 8.1 Average iterations in 1 000 trials for optimized β in the 4-replica divide-and-concur formulation of blank sudoku.

8.2 The Beast

Though your program easily meets the current needs of Pseudo-Co, the experience leaves you unsettled. That's because you are unable to extend Table 8.1 beyond $n = 5$. Instead of finding $n = 6$ solutions in a few thousand iterations, the algorithm shows no signs of even a near-solution when you allow millions of iterations!

You have no reason to suspect that sudoku solutions abruptly become scarce at $n = 6$. It's an interesting possibility, but more likely there's a bug in the program. When inspecting the code doesn't turn up anything, you rewrite the code from scratch being especially alert for effects that might kick in at $n = 6$. But alas, the results are unchanged.

It has not escaped your notice that the number of variables in one replica cube for 36×36 sudoku is

$$(6 \times 6 \times 6)^2 \,,$$

or that

$$1 + 2 + 3 + \cdots + 36 = 666 \,.$$

That your thoughts are drifting in these directions only exposes the depth of your frustration. Nothing in this book has prepared you for this.

Small β (flow limit) is no help. In fact, even your few successes for $n < 6$ evaporate when you decrease β below the values in Table 8.1. There are no obvious signs of problems in the gap, like "flatlining." But just in case there is a more subtle form of trapping, you give each of the 4×6^6 variables its own metric coefficient and use the auto-tuning heuristic. Alas, that also doesn't break the beastly curse.

By this time you are feeling reckless and increasingly motivated to go off-book. The most general update rule analyzed in Chapter 5,

$$x \mapsto x + \beta\Big(P_B\big((1 - \lambda)P_A(x) + \lambda x\big) - P_A(x)\Big) \,,$$

was a mathematical straw man constructed to motivate RRR, that is, the case $\lambda = -1$. Though some nice things happened for that choice of the parameter, the only settings for (β, λ) ruled out by Theorem 5.2 are the gray regions in Figure 8.1. Transverse contraction, essential for fixed-point-set convergence and also helpful for understanding how RRR avoids getting stuck, was only proved for the line segment of parameters $\lambda = -1$, $\beta \in (0, 2)$. But there was nothing in the proof that spelled disaster when λ was slightly above or below that line.

Breaking with the λ of RRR is easy because the program only needs to be edited in two lines: $(1 - \lambda)$ and λ replacing a 2 and a -1 in what formerly

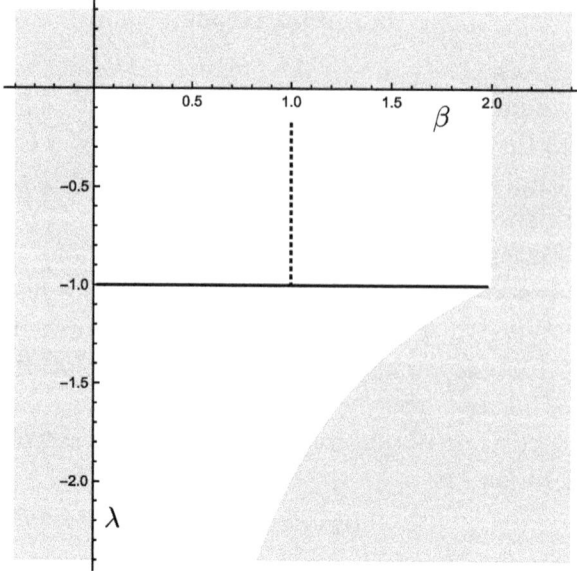

Figure 8.1 The RRR algorithm falls on the horizontal line segment in the (β, λ) plane. The gray region is ruled out because of expansive behavior in the transverse space. However, only the vertical departure from RRR, shown by the dotted line, is able to solve large blank sudoku puzzles.

was the reflection function, and adding λ to the input-parameter list. The simplicity of the change makes the new results in Table 8.2 all the more remarkable. Since β close to 1 worked best for RRR, for simplicity you've set $\beta = 1$ in the new experiments, optimizing only with respect to λ. You're not concerned about the deteriorating success rates for $n > 6$. The gap in the unsuccessful trials always reaches very small values before it flatlines. That's a clear sign of trapping by local constraints with high symmetry and you know a cure for that. Figure 8.2 shows the gap in two runs for $n = 10$, one has flatlined and the other is successful.

n	2	3	4	5	6	7	8	9	10
$-\lambda$	1.00	0.85	0.70	0.60	0.45	0.35	0.25	0.20	0.15
iterations	30	108	259	620	1 500	3 300	6 700	16 000	28 000
success (%)	100	100	100	100	100	97	91	86	60

Table 8.2 Average iterations for $\beta = 1$ and optimized λ parameter in the 4-replica divide-and-concur formulation of blank sudoku.

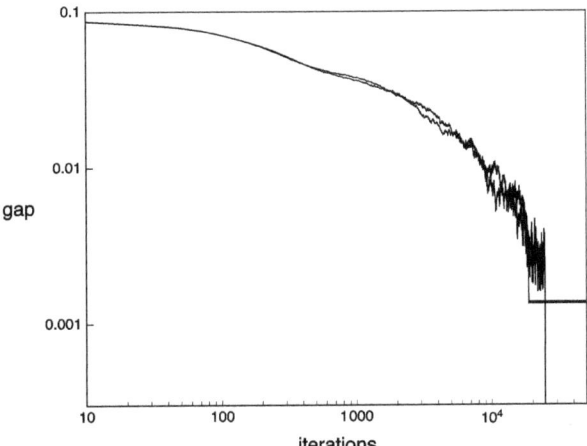

Figure 8.2 The gap in a successful and a trapped ("flatlined") run of the $\lambda = -0.15$ algorithm for solving blank 100×100 sudoku puzzles.

Your celebrations for solving 100×100 blank sudokus are short lived. True, it's amazing that you have a method that can handle a problem where one of the constraints has $40\,000$ one-out-of-100 selector variables. On the other hand, very few of the things you learned in this book can help you understand how this is possible! There is no proof of transverse contraction for the $\lambda > -1$ algorithm you are using. Troubling also is the fact that you cannot invoke flow-limit arguments, for your highly nonconvex problem, when your time-step $\beta = 1$ is very far from small. In fact, you discover that performance degrades dramatically when β is *not* close to 1. Finally, it seems likely that resolving all these questions cannot happen without also understanding why sudoku exhibits a transition at $n = 6$, at least for the $\lambda = -1$ algorithm.

If there is one constant for the brave souls who engage with nonconvex problems armed only with projections, it is this: paradigms are a bad bet. We can have faith that a paradigm will be useful only if there's a reasonable chance that some of its ideas extend to the intended area of application. Algorithms are a case in point. Obviously RRR is just a small generalization of Douglas–Rachford (DR), but the convexity paradigm that goes with DR does not explain why RRR works with nonconvex problems. In fact, you may recall that when we derived the RRR rule, it was only the much weaker property, of being locally affine, that mattered – not convexity.

The divide-and-concur chapter demonstrated that RRR is a versatile and practical tool in diverse settings. But apparently the paradigm we built up around this algorithm has its limits too. The shocking case of the 36×36 blank sudokus made that abundantly clear.

But let's not get too distracted by algorithms when it is the creative bipartisan formulations that make all the difference. This point has been made so many times we need to consult the index to help us recall all the contexts. Not only do nonconvex problems have a staggering variety, a given nonconvex problem can have quite different formulations. So if you're game for advancing the art of solving nonconvex problems with projections, it may be smarter to develop algorithms for particular kinds of problem formulations, rather than the impossible dream of an algorithm that can solve anything.

Notes

Chapter 1

1 The highly nonobvious pattern of bits in the solution is surprising in view of the perfect uniformity of the Fourier magnitudes in this hard instance of bit retrieval. Algebraic constructions of bit sequences with uniform Fourier magnitudes are known when the sequence length is prime, one less than a power of two, and, as in the example, the product of twin primes [7].

2 Though the language is unconventional, these are exactly the rules printed next to puzzles in newspapers. The choice of language brings out the fact that the two constraints have exactly the same structure.

3 Clues can be implemented as required occupations (by numbers), or as required nonoccupations (because of a direct conflict with a clue). The details of the implementation are covered in Exercise 4.14.

4 Both publications have the back-propagation formula, but alas the briefer and more highly cited *Nature* article omits the autoencoder study.

5 There is no better way to motivate students of multivariable calculus to study the chain-rule. The "back-prop" formula is nothing other than the chain-rule relationship among variables that live on a network.

6 There is uniqueness because the radii are designed to maximize their sum in the packing.

7 Working through the exercises at the end of each chapter is an essential part of learning the material, and this chapter is no exception!

Chapter 2

1 This particular anagram has a unique solution even if B is the Official Scrabble Players Dictionary with over 80 000 eight-letter words.

2 Thankfully the final ! is an exclamation, not a factorial.

3 All the wave-related math you will need, including Babinet's principle, is covered in the exercises.

4 Gravel, of course, used a divide-and-concur deconstruction, which is different from the deconstruction based on low-rank symmetric matrices. You will see divide-and-concur applied to periodic sphere packings in Section 7.2.

5 This is the appropriate distance when all the broadcast nodes have the same degree. The generalization is covered in Section 4.1.8.

6 Rounding to ± 1 is clearly distance minimizing, but may result in fewer than g good submatrices. Use the odd/even characterization of good/bad to show that bad can be

made good by flipping a single matrix element. Which one gets flipped, and on which bad submatrices?

Chapter 3

1 It's not that factorizations with symmetries cannot be worked out, just that the analysis gets complicated by all the cases that need to be considered.
2 Sometimes the expression "quantum state tomography" is used for reconstructions that treat the individual measurements, and not their expectation values, as the data.
3 There is no physical principle that says an object cannot be moved in space with finer precision than the resolution at which its image can be resolved. In ptychography with X-rays, this feat is accomplished with piezoelectric actuators.
4 **S**ymmetric, **I**nformationally **C**omplete, **P**ositive **O**perator-**V**alued Measures
5 The symbol L had already been taken.
6 If we also divide out the $25 \times 4 = 100$ translational and rotational symmetries, we find there are about $9\,000$ solutions. This is a small number when we realize there are about $20^3 = 8\,000$ choices to be made when placing just the second, third, and fourth tiles.
7 First find the three smallest square-triangular numbers and call them s_1^2, s_2^2, and s_3^2. Next find factorizations $s_i = p_i q_i$, $(i = 1, 2, 3)$ where (p_i, q_i) satisfy a very simple, two-term linear recursion relation. Use induction to show that the numbers $(p_i q_i)^2$ generated by the recursion are also triangular numbers. Assuming the recursion generates all possible square-triangular numbers, use its growth rate to answer the question. Alternatively, this is an instance of Pell's equation, if you are familiar with that.
8 If $f(z)$ is k_1-sparse and $g(z)$ is k_2-sparse, then $f(z)g(z)$ might not be $k_1 k_2$-sparse.
9 To analyze the large m, n behavior you will need this much of Stirling's approximation of the factorial: $k! \sim \sqrt{2\pi k}(k/e)^k$.

Chapter 4

1 Seasoned practitioners will sometimes remark, "What a fine set you are!".
2 In using the term *map*, we have already taken the position that P_A and P_B are not set-valued and can, from a practical perspective, be treated as maps.
3 They are absent from X-ray crystallography structures for the same reason.
4 Fienup was selected for this demonstration because he had no qualms about projecting to the nonconvex Fourier magnitude constraint. Permutation matrices are another highly nonconvex set.
5 It may help to think of the equilateral triangle as a probability simplex.
6 Indeed, based on the first number, Hamilton is a baseball fan.
7 Start with the computation of the Choleski decomposition, $G = yy^\mathsf{T}$. How is y, the "Choleski point" of A, related to any other element of A?
8 One of the sets should be obvious and is blind to nonnegativity. For the other, the one responsible for nonnegativity, consider a very minor generalization of the probability simplex constraint.
9 Use the method of Lagrange multipliers. Part of the solution involves finding the root of a polynomial of degree greater than two. Instead of invoking a general purpose polynomial root finder, show that the root in question is the unique root of a monotonic rational function, for which Newton's method works well.

Chapter 5

1 The most general solution to a linear recursion relation is the sum of a particular solution (consider the fixed point case) and the most general solution to the homogeneous relation.

2 A special case, described in [34], was called the "difference map." This term is now deprecated because of its widespread use in crystallography, where "map" refers to a terrain of atoms rather than a mapping in Euclidean space.

Chapter 6

1 Do not confuse the affine relaxation $\bar{P}_9$ with the convex relaxation G_9 (Garrett's polytope). The projection to $\bar{P}_9$ is much simpler than the projection to G_9 and is covered in Section 4.2.4.
2 The convex relaxation of P_9 is Garrett's polytope G_9 (doubly stochastic matrices). Projecting x to G_9, when x is far from the polytope's center, almost always results in one of its extreme points, that is, an element of P_9. This relaxation of one of the two sudoku constraints can therefore only have an effect if RRR keeps x near the polytope's center.
3 Puzzle `coly013` was posted by user `coloin` in 2009 on an online forum for sudoku programmers [24].
4 Equations marked $(*)$ should be treated with the same degree of skepticism as sports records with that annotation.
5 The intensity I is the rate of arrival of photons at the detector. Photon detection is a Poisson process in time, so by measuring for some fixed time one can only estimate I by one sample of a Poisson distribution with mean proportional to I (the number of detected photons). At the edge of the detector (high spatial frequencies), the rate of photon arrival is so low that only a handful of photons are detected over the course of the diffraction measurement.
6 There are two domains, comprising disconnected regions, of point-line flow. So it's just a matter of stitching together flows you already know.

Chapter 7

1 For Douglas–Rachford, see Theorem 4.5 of [9].
2 The 20 amino acids that comprise proteins have single-Latin-letter code names. The spelling of SEQVENCE is explained by the fact that U is *not* one of those codes.
3 Besides the translational nonuniqueness, hexeins have 12 rotation/reflection variants we will consider equivalent. When performing a depth-first search, we may place the first tile at the origin with its chain directed toward the east. We may also insist that the first turn is always toward the left.
4 Besides making data perceptible, dimensions reduction is also used for data compression, in which case the embedding dimension can be much larger than three.
5 From the **m**odified **N**ational **I**nstitute of **S**tandards and **T**echnology database.

References

[1] Amit, Daniel J, Gutfreund, Hanoch, and Sompolinsky, Haim. 1985. Spin-glass models of neural networks. *Physical Review A*, **32**(2), 1007–1018.

[2] Andersen, Hans Christian. 2008. *The Annotated Hans Christian Andersen.* WW Norton & Company.

[3] Anderson, Philip W. 1988. Spin glass III: Theory raises its head. *Physics Today*, **41**(6), 9–11.

[4] Aragón Artacho, Francisco J., Campoy, Rubén, and Tam, Matthew K. 2020. The Douglas–Rachford algorithm for convex and nonconvex feasibility problems. *Mathematical Methods of Operations Research*, **91**, 201–240.

[5] Ball, Keith. 1992. A lower bound for the optimal density of lattice packings. *International Mathematics Research Notices*, **1992**(10), 217–221.

[6] Bang, Thøger. 1951. A solution of the "plank problem." *Proceedings of the American Mathematical Society*, **2**(6), 990–993.

[7] Baumert, Leonard D. 1969. Difference sets. *SIAM Journal on Applied Mathematics*, **17**(4), 826–833.

[8] Bauschke, Heinz H., and Borwein, Jonathan M. 1993. On the convergence of von Neumann's alternating projection algorithm for two sets. *Set-Valued Analysis*, **1**, 185–212.

[9] Bauschke, Heinz H., and Moursi, Walaa M. 2017. On the Douglas–Rachford algorithm. *Mathematical Programming*, **164**, 263–284.

[10] Bauschke, Heinz H., Combettes, Patrick L., and Luke, D. Russell. 2002. Phase retrieval, error reduction algorithm, and Fienup variants: A view from convex optimization. *Journal of the Optical Society of America A*, **19**(7), 1334–1345.

[11] Best, M. 1980. Binary codes with a minimum distance of four (Corresp.). *IEEE Transactions on Information Theory*, **26**(6), 738–742.

[12] Birkhoff, Garrett. 1946. Three observations on linear algebra. *Universidad Nacional de Tucuman, Review Series A*, **5**, 147–151.

[13] Boyd, Stephen, Parikh, Neal, Chu, Eric, Peleato, Borja, Eckstein, Jonathan, et al. 2011. Distributed optimization and statistical learning via the alternating direction method of multipliers. *Foundations and Trends in Machine Learning*, **3**(1), 1–122.

[14] Bozóki, Sándor, Lee, Tsung-Lin, and Rónyai, Lajos. 2015. Seven mutually touching infinite cylinders. *Computational Geometry*, **48**(2), 87–93.

[15] Bruck, Yu M., and Sodin, L. G. 1979. On the ambiguity of the image reconstruction problem. *Optics Communications*, **30**(3), 304–308.

[16] Brunn, Hermann. 1887. *Über Ovale und Eiflächen*. Akademische Buchdruck-erei von R. Straub.

[17] Buhler, Joe, and Reichstein, Zinovy. 2005. Symmetric functions and the phase problem in crystallography. *Transactions of the American Mathematical Society*, **357**(6), 2353–2377.

[18] Burley, Stephen K., Berman, Helen M., Kleywegt, Gerard J., Markley, John L., Nakamura, Haruki, and Velankar, Sameer. 2017. Protein Data Bank (PDB): The single global macromolecular structure archive. *Methods and Molecular Biology*, **1607**, 627–641.

[19] Campos, Marcelo, Jenssen, Matthew, Michelen, Marcus, and Sahasrabudhe, Julian. 2023. A new lower bound for sphere packing. *arXiv preprint arXiv:2312.10026*.

[20] Candes, Emmanuel J., Strohmer, Thomas, and Voroninski, Vladislav. 2013. Phaselift: Exact and stable signal recovery from magnitude measurements via convex programming. *Communications on Pure and Applied Mathematics*, **66**(8), 1241–1274.

[21] Casselman, Bill. 2004. The difficulties of kissing in three dimensions. *Notices of the American Mathematical Society*, **51**(8), 884–885.

[22] Cohn, Henry. 2017. A conceptual breakthrough in sphere packing. *Notices of the American Mathematical Society*, **64**, 102–115.

[23] Cohn, Henry. 2024. *Sphere Packing*. cohn.mit.edu/sphere-packing.

[24] coloin. 2009. *The New Sudoku Players' Forum*. forum.enjoysudoku.com.

[25] Conway, John Horton, and Sloane, Neil James Alexander. 2013. *Sphere Packings, Lattices and Groups*. Vol. 290. Springer Science & Business Media.

[26] de Launey, Warwick, and Gordon, Daniel M. 2010. On the density of the set of known Hadamard orders. *Cryptography and Communications*, **2**(2), 233–246.

[27] Demaine, Erik D., and Demaine, Martin L. 2007. Jigsaw puzzles, edge matching, and polyomino packing: Connections and complexity. *Graphs and Combinatorics*, **23**(Suppl 1), 195–208.

[28] Dill, Ken A. 1985. Theory for the folding and stability of globular proteins. *Biochemistry*, **24**(6), 1501–1509.

[29] Dong, Junkai, Elser, Veit, Gyawali, Gaurav, Jee, Kai Yen, Kent-Dobias, Jaron, Mandaiya, Avinash, Renz, Megan, and Su, Yubo. 2021. Glass phenomenology in the hard matrix model. *Journal of Statistical Mechanics: Theory and Experiment*, **2021**(9), 093302.

[30] Douglas, Jim, and Rachford, Henry H. 1956. On the numerical solution of heat conduction problems in two and three space variables. *Transactions of the American Mathematical Society*, **82**(2), 421–439.

[31] Dudeney, Henry E. 1958. *Amusements in Mathematics*. Vol. 473. Courier Corporation.

[32] Dyson, Freeman. 2009. Birds and frogs. *Notices of the American Mathematical Society*, **56**(2), 212–223.

[33] Eckart, Carl, and Young, Gale. 1936. The approximation of one matrix by another of lower rank. *Psychometrika*, **1**(3), 211–218.

[34] Elser, Veit. 1986. The diffraction pattern of projected structures. *Acta Crystallographica Section A: Foundations of Crystallography*, **42**(1), 36–43.

[35] Elser, Veit. 2003. Phase retrieval by iterated projections. *Journal of the Optical Society of America A*, **20**(1), 40–55.

[36] Elser, Veit. 2008. Pauling's revenge. *Philosophical Magazine*, **88**(1883–1885).

[37] Elser, Veit. 2017. The complexity of bit retrieval. *IEEE Transactions on Information Theory*, **64**(1), 412–428.

[38] Elser, Veit. 2021. Reconstructing cellular automata rules from observations at nonconsecutive times. *Physical Review E*, **104**(3), 034301.

[39] Elser, Veit. 2023. Packing spheres in high dimensions with moderate computational effort. *Physical Review E*, **108**(3), 034117.

[40] Elser, Veit, and Gravel, Simon. 2010. Laminating lattices with symmetrical glue. *Discrete & Computational Geometry*, **43**, 363–374.

[41] Elser, Veit, and Millane, R. P. 2008. Reconstruction of an object from its symmetry-averaged diffraction pattern. *Acta Crystallographica Section A: Foundations of Crystallography*, **64**(2), 273–279.

[42] Fienup, James R. 1982. Phase retrieval algorithms: A comparison. *Applied Optics*, **21**(15), 2758–2769.

[43] Gamow, George. 1988. *One, Two, Three … Infinity: Facts and Speculations of Science*. Dover Books on Mathematics.

[44] Garris, M.D., Blue, J.L., Candela, G.T., Dimmick, D.L., Geist, J., Grother, P.J., Janet, S.A., and Wilson, C.L. 2024. *Optical Recognition of Handwritten Digits*. archive.ics.uci.edu/dataset/80/optical+recognition+of+handwritten+digits.

[45] Gilpin, William. 2019. Cellular automata as convolutional neural networks. *Physical Review E*, **100**(3), 032402.

[46] Golub, Gene H., and Van Loan, Charles F. 2013. *Matrix Computations*. JHU Press.

[47] Goscinny, René, and Uderzo, Albert. 1961. *Astérix le Gaulois*. Hachette Livre.

[48] Gravel, Simon, and Elser, Veit. 2008. Divide and concur: A general approach to constraint satisfaction. *Physical Review E*, **78**(3), 036706.

[49] Green, F. 1987. More about NP-completeness in the frustration model of spin-glasses. *Operations-Research-Spektrum*, **9**(3), 161–165.

[50] Groom, Colin R., and Allen, Frank H. 2014. The Cambridge Structural Database in retrospect and prospect. *Angewandte Chemie International Edition*, **53**(3), 662–671.

[51] Haas, Wolfgang. 2008. On the failing cases of the Johnson bound for error-correcting codes. *The Electronic Journal of Combinatorics*, **15**, R55–R55.

[52] Hales, Thomas C., Harrison, John, McLaughlin, Sean, Nipkow, Tobias, Obua, Steven, and Zumkeller, Roland. 2011. A revision of the proof of the Kepler conjecture. *The Kepler Conjecture: The Hales-Ferguson Proof*, 341–376.

[53] Hauptman, Herbert A., Guo, D.Y., Xu, Hongliang, and Blessing, Robert H. 2002. Algebraic direct methods for few-atoms structure models. *Acta Crystallographica Section A: Foundations of Crystallography*, **58**(4), 361–369.

[54] Hlawka, Edmund. 1943. Zur geometrie der zahlen. *Mathematische Zeitschrift*, **49**(1), 285–312.

[55] Hornik, Kurt. 1991. Approximation capabilities of multilayer feedforward networks. *Neural Networks*, **4**(2), 251–257.

[56] Hughston, Lane P., Jozsa, Richard, and Wootters, William K. 1993. A complete classification of quantum ensembles having a given density matrix. *Physics Letters A*, **183**(1), 14–18.

[57] Inkala, A. 2007. *AI Escargot: The Most Difficult Sudoku Puzzle*. Lulu.

[58] Johnson, John E., and Olson, Arthur J. 2021. Icosahedral virus structures and the protein data bank. *Journal of Biological Chemistry*, **296**, 1–11.

[59] Kamien, Randall D., and Liu, Andrea J. 2007. Why is random close packing reproducible? *Physical Review Letters*, **99**(15), 155501.

[60] Karle, J., and Hauptman, H. 1950. The phases and magnitudes of the structure factors. *Acta Crystallographica*, **3**(3), 181–187.

[61] Karp, Richard M. 1972. *Reducibility among Combinatorial Problems.* Springer US. Pages 85–103.

[62] Kingma, Diederik P., and Ba, Jimmy. 2014. Adam: A method for stochastic optimization. *arXiv preprint arXiv:1412.6980.*

[63] Knuth, Donald Ervin. 1997. *The Art of Computer Programming.* Vol. 3. Pearson Education.

[64] Koryo, Miura. 1985. Method of packaging and deployment of large membranes in space. *The Institute of Space and Astronautical Science Report*, **618**, 1–9.

[65] Krivelevich, Michael, Litsyn, Simon, and Vardy, Alexander. 2004. A lower bound on the density of sphere packings via graph theory. *International Mathematics Research Notices*, **2004**(43), 2271–2279.

[66] Kuhn, Harold W. 1955. The Hungarian method for the assignment problem. *Naval Research Logistics Quarterly*, **2**(1-2), 83–97.

[67] LeCun, Yann, Cortes, Corinna, and Burges, Christopher J.C. 2024. *The MNIST Database.* `yann.lecun.com/exdb/mnist`.

[68] Leland, McInnes, John, Healy, and James, Melville. 2018. Uniform manifold approximation and projection for dimension reduction. *arXiv preprint arXiv:1802.03426.*

[69] Lin, Keh Ying. 2004. Number of Sudokus. *Journal of Recreational Mathematics*, **33**(2), 120.

[70] Littlewood, John E. 1968. *Some Problems in Real and Complex Analysis, Heath Mathematical Monographs.* Raytheon Education.

[71] Luke, D. Russell. 2004. Relaxed averaged alternating reflections for diffraction imaging. *Inverse Problems*, **21**(1), 37.

[72] Marchesini, Stefano, He, H., Chapman, Henry N., Hau-Riege, Stefan P., Noy, Aleksandr, Howells, Malcolm R., Weierstall, Uwe, and Spence, John C. H. 2003. X-ray image reconstruction from a diffraction pattern alone. *Physical Review B*, **68**(14), 140101.

[73] Matoušek, Jiří. 2013. *Lectures on Discrete Geometry.* Vol. 212. Springer Science & Business Media.

[74] Matoušek, Jiří, and Gärtner, Bernd. 2007. *Understanding and Using Linear Programming.* Vol. 1. Springer.

[75] Mertens, Stephan, Mézard, Marc, and Zecchina, Riccardo. 2006. Threshold values of random K-SAT from the cavity method. *Random Structures & Algorithms*, **28**(3), 340–373.

[76] Miao, Jianwei, Charalambous, Pambos, Kirz, Janos, and Sayre, David. 1999. Extending the methodology of X-ray crystallography to allow imaging of micrometre-sized non-crystalline specimens. *Nature*, **400**(6742), 342–344.

[77] Millane, R. P., and Stroud, W. J. 1997. Reconstructing symmetric images from their undersampled Fourier intensities. *Journal of the Optical Society of America A*, **14**(3), 568–579.

[78] Minkowski, Hermann. 1910. *Geometrie der Zahlen.* BG Teubner.

[79] Moore, Cristopher, and Mertens, Stephan. 2011. *The Nature of Computation.* Oxford University Press.

[80] Musin, Oleg Rustamovich. 2003. The problem of the twenty-five spheres. *Uspekhi Matematicheskikh Nauk*, **58**(4), 153–154.

[81] Neutze, Richard, Wouts, Remco, Van der Spoel, David, Weckert, Edgar, and Hajdu, Janos. 2000. Potential for biomolecular imaging with femtosecond X-ray pulses. *Nature*, **406**(6797), 752–757.

[82] OEIS Foundation Inc. 2024a. *On-Line Encyclopedia of Integer Sequences.* oeis.org/A007299.

[83] OEIS Foundation Inc. 2024b. *On-Line Encyclopedia of Integer Sequences.* oeis.org/A206711.

[84] OEIS Foundation Inc. 2024c. *On-Line Encyclopedia of Integer Sequences.* oeis.org/A000609.

[85] Ðoković, Dragomir Ž., Golubitsky, Oleg, and Kotsireas, Ilias S. 2014. Some new orders of Hadamard and Skew-Hadamard matrices. *Journal of Combinatorial Designs*, **22**(6), 270–277.

[86] Ostergard, P. R. J., and Blass, Uri. 2001. On the size of optimal binary codes of length 9 and covering radius 1. *IEEE Transactions on Information Theory*, **47**(6), 2556–2557.

[87] Paley, Raymond EAC. 1933. On orthogonal matrices. *Journal of Mathematics and Physics*, **12**(1–4), 311–320.

[88] Pauling, Linus. 1987. So-called icosahedral and decagonal quasicrystals are twins of an 820-atom cubic crystal. *Physical Review Letters*, **58**(4), 365.

[89] Renes, Joseph M., Blume-Kohout, Robin, Scott, Andrew J., and Caves, Carlton M. 2004. Symmetric informationally complete quantum measurements. *Journal of Mathematical Physics*, **45**(6), 2171–2180.

[90] Rodenburg, John, and Maiden, Andrew. 2019. *Ptychography, in Springer Handbook of Microscopy.* Springer.

[91] Rumelhart, David E., Hinton, Geoffrey E., and Williams, Ronald J. 1985. *Learning Internal Representations by Error Propagation.* Tech. rept. University of California, San Diego, La Jolla Institute for Cognitive Science.

[92] Rumelhart, David E., Hinton, Geoffrey E., and Williams, Ronald J. 1986. Learning representations by back-propagating errors. *Nature*, **323**(6088), 533–536.

[93] Schattschneider, Doris. 1993. Crystalline symmetries: An informal mathematical introduction (Marjorie Senechal). *SIAM Review*, **35**(2), 335–336.

[94] Schütte, Kurt, and van der Waerden, Bartel Leendert. 1952. Das problem der dreizehn Kugeln. *Mathematische Annalen*, **125**(1), 325–334.

[95] Scott, Andrew James, and Grassl, Markus. 2010. Symmetric informationally complete positive-operator-valued measures: A new computer study. *Journal of Mathematical Physics*, **51**(4), 042203.

[96] Senior, Andrew W., Evans, Richard, Jumper, John, Kirkpatrick, James, Sifre, Laurent, Green, Tim, Qin, Chongli, Žídek, Augustin, Nelson, Alexander W. R., Bridgland, Alex, et al. 2019. Protein structure prediction using multiple deep neural networks in the 13th Critical Assessment of Protein Structure Prediction (CASP13). *Proteins: Structure, Function, and Bioinformatics*, **87**(12), 1141–1148.

[97] Shechtman, Dan, Blech, Ilan, Gratias, Denis, and Cahn, John W. 1984. Metallic phase with long-range orientational order and no translational symmetry. *Physical Review Letters*, **53**(20), 1951.

[98] Sheldrick, George M. 2008. A short history of SHELX. *Acta Crystallographica Section A: Foundations of Crystallography*, **64**(1), 112–122.

[99] Shen, Zuowei, Yang, Haizhao, and Zhang, Shijun. 2022. Optimal approximation rate of ReLU networks in terms of width and depth. *Journal de Mathématiques Pures et Appliquées*, **157**, 101–135.

[100] Sidorenko, Alexander. 1995. What we know and what we do not know about Turán numbers. *Graphs and Combinatorics*, **11**(2), 179–199.

[101] Sinai, Yakov Grigor'evich. 1963. On the foundations of the ergodic hypothesis for a dynamical system of statistical mechanics. Pages 1261–1264 of: *Doklady Akademii Nauk*, vol. 153. Russian Academy of Sciences.

[102] Smale, Steve. 1986. Newton's method estimates from data at one point. Pages 185–196 of: *The Merging of Disciplines: New Directions in Pure, Applied, and Computational Mathematics: Proceedings of a Symposium Held in Honor of Gail S. Young at the University of Wyoming, August 8–10, 1985. Sponsored by the Sloan Foundation, the National Science Foundation, and Air Force Office of Scientific Research*. Springer.

[103] Spence, John C. H., Howells, M., Marks, L. D., and Miao, J. 2001. Lensless imaging: a workshop on "New approaches to the phase problem for nonperiodic objects." *Ultramicroscopy*, **1**(90), 1–6.

[104] Springer, Jacob M., and Kenyon, Garrett T. 2021. It's hard for neural networks to learn the game of life. Pages 1–8 of: *2021 International Joint Conference on Neural Networks (IJCNN)*. IEEE.

[105] Stuart, Andrew. 2024. *SudokuWiki*. www.sudokuwiki.org/Main_Page.

[106] Thibault, Pierre, and Menzel, Andreas. 2013. Reconstructing state mixtures from diffraction measurements. *Nature*, **494**(7435), 68–71.

[107] Turán, Paul. 1961. Research problems. *Közl MTA Mat. Kutató Int*, **6**(417–423), 2.

[108] Turán, Paul. 1977. A note of welcome. *Journal of Graph Theory*, **1**(1), 7–9.

[109] Van der Maaten, Laurens, and Hinton, Geoffrey. 2008. Visualizing data using t-SNE. *Journal of Machine Learning Research*, **9**(11), 2579–2605.

[110] Vavasis, Stephen A. 2010. On the complexity of nonnegative matrix factorization. *SIAM Journal on Optimization*, **20**(3), 1364–1377.

[111] Vonnegut, Kurt. 1963. *Cat's Cradle*. Holt, Rinehart and Winston.

[112] Weakley, W. D. 1995. *Domination in the Queen's Graph in Graph theory, combinatorics, and algorithms: proceedings of the seventh quadrennial international conference on the theory and applications of graphs*. Vol. 2. Wiley.

[113] Zauner, Gerhard. 1999. *Quantum designs: Foundations of a non-commutative theory of designs*. PhD thesis, University of Vienna.

Index

For EU product safety concerns, contact us at Calle de José Abascal, 56–1°,
28003 Madrid, Spain or eugpsr@cambridge.org.